Animal Diversity of Pakistan

This book presents principles of animal biodiversity, its importance, sustainable utilization, and conservation. The introductory chapter explores the origins and evolution of biodiversity throughout time, shedding light on the remarkable developments that have taken place over the ages. The subsequent chapters of the book reveal the tools and techniques used to assess the status of biodiversity at any given stage. The book meticulously details the indicators and measurement methods employed to gauge the health and vitality of biodiversity. Furthermore, it elucidates the vital role biodiversity plays in serving humanity and underscores the dire consequences that may ensue if we fail to preserve it. It also introduces the nomenclature of species and explores their intricate phylogenetic relationships with both living and non-living entities. The spatial and temporal distribution of animals, unraveling the fascinating patterns and dynamics that govern their presence across the globe, is also discussed in the book. Additionally, the book unveils a repertoire of conservation methods aimed at preserving existing biodiversity and ensuring its sustainable future. Toward the end, the book offers valuable suggestions and highlights untapped avenues for future research, all geared toward safeguarding the current level of biodiversity and ensuring its prosperity in the years to come. This book is a valuable source for students and professionals in the fields of biology, ecology, and conservation biology.

Animal Diversity of Pakistan

Distribution and Conservation

Masroor Elahi Babar and Muhammad Ashraf

CRC Press
Taylor & Francis Group
Boca Raton London New York

CRC Press is an imprint of the
Taylor & Francis Group, an **informa** business

Designed cover image: Wikimedia Commons

First edition published 2025
by CRC Press
2385 NW Executive Center Drive, Suite 320, Boca Raton FL 33431

and by CRC Press
4 Park Square, Milton Park, Abingdon, Oxon, OX14 4RN

CRC Press is an imprint of Taylor & Francis Group, LLC

ISBN: 978-1-032-75009-5 (hbk)
ISBN: 978-1-032-75010-1 (pbk)
ISBN: 978-1-003-47200-1 (ebk)

DOI: 10.1201/9781003472001

Typeset in Times
by KnowledgeWorks Global Ltd.

Contents

About the Authors

Masroor Elahi Babar currently serves as the Vice Chancellor of The University of Agriculture, Dera Ismael Khan, KPK. Previously, he served as the Vice Chancellor of Gomal University and as the Dean of the Faculty of Science and Technology at Virtual University of Pakistan in Lahore. His extensive work focuses on functional genomics and molecular genetics of both domestic and wild mammal breeds. Dr. Babar has authored numerous books and papers in this field. In recognition of his contributions to the discipline of animal breeding and genetics, he was awarded the Tamgha-i-Imtiaz by the Government of Pakistan.

Muhammad Ashraf currently serves as a Subject Expert in Zoology at The University of Agriculture, Dera Ismael Khan, KPK. Previously, he held the position of Dean of the Faculty of Fisheries and Wildlife. Throughout his career, he has actively participated in various research projects related to fisheries and wildlife, serving as a Principal or Co-Principal Investigator. Dr. Ashraf has made significant academic and practical contributions to the advancement of the field of fisheries and wildlife. He is also the author of numerous scientific papers and books in these disciplines and has taught several courses at both undergraduate and postgraduate levels in Zoology.

1 What Is Biodiversity?

1.1 BIOLOGICAL DIVERSITY

"Biological diversity is variety of life that presently inhabits earth. It refers to all variation within and all species of plants, animals, and microorganisms and the ecosystems where these species are living and interact with them."

The millions of different animal and plant species that live on our planet, and the genetic differences within these species, all constitute biodiversity in a given area. Biodiversity also refers to the collection of different ecosystems where species from distinct groups interact with one another, as well as with the air, water, and soil. Biodiversity is not only present in animal or plant species but it is also found in genes present in different species, in species as a whole, and ecosystems which have different levels of biological organization. When we study biodiversity with reference to the number of species or taxa, three different dimensional scales are normally applied. Of these, Whittaker's scales are one of the most commonly applied, which include alpha, beta, and gamma diversity (de Vere, 2008).

If we develop this definition further then we can say that biodiversity is actually variation present among different organisms; they are either from marine, aquatic, or terrestrial resources or any other ecological complex where these animals belong. So, it can be inferred from these statements that biodiversity actually indicates the variations present within species, between species, and among the various ecosystems (Herrera Izaguirre, 2008). Therefore, "biodiversity" is the variety of life on earth, which includes all sorts of organisms, species, and populations. Biodiversity also covers the genetic variation present among these organisms either at species level or in a population ultimately expanding to complex groupings of these animals, their communities, and ecosystems. In addition to these differences, it also refers to the relationship of genes, species, and ecosystems and, in turn, their interactions with the environment.

1.1.1 Importance of Biodiversity

Biodiversity is an essential and basic feature of life on earth and covers all types of variations present in the living organisms (Wang et al. 2020). This variation may be genetic, species, and/or ecosystem variation. Like other countries of the globe, Pakistan is also under an inescapable risk of climate change which might further increase the different problems which can originate due to loss of biodiversity. Biodiversity is not important because it possesses some obscure and hidden values; however, it is important due to its ecosystem services, which a healthy biodiversity provides to all of us.

Kettunen et al. (2013) have categorized the services that biodiversity provides. In this context, they state that biodiversity provides us with food, clean water, renewable resources, and cultural values. When all these services combine together then different hydrological cycles consequently maintain climate. It was envisaged earlier that loss of biodiversity means erosion of climate. In 1962, Rachel Carson made a similar statement in her famous book *Silent Spring*, where she said that when there is persistent loss in biodiversity it weakens and undermines our ecosystem services which we depend on and means that the direct economic costs of these losses are very pronounced.

Science emphasizes the importance of biodiversity, although this is simply not enough because we also need to assess the social and economic impacts of biodiversity. Nevertheless, it cannot be concluded that nature is all good because some animal species have a negative impact on our daily lives; for example, agricultural pests and bacterial diseases, although part of biodiversity, negatively affect our daily life. Contrary to our previous positive statements, these organisms essentially need to be controlled and thoroughly checked by our functional ecosystem. Organisms considered as

DOI: 10.1201/9781003472001-1

nuisances in one ecosystem may be useful in another through interdependencies and evolutionary adaptation. It is generally accepted that a higher biodiversity will possibly suit and even benefit the stability of organisms and the natural environment. If the number of species is higher in a given area there is a possibility that the number of these species may increase and provide a similar function in an ecosystem in case of any exogenous change in that ecosystem (Vitsouek & Hooper, 1993; van Rensburg & Mill, 2010). This stable and useful function creates a buffer system and provides resistance against sudden change in a given ecosystem.

1.2 LEVELS OF BIODIVERSITY

Biological diversity can be examined at three levels:

- What is the genetic variation within a species?
- How can the species vary within a community?
- The organization and level of species in an area are divided into distinctive plant and animal communities. The three levels of biodiversity are genetic diversity, species diversity, and ecosystem diversity.

1.2.1 Genetic Diversity

Genes present in all organisms control the necessary information required for all life on earth. Genes are inherited from parents to offspring. They contain the information necessary to build and maintain cells, as well as to determine the required physical and biochemical features of each organism. Genetic diversity encompasses all the different genes present within a species. All the individuals in a species have their own particular genetic composition. Conservation of genetic diversity within a species demands conservation of different populations and helps the organisms to adapt themselves to environmental changes which are important for the survival of the species. Genetic diversity determines and tells us how the individuals differentiate and maintain their individuality, as well as other exceptional features it may contain. When DNA is transferred to different traits and expresses itself there, it hands over different capabilities to offspring; it may be tolerance to drought or frost, growth, and survival, or enhancing their capability to a changing environment. Based on the diversity in plants and animals, farmers select productive and tolerant species for their livestock and crop cultivation. Based on useful traits such as taste, height, color, and productivity, many varieties of corn, or maize, have been developed. When the question of conservation and sustainable use of unique animal breeds and crops is considered or their wild relatives are considered, they all actually genetically assure that animals and crops can adapt successfully to the changing climate. It further conveys the message how and in what manner they can meet the present as well as future needs of the farmer. In the light of salient features of biodiversity stated above, this is highly important to the food security of present and future generations.

1.2.2 Species Diversity

When organisms are morphologically similar and are able to interbreed and produce fertile offspring, they are called species. Thus, the definition of a species is actually a group of morphologically similar organisms that are able to interbreed and produce fertile offspring. There are many species of plants, animals, and microorganisms that inhabit our planet. The different plants cultivated and animals raised in agriculture are the result of human selection of these organisms for food, nutrition, and medicine. Human beings have further maintained these plants and animals for their future use and food security. Due to the differentiation of livestock species like cattle, sheep, chickens, goats and crop species like wheat, banana, cabbage, sweet potato, and ground nuts, they have been finally selected for further expansion and promotion.

Species diversity tells us how many different species are present in a certain region. Species diversity is a function of complex factors which are not well understood. At the global level, species diversity is not uniform nor is it evenly distributed across the continents. Scientists have identified thirty four hotspots all over the world. The total area of these hotspots is just 2.3% of the total land surface area available. These hotspots cover half of the total plant species and 42% of all terrestrial vertebrate species and they also host 75% of the planet's most threatened amphibians, birds, and mammals.

1.2.3 Ecosystem Diversity

Ecosystem diversity tells about different habitats, biological communities, and ecological processes going on in our environment. This diversity also tells us about different variations within and among individual ecosystems. Ecosystems vary in size and different species present in ecosystem are dependent on each other for their survival. A large stand of forest or a small pond serves as an ecosystem which is host of several organisms (Fig. 1.1).

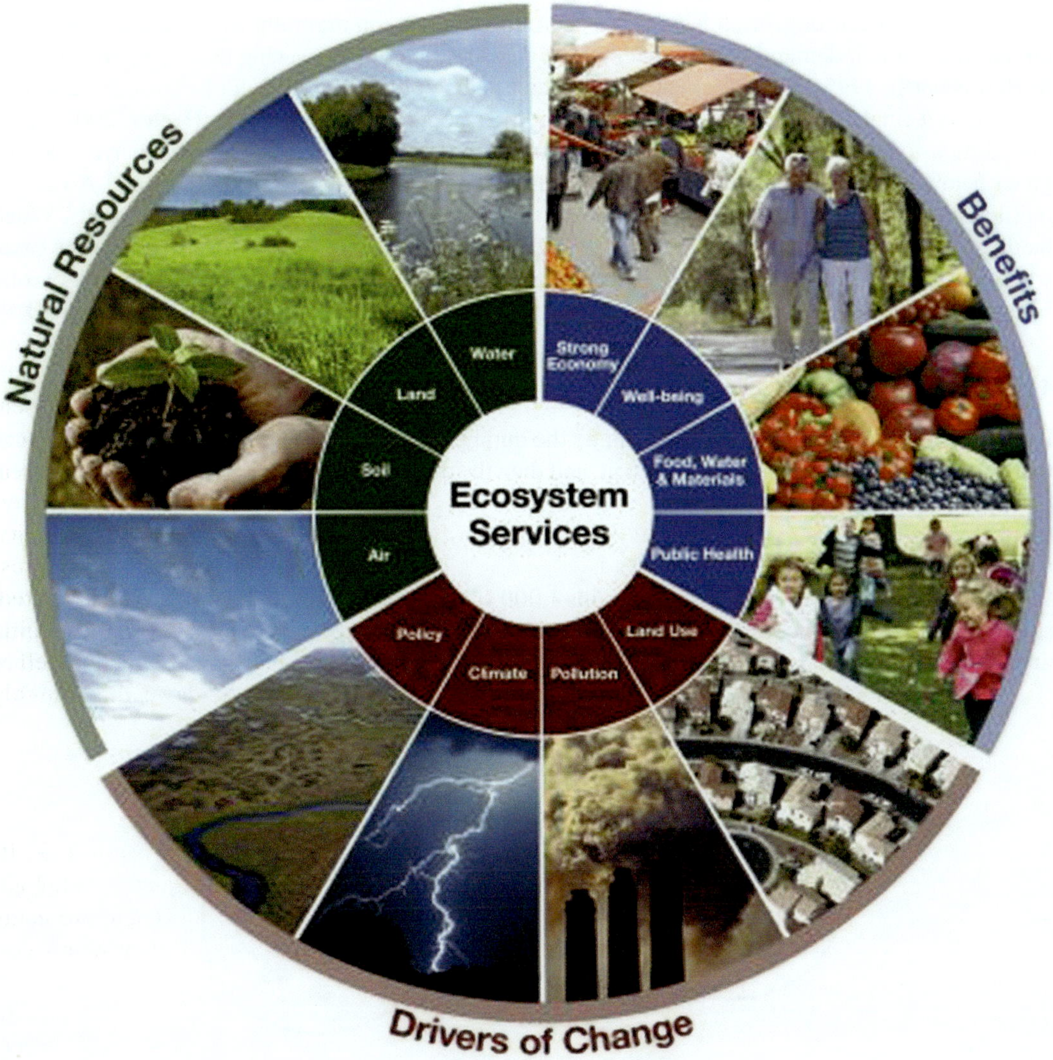

FIGURE 1.1 An ecosystem services. (Courtesy of EPA USA 2017.)

When we talk about ecosystem biodiversity it means types of different ecosystems present in a particular given place or an area. It all depends on the area of landscape: bigger areas hold many different independent ecosystems which are well integrated with each other. Conservation of an ecosystem is critical for the conservation of species and genetic diversity. An ecosystem is a functional unit and is a self-motivated composite group of plant, animal, and micro-organism communities that surround the environment where these organisms are living and are interacting with that environment. Wetlands, mountains, grasslands, forests, coastal areas, and lakes are examples of different ecosystems. The diversity of agro-ecosystems is partially a product of both agricultural and non-agricultural pieces of land utilization of water. Aquaculture systems, rice paddies, pastoral systems, and cropping systems are all examples of agriculture ecosystems. All of these smaller agriculture ecosystems are present in the broader ecosystems on which smaller ecosystems are based. When different elements of these systems are combined they make up combined or mixed systems.

1.3 MEASURING BIODIVERSITY

While discussing biodiversity a few questions come to mind. These are where biodiversity is found, at what rate is it declining, or, indeed, if it is declining how we can maintain it? These are fundamental questions which demand that biodiversity be measured before answering them properly (Harper & Hawksworth, 1995). Measurement of biodiversity is actually very difficult because it cannot be reduced to a single number because of its multidimensional spread (Purvis & Hector, 2000). This multidimensional concept can be further elaborated with the simple but comprehensive example given below. Suppose that there are two islands named Island A and Island B. Island A has six reptiles, one bird, and one mammal. However, Island B has four reptiles and four mammals. When biodiversity of both islands is considered, Island A is more diverse than Island B with three taxa, whereas Island B has only two (Fig. 1.2). Launchbaugh (2009) reported that biodiversity can be measured and monitored at several spatial scales which have been given on the next page (Fig. 1.3).

1.3.1 Numbers

If we take the number of any façade it can be the number of genes, the number of species present in that particular area, the populations present, and then the different taxa present in that particular area.

1.3.2 Evenness

In the context of evenness, if any site contains 1,000 species, there will not be much diversity at that site if 99.9% of the species are the same, in which case the site will be said to have evenness but little diversity. Simpson's and Shannon's diversity are two important biodiversity indices which ascertain

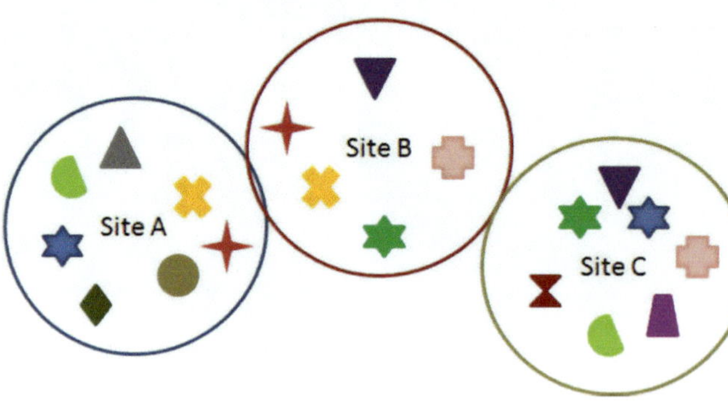

FIGURE 1.2 Measuring and monitoring biodiversity. (Courtesy of Launchbaugh, 2009; University of Idaho, College of Natural Resources USA.)

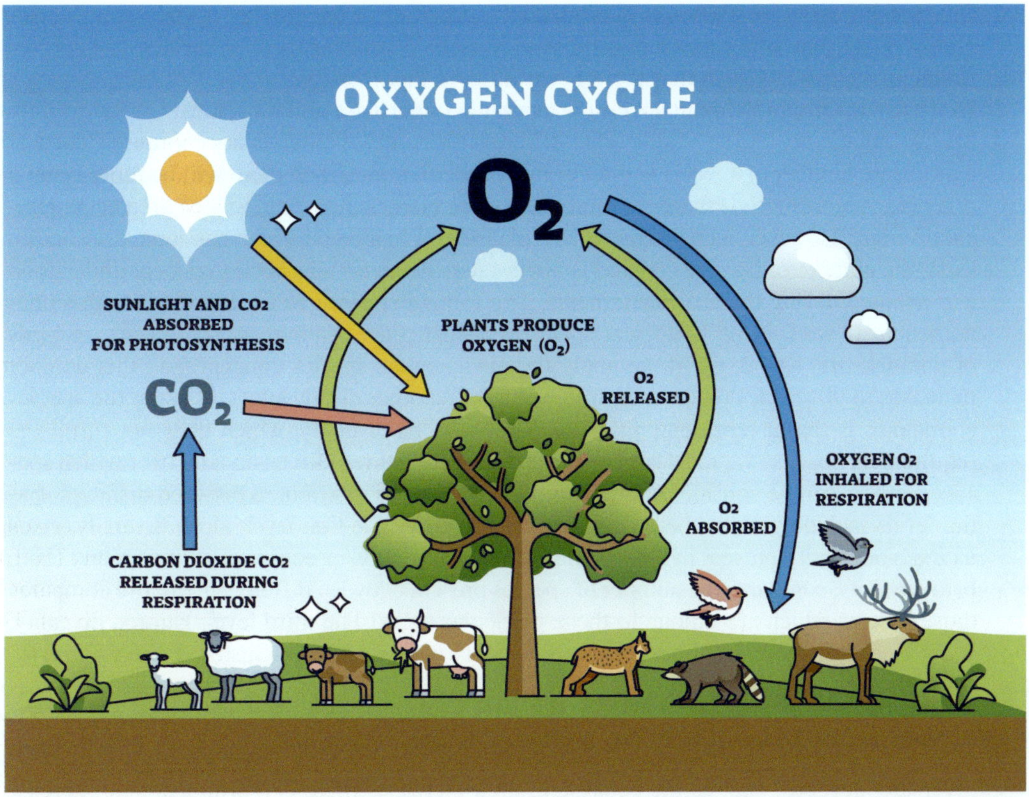

FIGURE 1.3 Stock-vector-oxygen-cycle-with-o-production-from-plants-absorbing-carbon-dioxide. (*Source*: Shutterstock.)

the degree of distribution of individuals among species. Parallel to that genetic diversity is measured by heterozygosity that covers number of alleles and their relative frequencies and/or occurrence.

1.3.3 Difference

It is generally observed that some taxa, allele, species, and populations may be comparable with one another, while others cannot be comparable, although they may be different from one another. For example, if within a population of the same species several individuals differ from each other, then they are named as units of evolutionary significance, sub-species, or management-units. Some differences among these individuals may be more significant than others, which may be of lesser importance. Among these differences, ecological variations among species are more important because they are essential for the smooth functioning of ecosystems. These different kinds of differences reflect phylogenetic diversity among various organisms. Intensity of this phylogenetic diversity is actually total length branches which emerge from an evolutionary tree (phylogeny) of organisms, which is a link between different organisms. Sampling of phylogeny at different places means presence of different things. There are many methods to measure biodiversity but the most commonly used one is species richness. This method of biodiversity measurement may be chosen for a number of reasons (Purvis & Hector, 2000; Gaston & Spicer, 2004), including the following:

1. Species have independent evolutionary routes and unique histories because they retain their own genes. This idea carries biological importance as to why we measure species richness instead of higher taxonomic grouping.

2. Considering different measures of biodiversity counting of the number of species is considerably easier than other measures sometimes used. Whenever there is variation in species richness, we can envisage that there is variation in biodiversity.
3. Extensive information about species is available in herbaria and museums.
4. Species abundance can substitute other different measures of biodiversity commonly used. If a moderate number of species is present in a particular area, then there will be more genetic diversity. Accordingly, if there are more habitats or biomes then there will be greater ecological diversity. However, using richness as a measure of biodiversity also has some drawbacks. First, the number of species counted is always dependent on the species concept understood and adopted during these measurements. The following example further clarifies this phenomenon. If we use the biological species concept, then in any country species of birds of paradise are 40–42, but if we apply the phylogenetic species concept then this number increases to 90 species (Gaston & Spicer, 2004). Another disadvantage of using the species abundance is that it represents just one element of biodiversity, which does not fulfill the requirement because we must consider this distribution three-dimensionally over which species richness can be considered, meaning that life can be distributed between different spatial levels including alpha, beta, and gamma diversity. The first level, alpha diversity, gives us the biodiversity present in a particular area, community, or ecosystem. The second level, beta diversity, compares the number of species present between different areas and compares those species which are unique to these particular areas. The third level, gamma diversity, is quite general and measures the overall diversity across a region (Gaston & Spicer, 2004).

1.3.4 Mapping of Biodiversity: Species Richness Estimation

It is universally accepted that as the ecosystem area increases there is a proportionate increase in the number of species present in that area. Mathematically, this area to species relationship is commonly expressed in the formula:

$$\text{Log } S = \log c + z \log A$$

where S represents the total number of species in that particular area, "A," "z," and "c" are constants and represent a particular relationship known as the Arrhenius relationship. This type of relationship tells us that there is about 50% variation in species richness between different areas with a slope of "z." This "z" slope ranges from 0 to 0.5 (most commonly 0.25 to 0.30). The mathematical relationship shows that if a given area decreases by up to 90%, then 50% of the species will become extinct and if this decrease is up to 99%, then 75% of the total species present in that area will become extinct. The constant "z" varies widely; nevertheless, this variation depends on whether the population is on an island or a continent, when latitudinal distribution as well as the size of any area where a particular population of species is present (Gaston & Spicer, 2004). It is clear that there is a direct relationship between the area and the number of species present: larger areas always have more species.

As discussed above, Myers et al. (2000) introduced the concept of biodiversity hotspots, that is, areas which are abundant in endemic species with habitat loss at a relatively accelerated rate. Biodiversity hotspots are important areas, which is obvious from their name; hence, their conservation is always a priority. Several conservation organizations like Conservation International suggest conservation of these sites on a priority basis. It is generally recognized that biodiversity hotspots hold at least 1,500 endemic plant species but, unfortunately, these hotspots have lost at least 70% of their original habitat area, with a concomitant loss of proportionate numbers of animal or plant species present there. Well known hotspots around the world include islands in the Indian Ocean, Madagascar, the coastal forests of Eastern Africa, islands in the Caribbean, and the Mediterranean Basin, with a total area of 2.3% of the earth's land surface collectively (Conservation International, 2024).

1.4 BENEFITS OF BIODIVERSITY

1.4.1 Direct and Indirect Use

Environment is very important to us because we use it daily, both directly and indirectly. Direct usage of the environment includes simple pleasures such as paying a visit to the countryside, walking in the woods, or hiking across hills. Ongoing natural processes in the environment, including improvement in water quality through natural filtration and regulation of climate through the absorption of CO_2 from the atmosphere, also provide beneficial services to human beings. These services are indirect in nature because most people are not aware that the environment is also helping them on a daily basis. Wetlands are a good example of human beings benefiting from them both directly and indirectly. These wetlands host a variety of birds and wildlife, consequently attracting large number of visitors.

Wetlands also purify water and help to control flooding through storage of excess water. If wetlands did not exist, considerable expenditure would be necessary to pay for purification of water and building the infrastructure necessary to control potential flooding. Although people rarely recognize benefits such as these, wetlands benefit humans indirectly on an ongoing basis.

1.4.2 Agriculture

Agriculture and livestock originated from biodiversity and maintenance of biodiversity will continue to enhance food production and other agricultural commodities, thereby ultimately ensuring food security. Biodiversity generates and promotes variations in agriculture and/or in livestock. In return, agriculture and livestock conserve biodiversity and promote their sustainable usage. Nevertheless, changes in both agriculture and livestock practices could result in drastic biodiversity loss if sustainable measures are ignored. Farmers and agricultural producers are the real owners of agricultural biodiversity and have all the required skills and knowledge for its sustainable management. Where sustainable agriculture promotes biodiversity, biodiversity in turn promotes agriculture. Sustainable agriculture uses water, land, and nutrients in such an efficient way that it impacts our society both economically and socially in a positive way (Brady and Weil, 1996). In the scenario given above, only then can agricultural producers actively and purposefully respond to consumer demands and government policies.

Although land hold the nutrients required for crops, artificial fertilizers and pesticides are still heavily used to enhance agriculture productivity. However, despite all these efforts the full potential of soil cannot be exploited without the provision of essential ecosystem services. This means that a huge variety of organisms present in soil are essential for the breakdown and recycling of nutrients present in soil. For example, earthworms can be named as a keystone species which fertilize the land by recycling nutrients. Insects pollinate a wide variety of crops including forage plants, and other grain producing crops. Honey bees (*Apis mellifera*) and bumble bees (*Bombus terrestres*) are vital for pollination. Bees collect nectar both for the benefit of an individual bee and for the colony and visit more flowers than any other insect. Robinson et al. (1989) have estimated that in the United States alone 31% of the value of production from agriculture is dependent on bees, whose very conservative value of production over €9 billion per annum, while a bolder stance is taken by Morse and Calderone (2000), who value agricultural production attributable to bees at a remarkable €14.6 billion per annum. Similarly, (Katumo et al., 2022) estimated that natural pollination benefits Canada to the tune of C$1.2 billion per annum. The benefits of biodiversity do not stop there. Biodiversity also helps in pest control, without which productivity losses would be far greater than normally envisioned. However, the excessive use of artificial inputs, non-native alien species, pollution, removal of semi-natural habitats, or the use of heavy machinery threaten these ecosystem services. When inappropriate farming methods destroy or impair balanced biodiversity, expenditure on artificial inputs increases and the prospects for sustainable agriculture are diminished. On the one hand, expenditure on agriculture increases in terms of external inputs to offset the effects of ecosystem imbalances and, on the other hand, it creates pollution due to excessive use of fertilizers and pesticides which can ultimately affect human health. Therefore, the value of biodiversity can be assessed where it assists in the moderate use of external inputs for agriculture, together with maximum output from sustainable ecosystems.

1.4.3 Forestry

Similar to agriculture nutrient recycling, pest control helps forests achieve their proper growth and survival. Forests are a habitat for many wild animals, which in turn can provide hunting opportunities. Forests also produce wild food in the form of fruit or fungi. Some forests are also a source of recreation for visitors. With the intensification of food production, the onset of pollution, and the introduction of alien species, the health of forest ecosystems can be threatened and attacks by serious pests are yet another threat to the survival of forests.

1.4.4 Fisheries

Oceans, rivers, and lakes provide an excellent ecosystem for the survival and nourishment of fish, as well as a source of food from fishing. Although fish are gathered directly from aquatic resources, they are not produced spontaneously in water; rather, the growth and production of fish depends on a healthy and functional ecosystem providing species for predators, nutrients for the growth of small organisms, habitats for living species, and good quality water for maintenance of the health and survival of fish. Pollution, over-exploitation of fishing permits, existing aliens species, and destruction of habitat for living and breeding of aquatic organisms threaten marine as well as freshwater biodiversity. It is hard to estimate the quantity of waste that aquatic life assimilates from rivers and sewerage outflows but they are certainly substantial. Although the benefits of fish biodiversity are huge, it has still not been possible to subsidize policy regarding the over-exploitation of fisheries. Although very little effort has been expended on the conservation of marine biodiversity, presumably the lack of real political effort and the determination of the people involved in this business have always been the principal constraints.

1.4.5 Water

In an aquatic environment, biodiversity recycles the nutrients and ensures and maintains the appropriate water quality for fisheries, agriculture, and human consumption purposes. In addition, aquatic flora and fauna assimilate human and animal waste, as well as industrial pollutants. Aquatic habitats thus mitigate floods, facilitate recreation, and provide many other amenities of life. Therefore, while depending on water we should be well aware of the impacts of any degradation through excessive human pollution. It is important to make a clear-cut distinction between the quality of ecosystem service and the intensity of external pollution.

1.4.6 Human Welfare

Biodiversity contributes a great deal to human welfare, either through eco-tourism or natural sightseeing or simply through the matching connotation between attractive biodiversity rich environments. Water sports and angling are some other roles which biodiversity provides.

1.4.7 Health

The relationship between biodiversity and human health is just the starting line to our understanding of this complex issue. It is well accepted that a functioning and healthy ecosystem with well-balanced biodiversity contributes to the provision of clean water quality and nutrient-rich food for human health and wellbeing. In addition, biodiversity also controls many diseases and their vectors inhibiting their further proliferation and outspread. Nevertheless, diseases have created havoc where the natural system has been disrupted by human interference. Bird flu can be quoted as an example. Biodiversity also contributes to the extraction and preparation of many drugs. Good health has a utility benefit that probably exceeds that of any other sector. The potential savings on health expenditure due to high environmental quality are equally great. The routes through which biodiversity contributes positively to health are too indirect and multidimensional to quantify. Nevertheless, at the same time they are certainly huge and deserve more attention (Fig. 1.3).

REFERENCES

Brady, N.C. & Weil, R.R. (1996). In: The Nature and Properties of Soil. University of Maryland College Park, Department Environmental Science, and Technology. Prentice Hall Inc. New Jersey, USA.

Conservation International (2011). Critical Ecosystem Partnership Fund. Crystal Drive, Suite 600 Arlington, VA 22202. Phone: 1.703.341.2400; Toll-free (within the US): 1.800.429.5660

de Vere, N. (2008). Biodiversity. National Botanic Garden of Wales Llanarthne Carmarthenshire SA32 8HG. UK EDIT Summer School 2008 Modern Taxonomy and Field Work. ndevere@gardenofwales.org.uk

Gaston, K.J. & Spicer, J.I. (2004). Biodiversity: An Introduction. 2nd Edition. Blackwell.

Harper, J.L. & Hawksworth, D.L. (1995). Preface. In: Biodiversity: Measurement and Estimation, ed. Hawksworth, D.L., pp. 5–12. Chapman and Hall, London.

Herrera Izaguirre, J. A. (2008). The 1992 United Nations Convention on Biological Diversity. Boletín Mexicano De Derecho Comparado, *1*(122). https://doi.org/10.22201/iij.24484873e.2008.122.4000

Katumo, D.M., Liang, H., Ochola, A.C., Lv, M., Wang, Q-F., & Yang, C-F. (2022). Pollinator Diversity Benefits Natural and Agricultural Ecosystems, Environmental Health, and Human Welfare. Plant Diversity 44(5), 429–435.

Kettunen, M., Brink, P.T., Dugley, N. & Stolton, S. (2013). The Social and Economic Benefits of Protected Areas: An Assessment Guide, eds. Kettunen, M. et al., Pp. 368. Taylor and Francis Publishers.

Launchbaugh, K. (2009). Why Measure Biodiversity? In: Principles of Vegetation Measurement and Assessment and Ecological Monitoring and Analysis. Lesson Series, University of Idaho, College of Natural Resources USA.

Morse, R.A. & Calderone, N.W. (2000) The Value of Honey Bees as Pollinators of U.S. Crops in 2000. Bee Culture 128, 1–5.

Myers, N., Mittermeier, R.A., Mittermeier, C.G., daFonseca, G.A.B. & Kent, J. (2000). Biodiversity Hotspots for Conservation Priorities. Nature 403, 853–858.

Purvis, A. & Hector, A. (2000). Getting the Measure of Biodiversity. Nature 405(6783), 212–219. https://doi.org/10.1038/35012221

Robinson, W.S., Nowogrodzki, R. & Morse, R.A. (1989). The Value of Honey Bees as Pollinators of U.S. Crops. II. American Bee Journal 129(7), 477–487.

van Rensburg, T.M. & Mill, G.A. (2010). Biodiversity Conservation in Managed Landscapes. In: A Handbook of Environmental Management, eds. Lovett, J. & Ockwell, D., Edward Elgar, pp. 471. Gloucester.

Vitsouek, P.M. & Hooper, D.U. (1993). Biological Diversity and Terrestrial Ecosystem Biogeochemistry. In: Biodiversity and Ecosystem Function, eds. Schultz, E.-D. & Mooney, H.A., Springer-Verlag, Berlin.

Wang, W-L., Daniel, D.O., Suman, O., Zhang, H-H., Xu, Z-B., Ma, F-Z. & Hu, S-J. (2020). Butterfly Conservation in China: From Science to Action. Insects 11(661), 2–29. DOI:10.3390/insects11100661

2 Biodiversity from Historical Perspectives

2.1 INTRODUCTION

If we want to interpret the present and future of biodiversity, it is essential to comprehend fully how biodiversity arose and what changes to it took place in the past. For comprehension of the changes in biodiversity, the first thing that we need to concerns is the different sources of information which help us in this comprehension and become the basis of this whole scenario. The next thing that we need to know is the history of life and the different patterns which took place in the history of biodiversity. What comes next is to understand the major activities that became the cause of these patterns, especially how biodiversity diversified and then became extinct. Finally, we need to know how many species are existing on the earth now.

2.2 SOURCES OF INFORMATION

There are two primary sources which provide knowledge about the history of biodiversity. These sources are, first, data from existing fossil records, and, second, the study of data which is available at molecular level about these species.

2.2.1 Fossil Records

A major part of the modern day scenery of life which can be seen on the planet earth originated from the past biodiversity, which left a rich fossil record following death and decay. This process has provided rich intuition and insight about the general history of life on earth. However, to fully understand the history of life on earth from the existing fossil is itself a challenge, for the following three reasons. First, when Darwin was justifying his theory of natural selection, the fossil record he collected and compiled did not show any precision or consistency. The record is variable in authenticity and is good for some periods while not reliable for others. Further, a number of species which have left this record range from a minimum of one and increases up to few percent of the total species that ever inhabited this earth (Sepkoski, 1992). Second, the information that is available from this fossil record is but a tiny fraction of that which has actually been explored. Third, whatever record is available is biased towards those animal species which are or have been more abundant, widespread, and are of long life. Take as an example soft body cnidarians (jellyfish, sea anemones). Seldom have any fossils of such animals been fossilized, contrary to the many vertebrates that have been fossilized (Fig. 2.1).

Soft-bodied groups of animals like flatworms, flukes, and tape-worms(*Platyhelminthes*) do not have any fossil record, while hard-bodied invertebrates such as the brachiopods and mollusks, echinoderms, and vertebrates(hard bodied), have been fossilized, although in essence that record is also not as perfect as it should be. From the total fossils collected so far the majority (95%) are from marine sources, while 5% (including both plants and animals) are terrestrial. Little information is available and those animals which can be studied are only those which have been well fossilized. Records from the 1990s show that 900–1,200 genera of dinosaurs were reported which had inhabited this earth (Dodson, 1990). Out of these, only 285 genera (336 species) were known from

DOI: 10.1201/9781003472001-2

FIGURE 2.1 Fossil of the first dinosaur ever found. (*Source*: catawiki.com from Shutterstock.)

fossilized individuals and the information regarding approximately half of the known genera has been collected from a single specimen.

It is has been further estimated that the total information collected so far about primates which existed in the past, 7% primates have been explored from fossil records (Tavaré et al., 2002). Although fossil records which are available now are far from perfect, they still provide useful pictorial information and marks on the rocks about the animals which lived on this earth in the past. Although there are several limitations in the fossil record it nevertheless helps us to understand the changes in biodiversity through geological time frames (Alroy et al., 2001; Smith 2001).

2.2.2 Molecular Evidence

Where the fossil record is helpful in terms of our understanding of the history of biodiversity, molecular evidence also has an integral role. When molecular data is compared it tells us about the relationships among species, genera, or populations and helps in the generation of new branches, ultimately establishing patterns of phylogenetic relatedness where sequences differ, this showing their divergence earlier in the evolutionary process. If fossil and molecular evidence are drawn closer together, normally they do not agree on the dates when different groups appeared on the Earth for the first time. There are contradictions between molecular and fossil evidence. Molecular evidence says that at least six animal phyla originated deep in the Precambrian period, more than 400 million years (Myr) earlier, while fossil record show the appearance of these phyla later on (Wang et al., 1999). To take another example, molecular evidence reveals that primates diverged from other placental mammals 90 Myr ago, while data taken from fossil records shows that the oldest known fossil of primates is only 55 Myr ago (Tavaré et al., 2002). Thus, it can be stated that fossil records always underestimate the actual appearance of life on earth because the possibility of fossilization of these individuals and their ultimate recovery is relatively low. It will now be important to interpret molecular divergence data and especially the form and dynamics of the molecular clock. However, when both fossil evidence and molecular data evidence are combined, they disclose many of the secrets of the origin and existence of life on earth in the past.

2.3 A BRIEF HISTORY OF BIODIVERSITY

If we take fossil records and molecular evidence together it, appears probable that all organisms discovered so far originated from a single ancestor. From the origin of life the first organism moved independently in different directions, which took place about 3.5–4 billion years ago, while the life of Earth is more than 4.5–5 billion years. This means that life has existed on earth in one form or another for almost all of its existence. Since the origin of life on earth this diversification has been continuous: life reproduces itself and then diversifies. In the very beginning, it appears this increase in life was slow, but with the passage of time it accelerated its pace, which happened with the appearance of multicellularity in organisms. These multicellular organisms possess different organs, tissues, and cells for the performance of different functions. From both fossil and molecular evidence it can be assumed that these multicellular organisms did not diversify until 1.4 billion years ago. At the time of their diversification, 60% of life on earth had already lapsed. Among multicellular organisms—specifically metazoans—there was no prominent diversification until about 600 Myr ago, when 80% of life on earth had already lapsed. Fossils of these metazoans indicate that large but soft-bodied metazoans were probably present in the late Precambrian (Vendian) rocks; the majority of them are present in Ediacaran fauna in Australia. Scientists say that they may be ancestral metazoans or those metazoans which could not maintain their life forms and became fossilized. This would only have been possible at the start of the Palaeozoic era (early life) and in rocks of the Cambrian period (550 Myr ago), where the sudden appearance of the first sizeable metazoans with hard parts can be seen (Gould, 1989). These metazoan fossils are frequently found in Canada (Burgess Shale fauna). There, we can find not only plentiful fossils but we can also see them with different body plans; some are presently experimental and appear to have lived for a shorter period of time, while others appear to have survived and can still be found today. It is generally presumed that if the metazoan had progressed—also called the Cambrian explosion until now—then 1,060 families of metazoans would have inhabited the oceans, instead of the 103 families which are actually considered and known so far (Sepkoski, 1997).

Phyla differing in their body plans means that they have conserved their own genetic makeup, which is very obvious in their homeotic genes, which regulate the expression of other genes in animal bodies—for example, the Hox genes which are present in nearly all organisms. These genes define the anterior–posterior identity of an organism in a very exact manner. If a Hox gene mutates or if a new Hox gene develops, it can drastically affect the morphological characteristics of organisms. Events like those mentioned above provide a mechanism for the early but fast evolution of body plans, which ultimately resulted in the diversification of phyla which happened on or around the Cambrian period. Further details can be seen in Figure 2.2, which shows the different timings and events which took place during the evolution of the Hox gene. These events have been charted onto a phylogenetic tree of metazoans and vertebrates. It is worth mentioning here that Cnidarians possess only anterior and posterior Hox genes (Fig. 2.2).

Additional central Hox genes can be seen in the rest of the animal phyla (the ecdysozoans or molting animals show further specialization), and further posterior Hox genes can be found in echinoderms and chordates. Vertebrates contain duplicate Hox genes. More than two Hox complexes are present in sharks and jawless fish, and teleost fish with five to seven complexes have been found in teleost fish, and four complexes are present in the tetrapods. The five-kingdom system of higher classification is the basis of the presence and distribution of Hox gene complexes, although a three-kingdom classification has also been suggested (Woese et al., 1990).

Viruses have not been included in these phyla because they are very minute and parasitic in habitat and are usually derived from others. Further, as the paleontological studies progress new phyla are continuously being discovered. In 1998 alone, some authorities reported more than 20 new divisions of bacteria at the phylum level (Fuhrman & Campbell, 1998). Nanoarchaeota, a nano-sized hyper-thermophilic microorganism, obtained from a submarine hot vent, has been the most recently discovered bacteria (Huber et al., 2002). When biodiversity began during the Cambrian period,

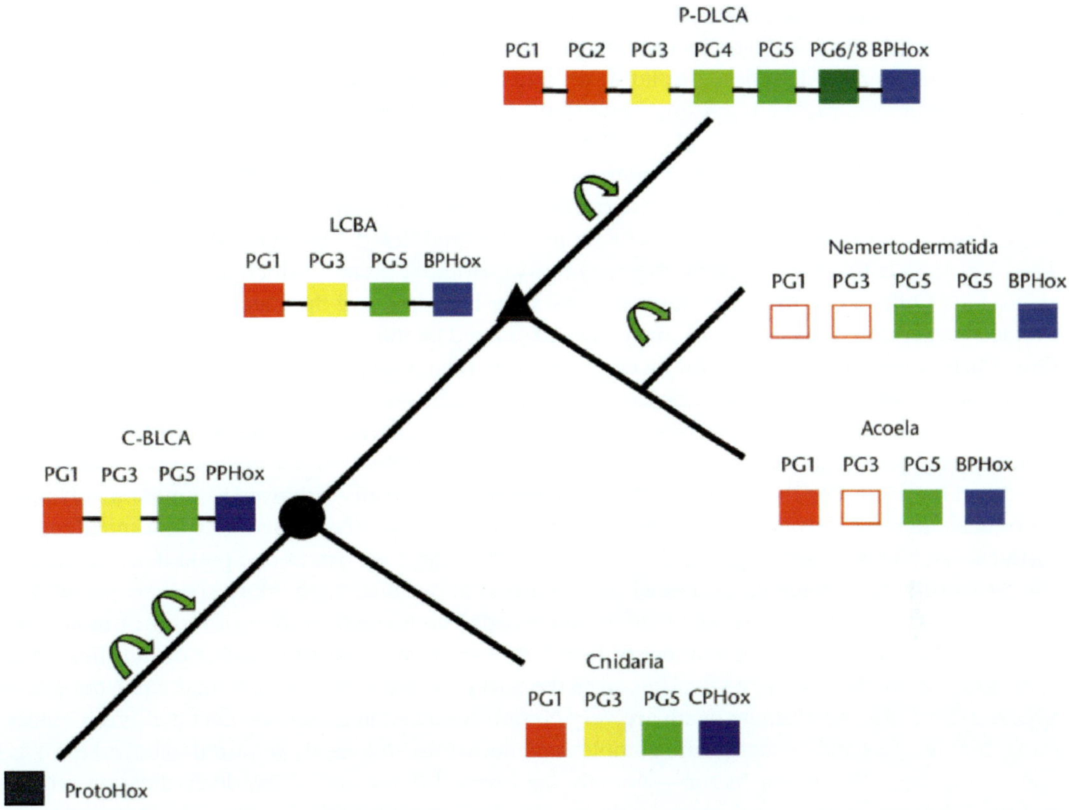

FIGURE 2.2 Hox cluster in metazoans based on recent phylogenetic studies in Cnidaria. (Courtesy of Moreno & Martínez, 2010.)

anatomical diversity had already reached its maximum (Gould, 1989). About 440 Myr ago, animals and plants colonized land before they diverged from each other, but this diversification was disguised behind the appearance of multicellular organisms in the oceans. Therefore, it can be opined that animal life appeared from fewer species in the Cambrian (early experimentation) period, having variable body plans, and continued up to the present day where there are many more species with body plans (Gould, 1989). Broadly speaking, we can say that in the Palaeozoic and early Mesozoic eras (although this has been a matter of some controversy (Signor, 1990)) there were relatively few species. However, there was a progressive and substantial increase in biodiversity starting just over 100 Myr ago but this biodiversity appeared short-lived and perished at the end of the Tertiary and beginning of the Quaternary (Pleistocene) period because there were more surviving species and higher taxa of animals and plants (both marine and terrestrial) than had been observed prior to this time period (Signor, 1990). About 5 Mye years ago the ancestors of the human family appeared from apes in Africa, and later (about 2 Mye year ago) the genus homo appeared, while the modern human being with unique structural modifications and evolution of specific features evolved relatively recently about 100,000–200,000 years ago. Presently, man inhabits in the quaternary (Holocene) period. Unlike the experience in the early evolutionary periods, there is not much diversity. This minimum change may largely be attributable to the start of human activities and onset of climatic changes. There is no agreement as to whether in a broader sense the path from a single species to many developed species can be elucidated by a simple mathematical model and, if so, what type of model can explain this problem. In part, it can be said that once a species started to diversify on a mass scale from the Cambrian period to the present day, the process was inconsistent.

Sometimes the process accelerated, while at other times it slowed down or stopped completely. Sometimes there was a dramatic increase in diversity, with periods of relative stagnancy in between (or at least no marked directional trend in diversity). Therefore, during the course of history biodiversity sometimes radiated and sometimes halted, and in between these periods, there were also mass mortalities (Signor, 1990; Sepkoski, 1992). Several families developed and grew in the marine environment and diversified in three main phases, known as the early Cambrian, the Ordovician, and then finally through the Mesozoic and Cenozoic eras. In the mid/late Cambrian, and through most of the Palaeozoic era, diversification more or less stabilized. Following these changes in biodiversity, mass extinction has been observed in five different periods which are hereby designated as the Late Ordovician, the Late Devonian, the Late Permian, and the Late Triassic, and the end-Cretaceous periods. This state of affairs can be explained by the sequential S-shaped logistic curves shown below, which show that initially the population increased slowly, then it speeded up, again it slowed down, and ultimately became an asymptote (Courtillot & Gaudemer, 1996). This phenomenon has also been clarified by a fundamental exponential curve, which states that the number of families vary greatly among themselves (Hewzulla et al., 1999). All organisms (both marine and continental)—including those families from continental organisms—showed a comparatively more continuous increase (Benton, 1995, 1997). In any case, whatever the biodiversity that inhabited the earth cannot be restricted. Supposedly, if diversification must persist, at one point it will definitely come to a full stop, although the reality seems different because many more species appeared on earth that survive today before any limitation ensued. The present number of families in any one period can be calculated as the number of families in the past+, plus the number of families developed now, minus the number of families from the past period which have perished. Such designs of appearance of life and ultimate death are very complicated and need several explanations to understand; their origins and extinction are complex and numerous studies are required to unveil the original symmetries. Two major factors—namely the internal dynamics of the diversification process and the influence of external factors—are driving these patterns and designs. The broad patterns of chronological change are reflected in local, regional, and global biodiversity. This scenario can be both informative as well as interesting. This means that as the biodiversity increases on a global scale it tends to do the same locally. The alternative to this scenario is that biodiversity remains constant at a local level but that differentiation occurs between the inhabitants of different localities giving rise to biodiversity. Now the question is why through time there is an overall increase in the pattern of biodiversity? The simple answer is that we do not have sufficient information to answer this question. It has been observed that a number of different factors is affecting these increased patterns of biodiversity, which may be attributable to external or internal factors. External factors are the change in climatic conditions, the breakup, and consequent drifting of continents. Internal factors are the occupation of more and more of the potential niche space by organisms, through evolutionary time, which is open to these organisms. Another factor of lesser importance may the further sub-division of the space occupied these organisms. When animals moved onto the land, it opened up more opportunities for these animals than they had experienced in the past.

2.3.1 Diversification

It cannot be said that the overall pattern of diversification has resulted from combined changes in the biodiversity of different groups of organisms. When they invaded habitats, some animal groups diverged differently during particular time periods or they became extinct altogether (Benton, 1997). Different animals and plants radiated in different directions and in different patterns and rates, including slow as well as accelerated rates. This can be interpreted by extrapolating from the current status of land plants and vertebrate tetrapods. The dominant primitive vascular land plants gave way to pteridophytes (ferns) and lycopsids (club mosses), which originated from primitive vascular land plants, which in turn produced gymnosperms (spore bearers), and finally angiosperms (flower bearers) replaced gymnosperms (spore bearers).

Similar to land plants, the number of successful reptile groups (including dinosaurs) originated from the vertebrate tetrapod, the early amphibians, and reptiles. The successful reptile groups simultaneously ended up in with the modern amphibians and reptiles, the birds, and the mammals. It will be really interesting to interpret how all these sequences of emergence can be translated into competitive replacement or improvement, in which one group was pushed out by the other when its numbers increased which ascended afterward. Several researchers are of the view that the rise and fall of diversity can fit in a model by modification of above-mentioned logistic model. In this model, different animal groups diversify first comparatively slowly and then speed up, and finally they attains a peak in richness at some point before declining and eventually reaching extinction. The generality of this model is not yet clear. Diversification may at any stage nevertheless be attributed to some groups of animals because not all animal groups diversify. There is a uniform pattern of diversification at all taxonomic levels. This can be further understood from the following complex examples and distributions. For instance, most of the animal species are in the kingdom Animalia. Most of the organisms though present in phylum Arthropoda fall under Animalia. Insect takes most of the organisms from phylum Arthropoda. The orders Diptera (flies), Hymenoptera (ants, bees, and wasps), and Coleoptera (beetles) take most of their species from the class Insecta. Similarly, order Rodentia (rodents) contain most of the animal species which fall under class Mammalia. Now the family Muridae takes most of the animal species which fall under order Rodentia and most of the animal species which fall under the *Muridae* are in the genu which is actually the largest in that group.

Presumably, there are three main explanations for this plotted pattern of diversity. First, the things developed in the past and during the classification of animals they were grouped considering any biological basis and background. Its authenticity is doubtful because in actuality differences among the organisms are real and reflect the evolutionary relationship among organisms. Secondly, there is a possibility that evolutionary patterns are happening by chance only. This matter of chance may result in many species in fewer groups or the same may happen in many groups which follow a specific evolutionary model, which is hereby designated as model of rapid speciation and extinction. Nevertheless, this model can be further elucidated by the following example. One family gives rise to one species which in turn produces four species. At all branching points of this splitting, there is one species which produces two offspring. This type of splitting is also called dichotomous divisions or separations. At the beginning, a parent species divides into two discrete species. In this splitting, possibly two or three species outcomes are expected. When these three species further form species, it can result in 6 four-species outcomes. This pattern of speciation is gradually repeated for bigger and bigger number of species. This shows that evolutionary trees grow very rapidly. If we follow the same rules of division and splitting, there can possibly be as many as 87,178,291,200 trees and ultimately 15 species. To understand the patterns of diversification, models of random speciation and extinction can help but they are universal. During this process, some groups develop and have disproportionately more species than can be the outcome of chance. This phenomenon leads us to the third possible reason for the pattern observed, why some animals disperse and diversify disproportionately. The reply is quite simply that some animal groups have some inherent unique features that predispose them to such conditions, which help them to diversify successfully in the given environment. Therefore, it is normally suggested that dispersal of animals from one environment to another helped the dispersal of vascular plant groups, the flying capability of insects helped them to diversify in different environments, and similarly the small body size of some birds facilitated their diversification. Although proposals are put forward in routine matters they are hard to testify. Gould and Lewontin (1979) used a phrase—"just-so" stories—which they borrowed from Rudyard Kipling's 1902 book of the same name, where they describe the reason that some animal species possess unique traits of their own kind which is either experimental or can no longer be examined or tested. They further tried to state why one group is more diverse than another, but they could not provide any sound empirical support. Nevertheless, there is a possibility that evolution of some traits provided opportunities for some groups to diversify more than others in a very disproportionate manner. Therefore, it can be said that adoption of phytophagy in insects ("plant eating") is the result of their disproportionate diversification (Mitter et al., 1988), while on the other hand the adoption of a carnivorous parasitic lifestyle is not attributed to this pattern of diversification (Wiegmann et al., 1993). Historically then, diversification can be called a specialization based either on pollination or dispersal and consumption (Thompson, 2002). If the pattern of biodiversity is followed by chronological changes which predict that speciation and extinction are occurring at different rates or if more species develop than are becoming extinct, this will increase the level of diversification. If the generation of species is equal to extinction, there will be pattern of stability. Stable diversity does not translate any change in population; however, over time individual outputs of different taxa will be on the rise. If the level of extinction is higher than the rate at which species are generated, then biodiversity will decline and if no more species are generated for a longer period of time then life may perish on the earth. It is estimated that approximately 99% of species have become extinct over the history of life on earth. The variety of groups from both marine and terrestrial sources witness that the average life span of a species is about 5–10 Myr (May et al., 1995). Raup (1994) used higher taxonomic units and observed that life spans of 17,500 genera of fossil marine animals have longer life spans and, when drawn in the form of a curve, their lifespans were strongly positively skewed or can be said to be right-skewed (Fig. 2.3).

This means that most of the genera survived for a shorter duration, while fewer survived for a longer period. The real pattern of this distribution may be more skewed because the animals which lived for only a very short duration are hard to record. Other than genera, this pattern can also be

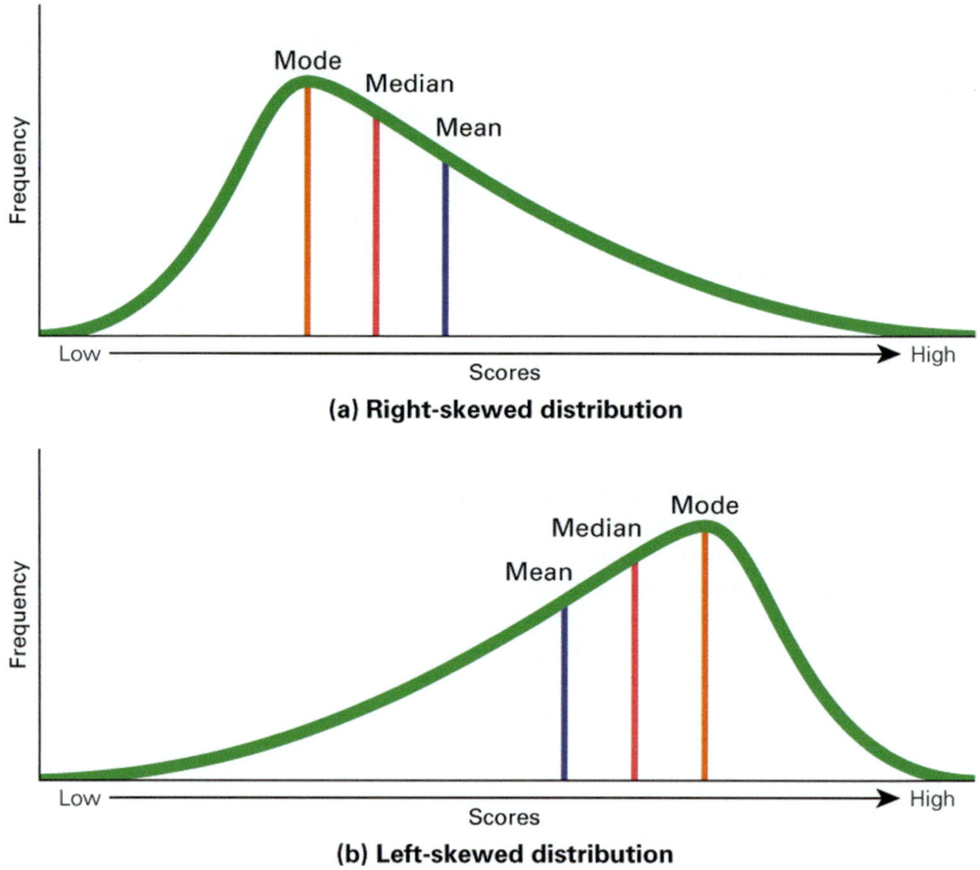

FIGURE 2.3 Curves with (a) right-skewed and (b) left-skewed patterns. (Courtesy of R. Gupta Published in Geek Culture November 9, 2022; What are right-skewed and left-skewed distributions? Statistics Interview questions.)

applied to species. When we compare life on earth, none of the genus survived for a longer period of time. The life of the longest-lived genus is approximately 160 Myr. This figure represents about 5% of the history of life. When different animal groups are compared for their survival, there is no hard and fast rule because some groups possibly have higher rates of extinction than others. Therefore, there is a great deal of variation regarding the life expectancy of different species present in different taxonomic groups. Therefore, it can be said that natural extinctions vary with the nature in which animals are presently placed. Hence, extinction is not uniform and is lower and disproportionate within species-poor groups. This shows that there is a loss of more genetic diversity than is further expected during the process of extinction. There are several other extinctions which are particularly attributed to human activities which can also be slow with major changes. This means that at present different birds like parrots and pheasants and mammals like primates are facing extinction without any definite proportion. These variations might be due to the prevailing external factors acting on the process of extinction and survival of different animals. So, in this context it has been observed from the available fossil records that marine groups have lower rates of extinction than do terrestrial groups. The probable reason of this variation can be the creation of shield-like structure for animals, which helps animals to resist changes emerging in the environment. An intrinsic factor may also be one which becomes the cause of vulnerability of some species over others. Now, the relationship of species with intrinsic factors and the possibility of extinction is very much dependent on extrinsic factors which are a persistent threat to the survival of the species. The rate and intensity of extinction always vary, being lower in some time periods and higher in others. Presently, it is estimated that in the short durations 75–95% of species which were alive in the past have now become extinct (Jablonski & Raup, 1995); nevertheless, the share of mass extinctions is only about 4% of all extinctions in the last 600 Myr (Raup, 1994) (Table 2.1). Their importance may therefore be attributed to the disruptive effect they have had on the patterns of development of biodiversity in a given area and time interval. These factors tell us that both terrestrial and marine animal life is not entirely resistant to environmental factors but can totally collapse on account of the occurrence of certain environmental stressors (Jablonski, 1994).

When biodiversity recovers, then an animal population becomes totally different at a given area other than that which was present when extinction started. Thus, the groups of animals which were considered very successful totally perished or significantly reduced in number. Mass extinction

TABLE 2.1
Different Taxons and Their Extinction Duration

Taxon	Species Duration
Marine bivalves and gastropods	10–14
Benthic and planktonic foraminifers	20–30
Marine diatoms	25
Trilobites (extinct)	>1
Ammonites (extinct)	Approximately 5
Beetles	>2
Freshwater fish	3
Snakes	>2
Mammals	1–2
Bryophytes	>20
Higher plants: herbs	3–4
Shrubs, hardwoods	27–34
Conifers, cycads	54

Source: Data from Levin (2000).

of these animals cannot be considered as a coincidence, and in fact the "big five" mass extinctions experienced previously had different reasons than scientists normally believe (Erwin, 2001; Hendricks, 2018).

2.4 HOW MANY EXTANT SPECIES ARE THERE?

The question here is if biodiversity was on the increase through evolutionary time then what is number of species living now. Then next question is how the biodiversity which is surviving now is distributed among the groups of different organisms and across the earth. It is generally accepted that an ideal method to know the number of species on earth which are currently existing and surviving are just to count them. Considering the volume and magnitude of biodiversity, counting animals will be a really difficult task. Despite the availability of resources and advancements in technology, it is still a formidable task to answer it properly. Some researchers are of the view that this task might be accomplished in decades but others do not believe it can be achieved. Considering the nefariousness of such a task, so many attempts have been made but all are indirect measures with several assumptions of type or another or another (Stork, 1997). To count the number of existing species on earth in large taxonomic groups or in all animal groups, five different methods have been used, which are based on extrapolations from:

1. **Canvassing experts**: This method is based on the opinion and estimates of those who have understood and studied the number of animals known to science. Although this method gives an overall estimation of number of species it is not well tested. However, despite this it is still considered to be a reliable estimate because it is based on the knowhow of experts who have worked on this task.
2. **Patterns of species description**: This method estimates the overall number of species based on future production of production and growth of taxonomically described species through time. It means that past patterns of growth predict the future growth of that particular species.
3. **Proportion of undescribed species**: This approach estimates the overall number of species by counting individuals from the ratio of previously known and unknown species. This method takes a large sample size and assumes that the samples taken are representative of the species being counted.
4. **Well-studied areas:** In this approach, species data obtained at global level or from bigger regions is extrapolated to smaller areas whose number of species are reasonably known. In this approach, it is assumed that areas which are inhabited by well-known numbers of species well represents those species that have not been counted before.
5. **Well-studied groups**: This involves estimated counting of overall numbers of species based on the assumption that ratios of numbers of unknown and known species remain reasonably constant across space, although all these methods consider several assumptions which are rarely met. Past and current estimates show that the estimated number of existing species is around 13.5 million and ranges from about 3.5 to 111.5 million species (Hawksworth & Kalin-Arroyo, 1995). Given the number of species (13.5 million), this is hard to imagine. At a global level one species is equivalent to 450 people. Nevertheless, it is always debatable if it helps to understand the extraordinary level of biodiversity. This counting of species for a particular taxonomic group including viruses, bacteria, fungi, nematodes, mites, insects, habitats, or biomes etc. is always a rough estimation. The estimated number of organisms in a given category of animals is set out below:
   - **Bacteria**: Standard species concepts cannot be applied to this taxonomic group and hence it is hard to estimate their number. The culture of bacteria, their proper identification, and then counting the huge quantity of these organisms is not an easy task to deal with. (The globally estimated number of prokaryotes is $4–6 \times 10^{30}$ cells.) It is worth mentioning here that their production rate is 1.7×10^{30} cells in year 1 (Whitman et al., 1998). Depending

on the estimation approach for bacteria, the number can vary several fold even if they are present in only a small area (Curtis et al., 2002). Globally, it is well recognized that the bacterial population is very diverse at present and is in the millions of taxa, rather than was thought in the past (Fuhrman & Campbell, 1998; Torsvik et al., 2002).

- **Protozoa:** In a very small area, there may be many species of protozoa. Some groups may be distributed in a wider geographic area and the number calculated in a local area may not be representative of global counting (Finlay et al., 2002). It is presumably true too because 85 ciliate species have been found in a volcanic crater-lake in Australia but interestingly none of this species was found to be an exclusive inhabitant of that continent (Finlay & Fenchel, 1999). Free-living ciliate species number just 3,000, while number of free-living protozoa number approximately between 12,000 and 19,000 (Finlay et al., 1998; Finlay & Fenchel, 1999).
- **Fungi:** From temperate studies it has been extrapolated that 1.5 million species of fungi are presented, which are also cited where required (Hawksworth, 1991). Researchers working in the tropical environment do not agree with this figure and say it is a substantial underestimation than actually present (Arnold et al., 2000). Nevertheless, in an area where fungus species have been fully explored, then it will be a significant overestimation. Fungal species are so diverse that even three individual plant leaves (two dicotyledonous and one palm leaf) from the Neotropics collectively may contain 108 foliicolous lichen species. This number is 25% of the total fungal taxa discovered from that area (Lücking & Matzer, 2001). Among fungal groups, only lichens maintain a mutualistic relationship between a fungus and an alga or cyanobacterium.
- **Nematodes:** Cobb (1914) opined that nematodes are so diverse that if we remove all the matter from the surface of earth we will still find that all the oceans, valleys, rivers, mountains, hills, and lakes will be covered with a film of nematodes. Recently, 1–100 million surviving nematodes have been reported, but some scientists consider this to be an extreme overestimation.
- **Mites:** For a long time, mites have been considered as a very diverse group of organisms and some say that their richness is comparable to taxonomic groups of insects (Walter & Behan-Pelletier, 1999). Less than 50,000 species of mites have been described, although the estimated number of mites falls in the hundreds of thousands (Walter & Behan-Pelletier, 1999).
- **Insects:** Like most other animals, insect counting is also overestimated which is not reasonable (Novotny et al., 2002a), although these estimations are honored in some quarters. This uncertainty is more prevalent on the numbers of insects present in the tropical rain forest canopy. These species of insects are very much restricted to this environment. Host specificity of herbivorous insects in these forests inflates the actual number of insects present there (Novotny et al., 2002a, b). It is still debatable which order of insects contains more species. Some researchers opine that Coleoptera (beetles) live in a tropical environment but experimental evidence is not available. It is further added that the majority of insect groups are parasites, which indicates that parasitism is a common lifestyle of living organisms on earth and hence the number of free-living species is far less than parasites (Windsor, 1998). Parasite biodiversity is not well understood because parasites are among the least explored organisms. It can further be clarified that human beings serve as a host to probably several hundred parasitic species, including insects, nematodes, and microbes.

2.4.1 Plants

Angiosperms (seed plants) have been counted to have up to 250,000 existing species, but most scientists do not agree with this figure and are of the view that these species may have as many as 300,000 species, while others say that the figure may be more than 400,000 species (Bramwell, 2002). Small areas may provide some reliable information but estimation in bigger areas still relies on assumptions and extrapolations. Sometimes the authenticity of such estimates becomes liable to two types of errors: (1) either different species are given the same name; or (2) the same species are given different names. The former is called synonymy, while the earlier is called homonymy. For example, out of 59 new species of mammals described—the Neotropics—between 1992 and 1998, two were Coendou *koopmaniand*, and *Thryoptera robusta*. These two species had already been named in 1997, and hence became the synonyms of those species which had already been named previously (Patterson 2000). Similarly, 20% of the names of the surviving known species are synonyms. Synonymy is very common in those insects which have distinct individual variations, are well studied, and have larger geographic ranges (Gaston & Spicer, 2004). Some groups of fungi and mollusks also experience the phenomenon of synonymy (Altaba, 1996; Aptroot, 1997; Bouchet, 1997). Due to this problem in North America, 24–31% of total currently accepted mammal names have been rejected (Alroy, 2002). Thus, for the Neotropics, since 1982 there has been an increase of only 150 species because 57 species were found to be synonyms, only 57 new mammal species have been described, and about 150 species have been revived from previous synonymy (Patterson, 2000). Other than species described earlier, 36 species are described every day with the constant over the past decades (Hawksworth & Kalin-Arroyo, 1995). There are 1,100 known species of viruses, 4,600 Monera (bacteria and similar forms), 74,000 insects, 290,000 plants, 23,000 algae, 30,000 protozoa, 66,000 fungi, and 280,000 other animals.

Nevertheless, a group of organisms a taxonomist may choose to study is just matter of chance or his/her personal preference or selection is driven by practical concerns; for example, the role of that particular organism in human affairs (agriculture, medicine) or researcher is seeking for research funding (Bouchet, 1997). Therefore, most of the animal species studied so far are comparatively more abundant locally or regionally, have a wide distribution, are of a large body, occupy a larger number of habitats, and derive a proportionate share from temperate zones, although some groups have collapsed (Bouchet, 1997). Previously, the description of the species did not ensure complete knowledge about them. This example elaborates this statement further. Indeed, 40% of described beetles were collected from a single locality (May et al., 1995), which reflects that they were captured many decades ago at one time(Bouchet et al., 2002).

There is always a gap between described and existing species, which is due to ignorance of small-bodied invertebrate taxa. Nevertheless, there is continuous discovery of new vertebrate species. For example, in class *Pisces*, lower vertebrates, every year between 130 and 160 new species are described (Berra, 1997). Similarly, in Africa 48 new species of birds were discovered between 1946 and 1995.

Taking this trend at a global level, 151 species of birds which became extinct between 1980 and 2002 have recently been described at an average rate of six or seven every annum; still others are awaited for description (van Rootselaar, 1999, 2002). Most of these species have been identified and described using traditional taxonomic approaches and techniques have been used for the description of the majority of these species. There are, however, prominent differences between morphological and molecular species descriptions. The number of morphologically described species is far lower than molecular techniques have revealed (Martin 1996; Price 1996). Molecular techniques say that probably 20,000 birds species are living now on the earth, a number which is double that described morphologically. Similarly, 16 new species of mammals per decade have been described from 1937 to the early 1990s (Pine, 1994). Those discovered and described so far include: four beaked whales (*Tasmacetus shepherdi*, *Mesoplodon ginkgodens*, *M. carlhubbsi*, *M. peruvianus*), a wild pig (*Sus heureni*), a peccary (*Catagonus wagneri*), four deer (*Mazama chunyi*, including gazelle (*Gazella bilkis*), "bovid" (*Pseudoryx nghe-tinhensis*), *Moschus fuscus*, *Muntiacus atherodes*, and *Muntiacus gongshanensis*), wild sheep (*Pseudois schaeferi*), the kouprey (*Bos sauveli*), and two porpoises

(*Lagenodelphis hosei*, *Phocoena sinus*). On the basis of historical patterns of species accumulation, Medellín and Soberón (1999) are of the view that that by the year 2032 an additional 247 mammal species will have been described, which will be above the total of 4,628 species counted in 1992. The majority of the new species will be small bodied (<100 g) and will belong to the order Rodentia, Chiroptera, and/or Insectivore. Considering the present rate of extinction, these species may perish before they are even described.

2.5 BIOGEOGRAPHIC DISTRIBUTION

2.5.1 Biological Diversity in Pakistan

Pakistan is located between the Palearctic, the Oriental, and the Ethiopian zoogeographical regions. In Pakistan, there are rapid changes in altitude that affect wildlife and vegetation. Protected Areas Management Proposal IUCN (1997) says that Pakistan has the world's rarest and unique plants and animals. The endangered snow leopard, *Uncia uncia*, famous for its spotted coat and survival skills, inhabits the mountainous region of the north which is. In the south is found an endemic threatened Indus dolphin, *Platanista minor.* However, its survival is now at risk due to construction of barricades on the Indus River.

Human settlements in Pakistan started a long time ago. Records of early human civilization have been found in Harappa and Mohenjodaro from around 300 BC. Before the start of the 20th century, human beings lived in relatively ecological harmony, which negatively impacted the environment very recently. At the turn of this century, many species which were common have now disappeared. In the past 50 years, there has been a tremendous increase in the population in Pakistan. This population has been forced to clear land for industry, agriculture, and human settlement. There are further developments in irrigation systems with construction of more dams for tapping water from the Indus River and its tributaries. A reduction in riverine swamps and forested areas in the flood plain and shrinkage in thorn scrub has been experienced due to the control of monsoon flooding. These activities have threatened many species because of overuse, loss of habitat, and the onset of pollution. For conservation of these species and to minimize further losses, an extensive protected area system has been developed to include in it all the sensitive areas from the wildlife point of view to protect further habitat degradation. Species-wise presence and distribution of different animals is briefly given below.

2.5.2 Mammals

Six mammal species are endemic in Pakistan. Of these, the Indus dolphin is currently endangered, while the woolly flying squirrel, *Eupetaurus cinereus*, is only found in the northern mountains of Chitral and Gilgit. The other endemic species of Pakistan are the four ungulates: goitred gazelle, *Gazella subgutturosa*, Marco Polo sheep, *Ovis ammon polii*, markhor, *Capra falconeri*, and urial, *Ovis orientalis*, Balochistan black bear, *Ursus thibetanus gedrosianus*, and the snow leopard. These mammals are also included in the list of threatened animals.

2.5.3 Birds

There is no bird species which is fully endemic to Pakistan. The Indian bustard, *Ardeostis nigriceps*, and two species of pheasant, the cheer pheasant, *Catreus wallichii*, and the Western tragopan, *Tragopan melanocephalus*, are presently in the endangered birds list. The Pallas Valley of Kohistan and the Neelum Valley of Azad Kashmir contain significant populations of Western tragopan. Pakistan is the second largest host to the white-headed duck, *Oxyura leucocephala*, during the winter, although during recent years there has been a considerable decline in its arrivals.

2.5.4 Reptiles

A total of 174 reptile species has been recorded in Pakistan, of which 18 (13 lizards and 5 snakes) are endemic to Pakistan. The Chagai desert contains the most distinctive heptrofauna, where there are six endemic species of reptiles. Four of these reptiles (the Olive Ridley turtle, *Lepidochelys olivacea*, the green turtle, *Chelonia mydas*, the gharial, *Gavialis gangeticus*, and the Central Asian cobra, *Naja Oxiana*) are now endangered.

2.5.5 Amphibians

In Pakistan, amphibian fauna is relatively scarce and contains only 16 recorded species. Among them all, only two species of frogs, *Rana hazarensis* and *Tomopterna strachani*, are endemic to Pakistan.

2.5.6 Fish

In the Indus River plain, the Himalayan foothills are rich in fish biodiversity. A total of 156 native fresh water fish species have been recorded so far and, of these, 17 are endemic to Pakistan.

2.5.7 Invertebrates

In all, 700 species of marine invertebrates have been recorded so far. Among them, a great deal of work has been done on mollusks and crustaceans due to their economic importance. The Karakoram, Himalayan, and Hindu Kush mountains host 41 species of endemic butterflies, including an important variety of Apollo butterflies of the genus Parnassius.

2.6 GEOLOGIC TIME SCALE

Proper recognition and comprehension of the geologic processes is important for understanding the concept of timescale, which is measured in billions of years. There are several terms used in geologic timescales which have been provided below with their brief definitions:

- Deep **time** mostly tells us about the history of earth before it was fully loaded with fossils.
- The **building blocks for life**: The chemical elements of water and heat, which are building blocks of life, appeared soon after formation of the earth about four billion years ago.
- Geologic **time.** It is estimated that Earth is 4.6 billion years old (4,600 million years). Geologic time, which is an important concept in earth sciences, actually tells us about the age of the earth, which is measured in time intervals of millions of years. Astronomical developments and advancements describe the vastness of this universe. Incomprehensible distances, which are far beyond the human experience, measure the vastness of the universe. Time, the fourth dimension, is a little different. In a geologic record that spans billions of years, we can distinguish essential and commonly used dimensions of time, such as minutes, hours, days, weeks, months, and years. The Earth's history is divided into four eons or ages, the most recent of which is the Phanerozoic eon, which has abundant fossils.
- The Phanerozoic has further three eras that have been divided on the basis of mass extinction events. Fossil records tell us about the changes in the biosphere over the most recent 600 million years of Earth's history, and also about the changes in the biosphere.

Both relative and absolute terms are used for time. For a couple of centuries, scientists used the concept of relative time to arrange geologic events in the correct sequence. Geologic time has four long eons or ages, namely the Proterozoic, the Archean, and the Hadean. There is no further subdivision

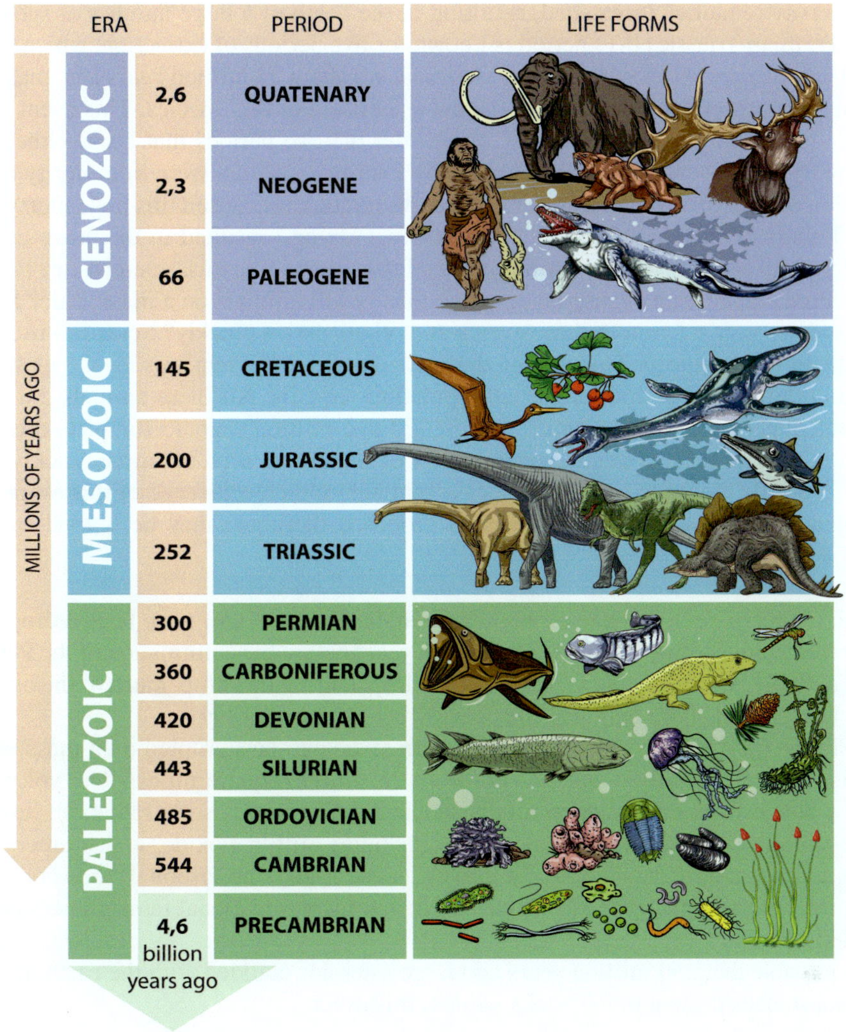

FIGURE 2.4 Geological timescale. (*Source*: Shutterstock.)

of these eons into shorter time intervals because of the non-existence of fossils. Early in this century, scientists used radio isotope methods to determine the actual age of some of the animals when they lived or were fossilized (Fig. 2.4).

At a global level, comparison of the relative ages of rocks is based on the presence or absence of fossils. In geologic time, eons are the largest intervals of time and their length is from 540 to 2,050 million years. Earth was a very hostile and inhospitable place to live in the first three eons. Due to this hostile environment, there was no life on earth at that time until the latter half of the Proterozoic eon, when preservation of a few fossils was observed. Prior to the Phanerozoic, geologists use the informal term Precambrian to refer to all geologic time.

2.6.1 Mass Extinctions

Biodiversity appeared within the biosphere at the start of Cambrian period (the oldest period). There was a progressive increase in the number of fossilized species from the Cambrian period onward and evolving organisms moved forward in time leaving behind more complex fossil forms. During this period, diversity remained threatened, resulting in the death of a large number of species (extinction). The Permian-Triassic (P-T) period is known for five periods of extinction, which significantly decreased biodiversity. At the close of the Mesozoic era about 65 million years ago, mass extinction of dinosaurs was observed, which led to the disappearance of this species. This event is known as the Cretaceous-Tertiary extinction or K-T event after the two periods that lie on either side of the time boundary. Dinosaurs were then replaced by mammals, which were in minority at that time. The collision of a large asteroid or comet with earth might have caused this huge mortality.

The P-T mass extinction, often called "The Great Dying," resulted in the death of 70% of all terrestrial species and 90% of all marine species. From events like this, scientists try to answer two questions. First, what factors cause mass extinctions or kill animals on a mass scale? Second, why are some species very tolerant to these events and others perish quickly? Scientists use P-T extinction to explain this scenario. They are of the view that the P-T event was on the global level and it affected both the marine and terrestrial populations equally. All these animals did not become extinct immediately; it took approximately 500,000 to 5 million years to complete this process. In response to these regional changes, some groups of organisms died off more quickly, while others survived a little longer. It could be the case that before the occurrence of this event some were already dying. Geologists have identified and proposed that there may be several causes of this global P-T extinction:

- With the increase in the size of world's ocean basins there may have been a drop of up to 100 meters in sea water level. A few million years before the culmination of the P-T extinction actually happened, sea water started to drop. During this drop, much of the ocean floor and the life inhabiting on it was exposed, which then died away.
- In Siberia (northeastern Russia), 251 million years ago lava erupted on a mass scale and covered over 1.6 million square kilometers (1 million square miles). The volcanic dust that resulted from this eruption totally blocked the incoming solar radiation, resulting in a short-term cooling effect. Some gases, including sulfur dioxide, which emerged from eruption might have resulted in acid rain on a mass scale. The carbon dioxide consequently emitted enhanced the greenhouse gas effect and increased global temperatures for longer periods of time.
- It is possible that 250 million years ago large asteroids collided with the earth; the chemical substances present in P-T rocks witness this event.

There is no single event which resulted in this mass extinction. However, mass extinction is clearly the result of several combined effects of several events which considerably altered the physical and chemical environments consequently killing huge animal population. Those species which could adapt themselves to these changes survived in the Mesozoic era that followed, while all others died and their remains became fossilized and became part of the fossil record for ever.

REFERENCES

Alroy, J. (2002). How Many Named Species Are Valid? Proceedings of the National Academy of Sciences, USA 99, 3706–3711.

Alroy, J., Marshall, C.R., Bambach, R.K., Bezusko, K., Foote, M., Fürsich, F.T., Hansen, T.A., Holland, S.M., Ivany, L.C., Jablonski, D., Jacobs, D.K., Jones, D.C., Kosnik, M.A., Lidgard, S., Low, S., Miller, A.I., Novack-Gottshall, P.M., Olszewski, T.D., Patzkowsky, M.E., Raup, D.M., Roy, K., Sepkoski, J.J. Jr, Sommers, M.G., Wagner, P.J. & Webber, A. (2001). Effects of Sampling Standardization on Estimates of Phanerozoic Marine Diversification. Proceedings of the National Academy of Sciences, USA 98, 6261–6266.

Altaba, C.R. (1996). Counting Species Names. Nature 380, 488–489.

Aptroot, A. (1997). Species Diversity in Tropical Rainforest Ascomycetes: Lichenized Versus Non-lichenized; Folicolous Versus Corticolous. Abstracta Botanica 21, 37–44.

Arnold, A.E., Maynard, Z., Gilbert, G.S., Coley, P.D. & Kursar, T.A. (2000). Are Tropical Fungal Endophytes Hyperdiverse? Ecology Letters 3, 267–274.

Benton, M.J. (1995). Diversification and Extinction in the History of Life. Science 268, 52–58.

Benton, M.J. (1997). Models for the Diversification of Life. Trends in Ecology and Evolution 12, 490–495.

Berra, T.M. (1997). Some 20th Century Fish Discoveries. Environmental Biology of Fishes 50, 1–12.

Bouchet, P. (1997). Inventorying the molluscan diversity of the world: What is our rate of progress? The Veliger 41(1), 1–11.

Bouchet, P., Lozouet, P., Maestrati, P. & Heros, V. (2002) Assessing the Magnitude of Species Richness in Tropical Marine Environments: Exceptionally High Numbers of Molluscs at a New Caledonia Site. Biological Journal of the Linnean Society 75, 421–436.

Bramwell, D. (2002). How Many Plant Species Are There? Plant Talk 28, 32–33.

Cobb, N.A. (1914). Nematodes and Their Relationships. US Department of Agriculture Yearbook, Washington, DC.

Curtis, T.P., Sloan, W.T. & Scannell, W.J. (2002). Estimating Prokaryotic Diversity and its Limits. Proceedings of the National Academy of Sciences of the United States of America 99(16):10494–10499. https://doi.org/10.1073/pnas.142680199.

Courtillot, V. & Gaudemer, Y. (1996). Effects of Mass Extinctions on Biodiversity. Nature 381, 146–148.

Dodson, P. (1990). Counting Dinosaurs: How Many Kinds Were There? Proceedings of the National Academy of Sciences, USA 87, 7608–7612.

Erwin, D.H. (2001) Mass Extinctions, Notable Examples of. In: Encyclopedia of Biodiversity, Vol. 4, ed. S.A. Levin, pp. 111–122. Academic Press, San Diego, CA.

Finlay, B.J., Esteban, G.F. & Fenchel, T. (1998). Protozoan Diversity: Converging Estimates of the Global Number of Free-Living Ciliate Species. Protist 149, 29–37.

Finlay, J., Monaghan, B.E. & Maberly, S.C. (2002). Hypothesis: The Rate and Scale of Dispersal of Freshwater Diatom Species is a Function of their Global Abundance. Protist 153(3), 261–273.

Finlay, B.J. & Fenchel, T. (1999). Divergent Perspectives on Protist Species Richness. Protist 150, 229–233.

Fuhrman, J.A. & Campbell, L. (1998). Microbial Microdiversity. Nature 393, 410–411.

Gaston, K.J. & Spicer, J.I. (2004). Biodiversity: An Introduction. Blackwell Publishing. P. 207.

Gould, S.J. (1989) Wonderful Life. The Burgess Shale and the Nature of History. Hutchinson, London.

Gould, S.J. & Lewontin, R.C. (1979) The Spandrels of San Marco and the Panglossian Paradigm: A Critique of the Adaptationist Programme. Proceedings of the Royal Society, London B 205, 581–598.

Hawksworth, D.L. (1991). The Fungal Dimension of Biodiversity: Magnitude, Significance, and Conservation. Mycological Research 95, 441–456.

Hawksworth, D.L. & Kalin-Arroyo, M.T. (1995). Magnitude and Distribution of Biodiversity. In: Global Biodiversity Assessment, ed. Heywood, V.H., pp. 107–199. Cambridge University Press, Cambridge.

Hendricks, J.R. (2018). Geological time. In: the Digital Encyclopedia of Ancient Life. http://www.digitalatlasofancientlife.org/learn/geological-time/

Hewzulla, D., Boulter, M.C., Benton, M.J. & Halley, J.M. (1999). Evolutionary Patterns from Mass Originations and Mass Extinctions. Philosophical Transactions of the Royal Society, London B 354, 463–469.

Jablonski, D. (1994). Extinctions in the Fossil Record. Philosophical Transactions of the Royal Society B 344, 11–17.

Jablonski, D. & Raup, D.M. (1995). Selectivity of End-Cretaceous Marine Bivalve Extinctions. Science 268, 389–391

Levin, D.A. (2000). The origin, expansion and demise of plant species. Oxford: Oxford University Press. 230 pp.

Lücking, R. & Matzer, M. (2001). High Foliicolous Lichen Alpha-Diversity on Individual Leaves in Costa Rica and Amazonian Ecuador. Biodiversity and Conservation 10, 2139–2152.

Martin, G. (1996). Birds in Double Trouble. Nature 380, 666–667.

May, R.M., Lawton, J.H. & Stork, N.E. (1995). Assessing Extinction Rates. In: Extinction Rates, eds. Lawton, J.H. & May, R.M., pp. 1–24. Oxford University Press, Oxford.

Mitter, C., Farrell, B. & Wiegmann, B. (1988). The Phylogenetic Study of Adaptive Zones: Has Phytophagy Promoted Insect Diversification? American Naturalist 132, 107–128.

Moreno, E. & Martínez, P. (2010). Origin of Bilaterian Hox Patterning System. In: Encyclopedia of Life Sciences (ELS). John Wiley & Sons, Ltd, Chichester. https://doi.org/10.1002/9780470015902.a0022852.

Novotny, V., Basset, Y., Miller, S.E., Weiblen, G.D., Bremer, B., Cizek, L. & Drozd, P. (2002a). Low Host Specificity of Herbivorous Insects in a Tropical Forest. Nature 416, 841–844.

Novotny, V., Basset, Y., Miller, S.E., Drozd, P. & Cizek, L. (2002b). Host Specialization of Leaf-chewing Insects in a New Guinea Rainforest. Journal of Animal Ecology 71, 400–412.

Patterson, B.D. (2000). Patterns and Trends in the Discovery of New Neotropical Mammals. Diversity and Distributions 6, 145–151.

Pine, R.H. (1994). New Mammals Not so Seldom. Nature 368, 593.
Price, T. (1996). Exploding Species. Trends in Ecology and Evolution 11, 314–315.
Protected Areas Management Proposal IUCN (1997), Islamabad. L E G E N D Adapted from: Roberts, T.J. 1991. The Birds of Pakistan, Vol.1. Oxford University Press, Karachi. Pages 19.
Raup, D.M. (1994). The Role of Extinction in Evolution. Proceedings of the National Academy of Sciences, USA 91, 6758–6763.
Sepkoski, J.J. Jr (1992). Phylogenetic and Ecologic Patterns in the Phanerozoic History of Marine Biodiversity. In: Systematics, Ecology, and the Biodiversity Crisis, ed. Eldredge, N., pp. 77–100. Columbia University Press, New York.
Sepkoski, J.J. Jr (1997). Biodiversity: Past, Present, and Future. Journal of Paleontology 71, 533–539.
Signor, P.W. (1990). The Geologic History of Diversity. Annual Review of Ecology and Systematics 21, 509–539.
Smith, A.B. (2001). Large-scale Heterogeneity of the Fossil Record: Implications for Phanerozoic Biodiversity Studies. Philosophical Transactions of the Royal Society, London B356, 351–367.
Stork, N.E. (1997). Measuring Global Biodiversity and Its Decline. In: Biodiversity II: Understanding and Protecting Our Biological Resources ed. Reaka-Kudla, M.L., Wilson, D.E. & Wilson, E.O. (eds.). Washington DC: Joseph Henry Press.
Tavaré, S., Marshall, C.R., Will, O., Soligo, C. & Martin, R.D. (2002). Using the Fossil Record to Estimate the Age of the Last Common Ancestor of Extant Primates. Nature 416, 726–729.
Thompson, J.N. (2002). Plant–animal Interactions: Future Directions. In: Plant–Animal Interactions: An Evolutionary Approach, eds. Herrera, C.M. & Pellmyr, O., pp. 236–247. Blackwell Publishing, Oxford.
Torsvik, V., Øvreås, L. & Thingstad, T.F. (2002). Prokaryotic Diversity – Magnitude, Dynamics, and Controlling Factors. Science 296, 1064–1066.
van Rootselaar, O. (1999). New Birds for the World: Species Discovered During 1980–1999. Birding World 12, 286–293.
van Rootselaar, O. (2002). New Birds for the World: Species Described During 1999–2002. Birding World 15, 428–431.
Walter, D.E. & Behan-Pelletier, V. (1999). Mites in Forest Canopies: Filling the Size Distribution Shortfall? Annual Review of Entomology 44, 1–19.
Wang, D.Y.-C., Kumar, S. & Hedges, S.B. (1999). Divergence Time Estimates for the Early History of Animal Phyla and the Origin of Plants, animals and Fungi. Proceedings of the Royal Society, London B266, 163–171.
Whitman, W.B., Coleman, D.C. & Wiebe, W.J. (1998). Prokaryotes: The unseen majority. Proceedings of the National Academy of Sciences of the United States of America 95(12), 6578–6583. https://doi.org/10.1073/pnas.95.12.6578.
Wiegmann, B.M., Mitter, C. & Farrell, B. (1993). Diversification of Carnivorous Parasitic Insects: Extraordinary Radiation or Specialized Dead End? American Naturalist 142, 737–754.
Windsor, D.A. (1998). Most of the Species on Earth Are Parasites. International Journal of Parasitology 28, 1939–1941.
Woese, C.R., Kandler, O. & Wheelis, M.L. (1990). Towards a Natural System of Organisms: Proposal for the Domains Archaea, Bacteria and Eucarya. Proceedings of the National Academy of Sciences, USA 87, 4576–4579.

3 Biodiversity Indicators, Assessment, and Measurement

3.1 INTRODUCTION

It is difficult to maintain biodiversity by merely assessing its condition following audition. Monitoring the actual current status of biodiversity—whether abundant or declining— tells us whether it is protected or not and how effectively it is managed (Hunt et al., 2006). Monitoring is actually repeated the measurement of biodiversity just to get an idea which species are at risk and what decisions need to be taken to ensure its integrity. Desired outcomes of natural resource management (NRM) plans are hard to achieve without the organized activity of biodiversity monitoring. Sakar and Margules (2002) stated that monitoring carries several challenges but an important one is that it is not possible to detect changes and know each and everything about biodiversity everywhere. They further stated that not all biodiversity present at different places on earth needs assessment.

3.1.1 NRM or Natural Resource Management Planning

Audition of biodiversity uses scientifically reliable indicators and measures different substitutions in biodiversity. Smyth et al. (2003), James (2004), and Hunt et al. (2006) say that monitoring of diversity actually involves production of sets of indicators but the development of methodology for the selection of indicators has been poorly worked on. However, sufficient research has been done for assessing the condition of biodiversity for making decisions for its management (Carignan & Villard, 2002; Read et al., 2005). The qualities of biodiversity indicators should be sustainable, natural, and resilient (Andreasen et al., 2001). Useful selection criteria have been developed for identification of scientifically meaningful indicators for assessment of biodiversity, which is one outcome of continuous research.

Despite all these efforts, the selection of a reliable and credible set of indicators is still a challenge. Other than that biodiversity which is regionally at risk, there is another constraint hindering biodiversity monitoring (Smyth & James, 2004). Different commotions which affect biodiversity may overlap in some areas, confounding the effect of others; however, depending upon the resilience of the ecosystem there is substantial variation in the intensity and degree of these disturbances. Such indicators will be specific for particular regions and, accordingly, they will predict the nature of biodiversity as well as the effects of different disturbances on biodiversity on existing biodiversity in that certain area (Suter, 2001). This prediction becomes not only difficult but impossible in those ecosystems which are dynamic and self-motivated. In such ecosystems, scientists may ponder about complex biophysical factors which hamper the process of monitoring and assessment rather than obtaining information regarding decision-making about biodiversity. Rather than getting non-understandable and non-predictable knowledge, under present ecological theory scientists make NRM decisions on the overall integrity of ecological systems (Fig. 3.5). In this scenario, those indicators are recommended for measurement of the condition of biodiversity which have known and predictable responses to environmental stressful factors and in a general sense have known levels of confidence (Suter, 2001). This criterion, however, is slightly different for rangelands which do not require having predictive qualities. When rangelands are monitored for biodiversity, two types of indicators are generally used, known as response and pressure indicators (Smyth et al., 2003; Hunt et al., 2006). Niemeijer and de Groot (2007) say that these indicators, including pressure, driving force, impact, and response, represent different links of series

DOI: 10.1201/9781003472001-3

of mortalities in disturbed ecosystems. Contrary to this, in modified ecosystems societal effects like land uses are more important, mediating different pressures on biodiversity. Societal effects change the habitat structure of animals, which may or may not influence biodiversity. Rather, society here plays an important role in protection of key animal species. Niemeijer and de Groot (2007) are of the view that each of the indicators should be used to assess and measure the condition of biodiversity. The monitoring and measurement of biodiversity against resource condition targets is an important activity for regional NRM plans (Hunt et al., 2006). Bioregions will have different conservation issues, outcomes to achieve, and management objectives when values of biodiversity and consequently their stressors are specific for different regions. Each result has its own peculiar indicators for measurement and sets of biodiversity values. Starting with an "An Index of Biological Integrity" (Majer & Beeston, 1996), also called simple aggregates, next is the biodiversity significance score of biodiversity incentive metrics (Oliver & Parkes, 2003), and graphical representations like radar diagrams (Suter, 2001) and Biohyets (Read et al., 2005), before gradually moving toward more sophisticated multivariate mathematical treatments (Andreasen et al., 2001). These are some of the methods and strategies that can be utilized for measurement of the condition of biodiversity.

3.2 BIODIVERSITY INDICATORS

Biodiversity indicators are actually procedures or measures to recapitulate complex data into simple, standardized, and communicable form. Due to the presence of complex concepts and processes, it is hard for people to understand them fully. In the previous chapters, it has been explained in depth that biodiversity actually tells us about different variations among life forms present both in terrestrial and marine environments, as well as in other the aquatic systems and ecological complexes these organisms come from. This diversity normally has three different forms, including ecosystem, species, and genetic diversity. Biodiversity strengthens those items which are vital for the betterment of human beings, for example, in food production. At a global level, all governments realize the importance of biodiversity and its conservation is always their priority. As well as goals and sub-targets, all of these the governments focus on seven focal areas from the point of the importance of biodiversity and its conservation.

3.2.1 Threats to Biodiversity

This area observes some of the pressures like pollution, invasive species, and climate change on biodiversity (Fig. 3.1).

3.2.2 Ecosystem Integrity and Its Services to the Society and Goods for the Society

This detects the major changes which indicate how the ecosystem is working.

3.2.3 Sustainable Use

This deliberates a sustainable use of biodiversity and its extent.

3.2.4 Access and Benefit Sharing

This monitors the mechanism of the sharing of biodiversity benefits equitably.

3.2.5 Status of Transfer of Resources and Their Utilization

This sets out the level of resources that have been reserved for biodiversity.

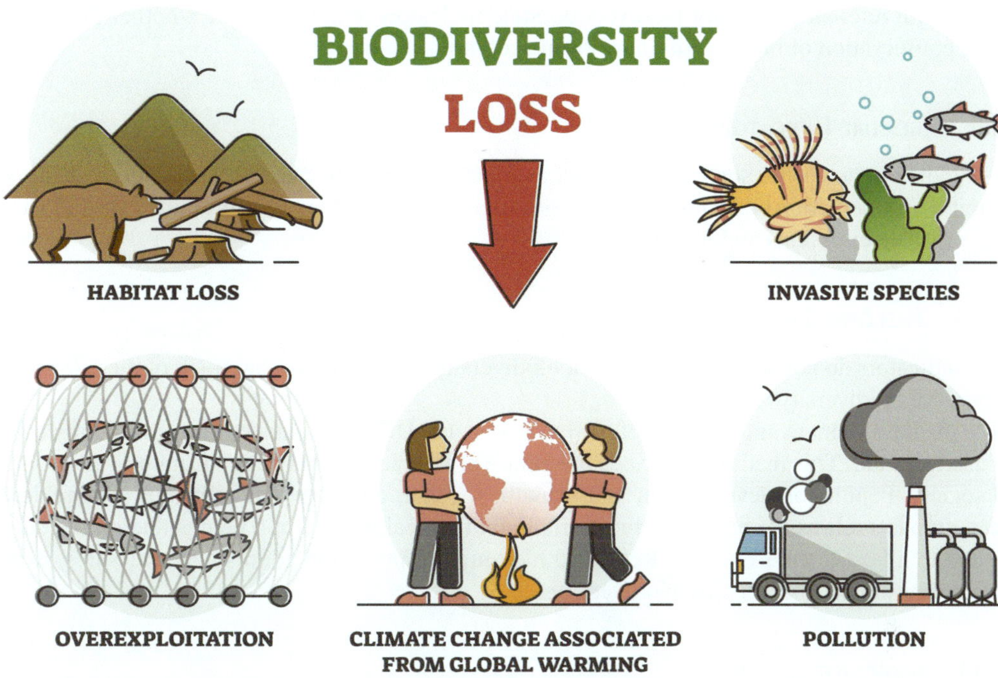

FIGURE 3.1 Biodiversity loss: issues and causes. (*Source*: Shutterstock.)

3.2.6 Status of Conventional Knowledge Innovation and Practices

This sets out the status of conventional knowledge that is currently available about biodiversity.

Public is another important component which takes interest in utility of biodiversity and its conservation. Other than public at global level all the countries develop goals and strategies for conservation of biodiversity with allocation of proper time durations and alleviation of constraints which can hinder the accomplishment of objectives and goals. Conservation of biodiversity is equally beneficial for all animals and human beings. It helps in poverty alleviation for human beings. At world level, each country focuses on its own target areas, but a list contains seven topics that are vital, including: agriculture and biodiversity; forests and biodiversity; ecological networks; invasive alien species; financing biodiversity; monitoring and indicators; and public participation and knowhow. All of these target areas focus on management of natural resources and their sustainable use. They tend to stop and totally eliminate the possibility of biodiversity decline at each and every level. Now, we will consider these biodiversity indicators and whether they are sustainable, are on the rise, or declining.

3.2.7 Types of Indicators

Indicators provide a variety of information on biodiversity. Every country has its own set of indicators, e.g., the European Environment Agency (EEA) classifies the EU indicators in the driving force-pressure-state-impact response (DPSIR) framework. Broadly speaking, however, biodiversity indicators can be placed in three main categories, namely state indicators, pressure indicators, and response indicators.

3.2.8 State Indicators

Each State sets out the quantitative and qualitative condition of the environment and indicators measure, how the abundance of species changes through time, setting out the condition of important habitats with reference to extant biodiversity. State indicators tend to indicate whether different targets of conservation of biodiversity have been accomplished or not.

3.2.9 Pressure Indicators

These indicators measure some of the factors which are currently causing loss to biodiversity such as habitat fragmentation or pollution. Changes to the environment result in pressures such as various kinds of waste emissions and land use.

3.2.10 Response Indicators

These indicators do not describe the state of biodiversity. Rather, they keep track of the efforts made to conserve biodiversity.

For the development and analysis of the indicators, the DPSIR framework is a commonly applied and used method. This method describes the nature of interactions between the natural environment and society. Population development and growth, societal, and economic changes are the driving forces. Impacts tell us the effects of biodiversity on the health of human beings and on the ecosystem due to a hostile environment. Response measures respond and address the pressures, drivers, state, and/or impacts on existing biodiversity.

3.2.11 Indicator Levels

Currently, many diversity indicators are used. Many NGOs like "WWF Living Planet Index" have developed and established their own indicators for the assessment and monitoring of biodiversity. Specific indicators work better at different scales. For instance, indicators used for species extinction are more relevant and appropriate at global levels and indicators which cover protected areas are more relevant at national levels. Every country has different indicators to assess biodiversity and these should not be the same. This flexibility allows the utilization of existing indicators and available data. It is further added that various countries have climatic, ecological, and geophysical differences which prevent them from using the applications of the same indicator. Presumably, it will be not possible to have similar indicators at different levels and scales but whatever indicators are used should certainly match these indicators and be well matched in a way that the measurements done with these indicators are comparable with other indicators at different levels and scales.

3.2.12 Targets and Reporting of Their Progress

At global level, each country has its own reporting procedures and protocols which differ from those of other countries. For example, the UK uses its "Joint Nature Conservation Committee (JNCC)" website and "Biodiversity Indicators in your Pocket." This was published in 2007 and is now regularly updated annually. Recent data collection and analysis shows that biodiversity in UK biodiversity is still under pressure of extinction and measures taken so far for conservation of biodiversity have succeeded only selectively. The global set of indicators was published in "The Global Biodiversity Outlook2" in 2006 and, as required, changes and developments are regularly updated.

3.2.13 Issues

Regarding biodiversity coverage, the set of indicators provides the best available information. Hence, the creation of a set of indicators for any biodiversity assessment is a real challenge because of the following four key factors:

- **Vagueness of targets:** If the targets for biodiversity conservation and sustainability are vague, then indicators are difficult to define.
- **Time available:** This factor tells us how much time is needed to ensure the indicator is fully developed and becomes fit-for-purpose.
- **Lack of knowledge:** This becomes an issue if we are not fully aware of the different mechanisms which make changes in some state biodiversity indicators.
- **data availability:** As the data on these indicators is often collected from academic projects or volunteers, they have limited applicability because they are based on an existing data set.

3.2.14 Choice of Indicators

It is essential that at all indicator levels need to be relevant and should be scientifically comprehensive. Currently, the best available indicators should always be used, although they have inherent strengths and weaknesses. Some already extremely useful and well established indicators are used as instruments for formation of policy. For example, the Common Bird Index indicator has wider public qualities and is scientifically accurate. Nevertheless, when we use heterogonous sets of indicators, then usefulness of these indicators becomes restricted.

It is essential for policy and management reactions that state indicators should directly relate to pressure indicators, often not currently the case. In addition, some indicators can be inappropriate for the given targets for the assessment of the condition of biodiversity. For instance, a well-known indicator known as the Marine Trophic Index (MTI) is not sufficiently sensitive and does not respond to important changes taking place in biodiversity. Further, this indicator cannot be applied on a three-dimensional scale. For example, the MTI and other analysis of trawl survey data may not show the effectiveness of different measures adopted for conservation of marine resources.

3.2.15 Common Bird Index

This Index is currently used in national policies at country level. The Index demonstrates the richness and abundance of birds over the specified time period. Here, birds act as indictors because of their widespread distribution and their sensitivity to disturbances in an ecosystem. They can therefore act as a barometer for assessing the health of the system. Success of this Index depends on the comprehensive understanding of the ecological processes, accessible data which are gathered by a network of professionals and amateurs, and its relevance to the policy and ease of its interpretation. Links between these indicators and the gaps present in them are being investigated.

3.2.16 The Marine Trophic Index (MTI)

Trends in fish catches (reported to the United Nations Food and Agriculture Organization (FAO)) is the basis of the MTI. This index depicts the rank of species in the food chain, from primary producers to large predators of the catches (Piacenza et al., 2019). This indicator discloses the changes in the average trophic levels of fish catches through time, and hence this indicator is considered as an indicator of health and diversity of an ecosystem. However, the indicator has several issues: The major one is that it is hard to reflect the state of biodiversity from the extent of fisheries catches. For estimation of the health of marine ecosystems, this indicator should be interpreted with other indicators.

3.2.17 Number of Indicators

How many indicators we currently have is a debatable issue. Decision-makers will prefer fewer indicators, while scientists prefer more and more indicators for better assessment of biodiversity. The Conference of the Parties (COP) to the Convention on Biological Diversity (CBD) and the European Environmental Agency (EEA) streamlined the indicator assessment operation and reduced the number of indicators from about 200 to the current number of 18 and 26, respectively. Dealing with huge of number of indicators is risky and creates confusion. A combination of these indicators is also not possible and hence, in order to obtain clear picture, reduction in biodiversity is hard to achieve. Careful interpretation of indicators is essential to get an overall picture. Nevertheless, when viewed in the form of a data set in the DPSIR framework, these indicators are very instrumental in policy formation and facilitate answering of different questions.

3.2.18 Value of the Biodiversity Indicators Used for Biodiversity Assessment

Indicators measure outcomes of biodiversity and define targets for national policy of any country. In addition, they focus on research and inform about specific questions and act as a communication tool by providing a basis for discussion. They are meaningful tools, complement other sources of information, and measure targets of biodiversity. Mechanisms for driving change of some indicators and their link to biodiversity are well understood. Hence, such indicators have shown more usefulness in making policy about biodiversity conservation and management.

Due to the lack of sufficient knowledge about indicators, some of the newer indicators—specifically those relating to climate change or invasive species—suffer if we do not have sufficient knowledge about them. For the evaluation of the usefulness of such indicators, more research and time is required. For other areas of environmental sciences, e.g., air pollution, a single monitoring program can work, but in the case of biodiversity several monitoring programs are applied and tested. Limited and scattered data are available with non-governmental organizations (NGOs) and some institutions can benefit each other by adopting common standards and with better coordination. The Biodiversity Research Advisory Group (BRAG) and the Environment Research Funders' Forum (ERFF) emphasize addressing this area, but a lack of coordination and the shortage of required funds is always a problem. At present, there is sufficient data for some indicators, e.g., birds but for some others, e.g., the Red List Index, only rough data is to hand for some animal species at a regional or global level. In general, data on biodiversity is very scanty in many areas, e.g., genetics. Accordingly, required information and knowledge of different ecological systems is also poorly collected and comprehended.

3.2.19 Future Challenges

It is expected that role of biodiversity indicators will increase in ecosystem service assessments, economics, and climate change. The Millennium Ecosystem Assessment (2005) has recognized the CBD adoption of the principle of the ecosystem approach, which is a useful framework and is extensively used for the assessment of biodiversity and ecosystems services (Fig. 3.2). This principle also helped in the evaluation and implementation of the potential responses from these indicators. It is expected that with the gradual increase in the value of biodiversity, biodiversity indicators will be more closely linked and will favor the ecosystem approach in their measurement and assessment functions. This activity will be directly related and will be in a fashion similar to the EcoQoS(Eco Quality of Service) in the Convention for the Protection of the Marine Environment of the North-East Atlantic (OSPAR Convention).

3.2.20 Ecosystem Approach

The Convention on Biological Diversity (CBD) suggests that for any action on biodiversity, the ecosystem approach is the primary framework to apply. For the integrated management of water, land, air, and living resources, the ecosystem approach works as a policy for the integrated management of water, air, land, and living resources and their impartial exploitation and consumption sustainably. DEFRA (the Department of Environment, Food, and Rural Affairs) says that the ecosystem approach is a model that helps in the promotion of adaptive management of the environment and

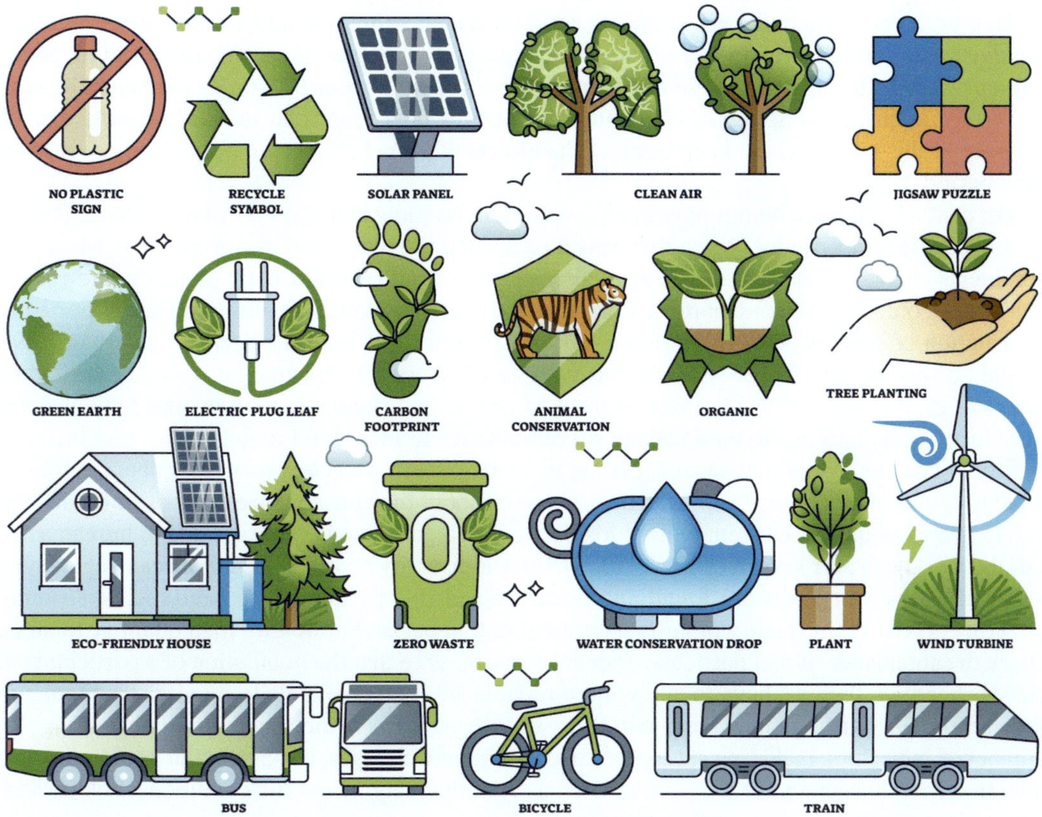

FIGURE 3.2 Mitigating future biodiversity loss. (*Source*: Shutterstock.)

reflects the value of ecosystems in decision-making. This model or approach further suggests that policy-making and delivery ensures the maintenance of a healthy ecosystem.

3.3 THE NEED FOR BIODIVERSITY ASSESSMENT

Biodiversity in the natural world on the earth is constituted and organized in such a way that its understanding in its true sense is hard. We want to monitor life in a meaningful manner by using a simple model system. Application of this model requires the selection of a few areas which are sensitive to global change and species and immediately describe the status of the system (Schipper & Rovero, 2018). In understanding ecological indicators, advancements in technology have helped considerably, which in turn has resulted in an increase in advancements in technology. These enhance our understanding of these ecological indicators and the increase the range of species distribution and the processes currently used for their measurement. Although considerable effort has been expended in the collection of knowledge regarding changes in and around the world, we are nonetheless still "counting the books while the library burns" (Lindenmayer et al., 2013). In other words, we need to devise better methods for the measurement of life and the systems which support it on earth. Ultimately, we should use this information to alleviate human impacts on the extinction of biodiversity now and in the future. We face species loss every day and some species are not even described before they become extinct. Species are perishing at such a high rate that it is hard to note them all. In this scenario, monitoring is not only required to report the extinction of species (Lindenmayer et al., 2013), but we also need to devise a reliable and universally available early warning system (Rovero & Ahumada, 2017) which can help us to detect which species are on decline or, indeed, are endangered and in need immediate attention and which natural resources are declining and are thus in need of new management strategies for their conservation. Simply put, our global life support system needs more advanced and better organized indicators to assess and measure change (Fig. 3.1). This change will tell us how important our assessments are. The purpose of this exercise is to keep track of the effects of global change on biodiversity over longer time periods.

For centuries, human beings have been involved in the study and assessment of the natural world, be it species or ecosystems. When they travelled around the world and discovered new islands and continents early explorers observed animals and plants and reported their presence and relevant habitats accordingly. When Charles Darwin visited the New World, he observed and reported the distribution of species. This observation led to his theory of "natural selection" (Darwin, 1896). While he was working in SE Asia, Alfred Wallace (Wallace, 1869) came up with similar findings (Darwin & Wallace, 1858). Animals started to perish with human activities and human beings introduced more animals to compensate for these losses to meet food requirements and their own ventures. The Dodo— which became extinct in 1690s—sends us a clear message about the fragility of islands and other ecosystems. It further warns us about the consequences of loss of biodiversity due to overexploitation. Current animal losses are the largest extinction crisis due to human activities the world has ever known (Dirzo et al., 2014). Biodiversity assessments are measuring numbers which tell us what species are found and/or distributed in a particular area and their distribution pattern. As well as the quality of animals present in that particular area we then quantify them and check the abundance of that particular species. If we observe that the population of a particular species is changing, then we have to apply more reliable and sensitive tools to monitor its population. We can obtain warning signs from our early warning system if the population is changing. The number of indicators, including status, the extent of habitat changes, trends of wild populations, and the effective management of the protected areas quantitatively assess the condition of biodiversity and achievement of its conservation over time.

3.3.1 Biodiversity Monitoring and Assessment

It is important to understand biodiversity and its principles. Equally important is to devise the means and methods for measurement of its complex and intricate processes. Moreover, it will be worth understanding the ever-increasing need for biodiversity monitoring and assessment. Humans never spared biodiversity and left its impacts in and on every piece of the globe where diversity existed even in the Anthropocene era (Crutzen, 2006). This was the era when humans ruled this earth with full power and strength. However, effective biodiversity assessment and monitoring is very helpful and can warn us if different systems which are supporting biodiversity in one way or another are swiftly switching off and that biodiversity is perishing at an accelerated pace (Fig. 3.3).

Typical warning systems inform the general public about the future and lead the public to prepare themselves to avoid the danger or to mitigate its effects. Recently, Rovero and Ahumada (2017) proposed that, operationally, an early warning system for biodiversity should comprise the following chain of functions: (1) monitoring biodiversity along with potential drivers of changes; (2) integration of that data into models that can then be used to simulate biodiversity change under different scenarios of deforestation, poaching, climate change, etc.; (3) sharing these data, models, and predictions with managers covering protected areas, government agencies, the general public, and local communities; and (4) finally providing recommendations based on the changes which have been identified and predicted for the future applications before initiating and promoting activities for the management of conservation. This early warning system will also tell us about the effects of global changes on species and communities and especially about the progress in extinction of animals which demands further exploration. Earthquakes, tsunamis, hurricanes, catastrophic fires,

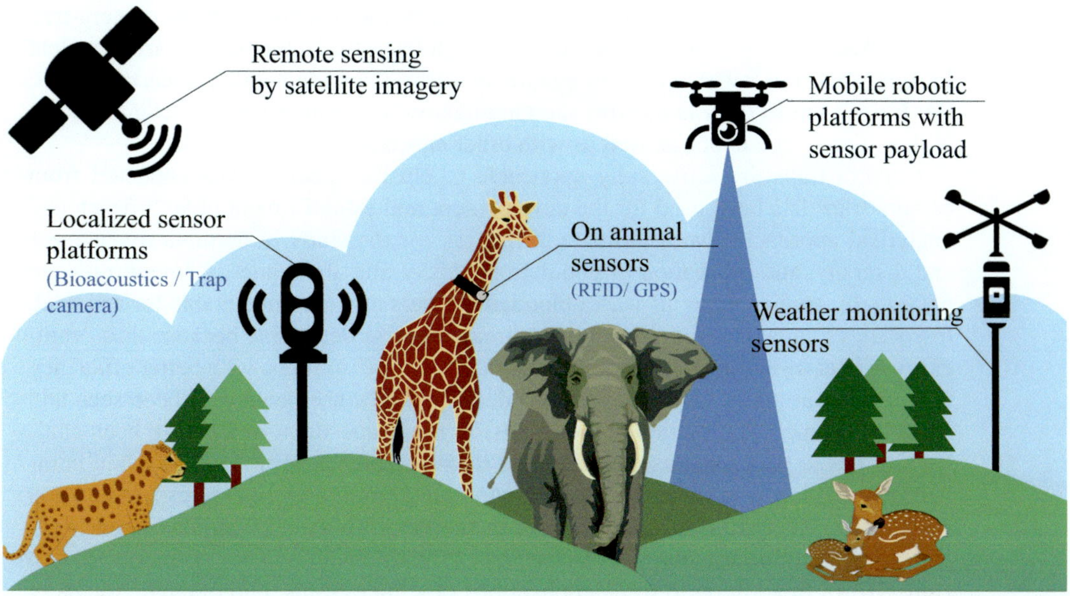

FIGURE 3.3 Existing methods for biodiversity analysis using sensor set-ups. (Courtesy of Panigrahi et al., 2023; PMC Free Full Text.)

hurricanes, and volcanos are well-established early warning systems for a number of natural phenomena. Persistent monitoring and assessment is needed to establish mechanisms so they can predict and forecast changes in biodiversity.

3.3.2 Tools for Biodiversity Assessment

Some of the tools used for monitoring and assessment are:

For monitoring and assessment of species and ecosystems there are a variety of tools and techniques, a few of which are used for monitoring larger vertebrates and those moving on the ground, including:

a. Camera-trapping. It is said that "a picture is worth a thousand words" because when we collect pictures of an animal for data purposes we do not simply obtain an unambiguous record of its presence at a particular time but we also receive a great deal more information from the photograph of an animal (Fig. 3.5). Camera traps are important tools and are extensively used for monitoring and assessing larger animals and their habitats. Using this method, scientists collect data from a TEAM (tropical ecology assessment and monitoring) monitoring site.

   In addition to animal photography we can also observe various weather events, follow different phases of the moon, observe the condition of an animal, the fruiting patterns of trees, animal behavior, as well as check the different resources and patterns of different activities in animals and plants. Camera traps and small automatic cameras can capture images of those animals moving around. For monitoring and detecting terrestrial animals and birds, cameras are very efficient (Rovero et al., 2013). Bigger mammals can also be good indicators of biodiversity assessment because they are very sensitive to area, habitat conversion, hunting, etc. (Schipper et al., 2008). Cameras are normally installed in larger areas and are left unattended to collect data, e.g., (A) the Giant anteater (*Myrmecophaga tridactyla*) shown below with its baby on its back conveys us information about its reproductive status, as well as the time of the year; (B) Jaguar (*Panthera onca*) behavior suggests the invisibility of the camera-trap (C) the Melanistic jaguar (aka "black panther") tells us that these big cats are prevalent in some areas but fewer in number elsewhere due to their adaptability to certain areas; (D) Baird's tapir (*Tapirus bairdii*) showing its nose as an instrument to communicate with its family members, as well as with other animals.

Camera traps are universally applicable to elusive species. Data collected from camera traps has been used for the development and establishment of well-developed analytical approaches with ultimate utilization for the study of community richness and diversity, and estimation of population density and abundance. Camera traps are not equally applicable to all warm-blooded animals and are not suitable for arboreal, fossorial, or aquatic animals. Smaller animals are less detectable because these cameras use heat-in-motion activity and so can detect larger animals with better efficiency than smaller animals. Over the past decades, great advancements have been seen and nowadays camera traps can be used to answer those questions which have been challenging in the past, such as evaluation of the structure of animal and/or plant communities and levels of diversification of mammals living in tropical forests (Ahumada et al., 2011). Camera traps also help us in the determination of model habitat preferences of certain animals, with subsequent estimations of species richness (Rovero et al., 2014). In addition to these applications of camera traps, photographic intensity can be used as an animal density index (Rovero & Marshall, 2009).

b. USGS: In this method, applications of high-precision GPS and terrestrial LIDAR scanners record the changes in the structure of the earth and restoration of vegetation on it.
c. USGS scientists and their helpers evaluate the structural restoration effects on the abundance and composition of species within the channel at the Deep Dirt Farm Institute in Patagonia, Arizona (Fig. 3.4).

3.3.3 Biodiversity Monitoring: Approaches and Challenges

There are several purposes of biodiversity monitoring. It provides basic data about types of evaluable changes, e.g., climate changes. Monitoring also determines the level of human activities with

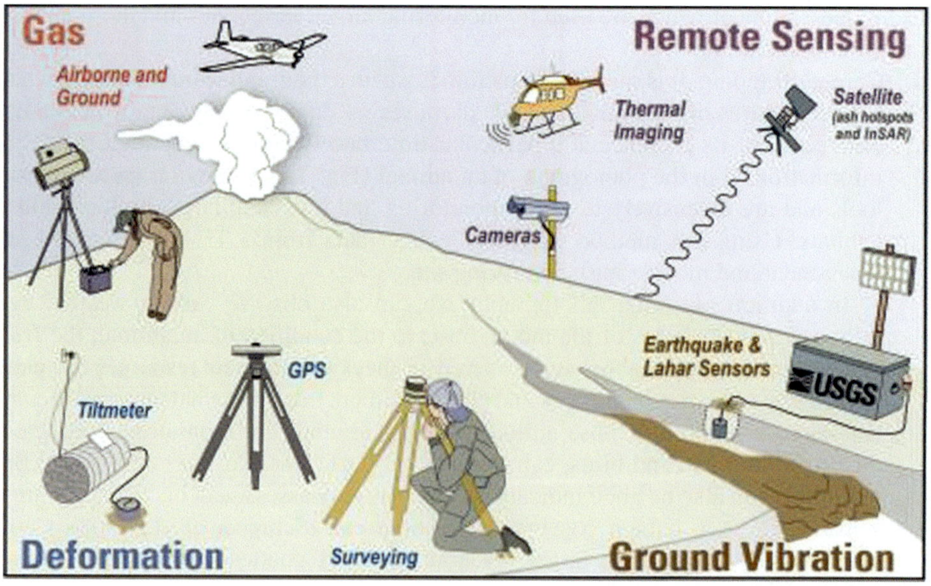

FIGURE 3.4 Volcanic monitoring types and methods employed by the USGS Volcano Hazards Program. www.shutterstock.com.)

the disturbances in the ecosystems and, in turn, their responses. It also assesses the effects of management teams working on the conservation of biodiversity and recognizes different changes developing in the forest and animal production systems, in turn detecting alterations in the structure and function of ecosystems. Finally, monitoring provides long-term data for assessment of ecological theories and models and many others. Lindenmayer and Likens (2010) have categorized monitoring into three main types:

- **Curiosity-driven, or passive monitoring**: This type of monitoring has little relevance to science because it is not based on any scientific hypothesis and hence it is not able to address environmental problems. It means that this lacks management interferences which are normally hypothesis-driven with scientific understanding of the system under monitoring.
- **Mandated monitoring:** This method of biodiversity is used according to the requirements of governments for their legislation or by political directive. This monitoring is practically oriented and is of a more applied nature and it usually deals with understanding of post hoc changes, their description, and understanding the drivers of these changes.
- **Question-driven monitoring**: This type of monitoring is based on well-planned scientific studies. It addresses particular questions and/or problems, and hence we can say it hypotheses or is question-driven. It uses robust experimental designs and has the capability to address suitable questions and relevant problems.

This classification is relevant and important because most monitoring programs fail and cannot deliver useful outcomes because of their inability to focusing on specific problems. Therefore, they are unable to diagnose or ascertain the causes of specific changes during monitoring. Improper sampling design is among one of the specific changes, although often it is not clear what species should be monitored or indeed why a species should be monitored (Lindenmayer & Likens, 2010). Landscape, organism, and ecosystem-specificity are the few complex processes which actually tell how the indicator works. Most probably, one indicator species working in one place or an ecosystem may not work in another place or ecosystem. Wolves (*Canis lupus*) in Yellowstone function as structuring agents and control herbivores in that area and also sometimes prey on meso-carnivores. However, the opposite has been observed in eastern United States due to its very low wolf population, with high human/wolf conflicts and hence humans are taking over their "natural" ecosystem function (Sergio et al., 2008). In this context, another issue is performance of each species which play differently in different ecosystems and accordingly responds differently to changes. Therefore, a response of one species cannot be used to account for other species. Further, the relationship of indicator species and those entities they are working as indicators for are poorly understood, pollution being but one example. Another aspect of monitoring that is worth mentioning is that monitoring should be effective for the long term. The current scientific community is not in the favor of long-term research and hence it is difficult to generate results from this process and publish them in time. Additionally, funding agencies provide funds for short-term projects with quick results; hence, provision of sustained funding is always a constraint to this process of monitoring (Lindenmayer & Likens, 2010).

3.3.4 Things to Know Before Starting to Monitoring

For assessment and monitoring of biodiversity is to frame the question and then to look for experimental design options. Then comes methodology: what equipment is necessary and what analytical tools must we use. This might seem overly naïve, but we must been sure that all the confounding factors have been fully covered, which can restrict the applicability of the data collected from these studies. Numerous protocols exist for monitoring species, e.g., tigers (Karanth et al., 2006) or species assemblage, e.g., terrestrial vertebrates (Ahumada et al., 2011). Much will depend on whether we are starting to monitor from a small scale and gradually moving toward higher structures like wildlife reserves (Simcharoen et al., 2007) or whether our monitoring is at the national level—which means working within the country (Karanth et al., 2011)—and/or we whether are covering the whole globe for monitoring purposes (Fegraus et al., 2011). It is well established that detection of species varies not only among different ecosystems but there is also variation among species when we move from site to site. Keeping all these factors in mind, we should ponder the statistical power which is always dependent on the collection of sufficient data over the appropriate scale. If we assess population of jaguars (*Panthera onca*) over just a 1 km^2 area, this monitoring will be meaningless because the area is very small. Similarly, if we monitor jaguars for one year it will not give us fruitful information due to the short duration of monitoring. To answer any question, study, design, and appropriate methods of investigation are very important. Defining the limitations of the ecosystem is also very important for delineating the required methodology. Sometimes animals may be very scarce and we cannot see or quantify them. Sometimes we consider and measure single a feature and ignore something less important, which might in fact be a more important indicator for species description. Taking for example estuaries where sub-aquatic vegetation, macroinvertebrates, and/or even blue crab (*Callinectes sapidus*) in the Chesapeake Bay can be much better indicators than some others that are normally conceived. As ecosystems, species, and ecological processes all operate at fundamentally different scales of space and time, hence they bring rapid change in biodiversity (Rosenzweig, 1995). Therefore, for monitoring and measuring of biodiversity temporal and spatial scales should be given due importance. Increases in species diversification have been observed when we approach the equator along both latitudinal and elevational gradients, but there is a proportionate decrease in biodiversity when we move from rural to urban areas due to a rise in population and development of wild areas. Biodiversity is heavily affected by humans, and therefore spatial scale is most important for the determination of the appropriate scale of the project. Logistic problems (if they exist) should be considered at the outset of the studies for adequate measures to maintain integrity of the data. Some areas may be totally unsafe, including miner treats(crafted food item that provides a temporary boost to mining and magnetism, making it easier to find and collect ores, gems, and other treasures buried in the earth.), borders, guerillas, and war zones, and some areas may not accessible at all like cliffs, mountains, rivers, ice, etc. Therefore, consideration of these constraints is important at the time of formulation of this hypothesis. During monitoring the temporal scale is also necessary to consider because it is a primary component of monitoring. In addition, the frequency of monitoring and time duration of the monitoring should also be considered. Both these parameters are very closely linked to the seasons of the year. Weather parameters, including monsoons, hurricanes, fire cycles, and behavioral parameters of animals/plants like hibernation, migration, and fruiting patterns, as well as other biotic and abiotic factors, can affect the probability of the presence of species at a given place and time of the year. Therefore, during a biodiversity assessment it will be important to consider which season is most suitable for monitoring and to search for appropriate answers to questions regarding which species is to be monitored. Sometimes due to various unexpected factors like storms and/or fire cycles (natural parts of the ecosystem), multiple sampling is required. Nevertheless, segregation of natural factors from human activities is still a challenge (Fig. 3.5).

3.3.5 Things to Monitor

It is important to decide which species to focus on. As everything is hard to monitor, we should select species which can be detected, which serve as indicators for the functioning of the ecosystem, and for which sampling and analytical tools and methods are available. Depending on the potential function of species, Caro (2010) classifies different species as follows:

- **Ecological "indicator species":** These species are ecologically sensitive species. They immediately detect and inform us of any disturbances in the ecosystem. Arboreal primates in a rainforest may be a good example (Fig. 3.6).

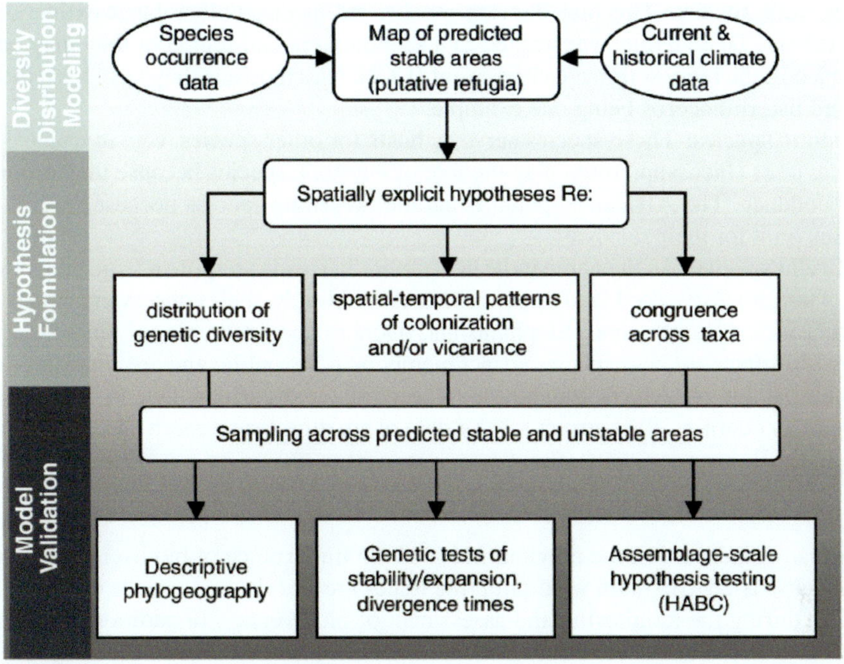

FIGURE 3.5 Proposed method of biodiversity prediction. (Courtesy of Carnaval; Hickerson et al., 2009; www. shutterstock.com.)

FIGURE 3.6 Arboreal primate: an indicator species. (*Source*: pxhere.com; Free public domain.)

- **Vulnerable species:** This includes rare species and species vulnerable to extinction. These species may be endemic or restricted in a specific area and region or they may have been included in the highest threat categories of IUCN, which means they have been included in the red list, rhinoceros being one example.
- **Keystone species:** These species serve as hosts for other species, e.g., in tropical rainforests fig trees (Ficusspp.) often play the role of keystone species because they produce fruit in abundance. The African elephant is another keystone species because several animals depend on it for their survival.
- **Umbrella species:** As the name indicates, umbrella species require a large area in which to live. They are also called landscape species. Accordingly, their conservation also demands conservation of larger areas. Elephants and tigers re good examples of umbrella species.
- **Flagship species:** These species attract members of the public and are considered as icons of their habitat, e.g., tigers, pandas, snow leopards, and gorillas. Due to their charismatic (magnetic) qualities, they attract a great deal of funding for research and conservation.

3.3.6 Components of Effective Monitoring

We have already discussed in the previous sections the importance of biodiversity and its monitoring. We have also tried to explain what sampling issues are and the importance of chronological and spatial scales during the monitoring and assessment of biodiversity. In addition, we also discussed the successful monitoring network and efficient monitoring with examples. In the light of the above, effective monitoring can be summarized as follows:

- Biodiversity monitoring should be question-driven, and with scientifically sound footings.
- Monitoring should be based on appropriate study design which can produce a healthy data set possessing sufficient statistical power to draw meaningful inferences on the study system adopted in space and time.
- The monitoring process should be clear, and it should be well known what is to be monitored and why it is to be monitored. If the target unit is a biodiversity indicator, then it should be very clear what this indicator is for, as well as the terms and conditions which make it an indicator.
- The protocol developed for the monitoring process should be based on a simple, repeatable, and consistent data collection protocol.
- It is worth mentioning here that the monitoring process should be standardized across all its phases, and daily procedures should be very clear for data management and analysis.
- The monitoring process should have sufficient funding support so the process can be persistently continued until accomplishment of the desired targets.
- The monitoring process should be initiated and pursued by those researchers/agencies that have sufficient patience and have long-term commitment.

3.4 THE ROLE OF ASSESSMENT IN BIODIVERSITY

See Section 3.2 above.

3.5 ISSUES OF SCALE

At a global level, biodiversity loss is a major economic, cultural, and ecological problem. Urgent action is then required for biodiversity assessment and monitoring to evaluate the policies developed and their level of success for conservation of nature and to protect it from further decline. Scientific work has been expanded many time over on biodiversity and its assessment. Thus, it can be envisioned that an assessment of biodiversity may provide the basic data for developing hypotheses on the processes underlying diversity patterns, controlling the effectiveness of agri-environmental schemes, and designing the networks of nature conservation areas. Obviously, when assessing biodiversity there are several scale-related problems. Therefore, the methodology implied to explain the condition of biodiversity will mainly depend on the complexity of the study area and the extent of its study, which is in addition to other factors. Grain size may be critical at the time when we study and elucidate spatial patterns within that particular study area. Keeping these deliberations in mind, the following questions of scale are relevant:

1. What are biodiversity indicators and their scaling problems?
2. What are patterns of biodiversity and their scaling problems?
3. How is biodiversity measured and predicted in large areas?

3.5.1 Biodiversity Indicators

Biodiversity indicators are cost-effective methods to assess and identify species richness. There are specific relationships between types of biodiversity and where they are living. These relationships include:

1. The relationship between species richness and environmental variables.
2. The relationship between the richness of different and unrelated taxa, such as birds and plants.
3. The relationship between numbers of species and numbers of supra-specific taxa, such as genera or families.

3.5.2 Patterns of Biodiversity

Patterns of biodiversity depend on the extent of the study area and grain size. For example, the famous latitudinal gradient of species richness can only be seen if studies are performed on several biogeographical zones. However, the problem here is that, despite the extensive area under study, it is still not clear whether the gradient would emerge only very small plots were used, e.g., only 1 m^2. It should further be added that grain size is critical if we want to recognize trends in biodiversity and it will further differ depending on whether we will measure trends in rare or common species. Furthermore, very large plots are not acceptable either. For example, if we use indicators to detect changes in the sub-continent, sensitivity would be poor if we considered the whole of the sub-continent as one large plot (Fig. 3.7).

FIGURE 3.7 Global patterns of biodiversity. (*Source*: www.shutterstock.)

If we compare biodiversity in regions with different sizes, scale-related problems may emerge. Nevertheless, if biodiversity is further subdivided into alpha and beta components over multiple spatial scales, this exercise may provide an option of the applications of hypothesis which are testable on the different processes which cause the patterns. Proper quantification of variability in chosen response variables at a hierarchy of scales is required if we want to understand the patterns of biodiversity.

3.5.3 Measuring and Predicting Biodiversity

It is still a challenge how to assess species richness in large landscapes. If we want to assess species richness directly, then in most cases sampling is necessary. Nevertheless, it is the sampling effort which determines the number of species (Fordham et al., 2016). In this scenario we have two options, namely whether to go for standardization of sampling efforts or develop such methods that can work without an application of sampling effort. Finally, results or opinions on biodiversity assessment depend on the organisms concerned and on the ecosystems or landscapes where we have planned our studies. Under these conditions, we expect and hope for the contributions of scientists from a variety of backgrounds.

3.6 GRADIENTS IN BIODIVERSITY

It is well known that there is a rise in species richness when move from the poles to the tropics, usually known as the latitudinal gradient in diversity of species. This pattern of species distribution is a very widely accepted pattern in ecology. To say it another way, localities at higher altitudes have a lower number of species while the localities at lower altitudes have a higher number of species. It is a great challenge for macro-ecology how to explain the latitudinal diversity gradient appropriately (Willig et al., 2003; Pimm & Brown, 2004; Cardillo et al., 2005).

3.6.1 Hypotheses for Patterns in Biodiversity

For the exploration of the latitudinal diversity gradient, there are several hypotheses which are closely related and hence interdependent. The majority of these hypotheses are generally divided into four general categories: spatial/area, energy/climatic, evolutionary/historical, and biotic.

3.6.2 Spatial/Area Hypotheses

In this hypothesis, the mid-domain effect has a pivotal role in its elaboration and explanation, which will be discussed here to explain this hypothesis.

3.6.3 Mid-domain Effect (MDE)

Colwell and Hurtt (1994) and Willig and Lyons (1998) stated that ranges of different species will overlap towards the center of a shared domain if it is bounded by geometric domains. The peak of species richness will appear toward the center of the domain. Colwell and Lees (2000) called this random distribution (stochastic) phenomenon and tilting of species toward the center of the domain, the mid-domain effect (MDE) Finally, they came up with the hypothesis that MDE might contribute to the latitudinal gradient in species richness. Models developed so far are called null models because, unlike null models, pure models try to make the species richness model free of evolutionary and environmental aspects (Colwell et al., 2004, 2005). If latitudinal gradients of species richness are purely determined by MDE (Colwell & Lees, 2000), then the observed richness patterns at the biogeographic level would not be distinguishable from patterns produced by random placement of the observed ranges. Other scientists say that as the MDE models do not exclude the role of the environment at the population level, as well as when it sets domain boundaries, then these models cannot be considered as null models (Zapata et al., 2003, 2005; Hawkins et al., 2005). Rahbek et al. (2007), Storch et al. (2006), and Currie and Kerr (2007) say that MDE is a controversial model. Others say that in wide-ranging species MDE has a potential role in latitudinal gradients of species richness (Romdal et al., 2005; Storch et al., 2006; Lees & Colwell, 2007; Rahbek et al., 2007; Dunn et al., 2007). Bokma and Monkkonen (2001), Currie and Kerr (2007), Hawkins and Diniz-Filho (2002), and Kerr et al. (2006) contradict this phenomenon and suggest there is little relationship between predicted and observed latitudinal diversity patterns.

3.6.4 Geographical Area Hypothesis

Geographical area hypothesis (Terborgh, 1973) claims that as the tropics have the largest biomes hence they host more species. Larger tropical areas support larger species population sizes with wider distribution. Therefore, species found on larger ranges and areas normally have lower extinction rates (Rosenzweig, 2003). Additionally, there is also a possibility that species living in these larger areas can increase the rate of speciation by undergoing allopatric speciation (Rosenzweig, 2003), ultimately promoting the higher levels of species richness in the tropics. If we accept that the geographical area hypothesis is correct, then the same level species richness should be present in all these areas but it does not happen like this. For example, temperate regions host more species than polar regions (Gaston & Blackburn, 2000). Rosenzweig (1992) elaborated this phenomenon further and opined that to the north of the tropics the species richness gradient will disappear if we exclude the parsley distributed species from this area. Blackburn and Gaston (1997) tested and found a relationship between available land area and species richness in a given biome latitudinal patterns in avian species richness in the New World. Some bio-geographers do not accept the terrestrial tropics as the largest biomes and hence cannot reliably explain the latitudinal species diversity gradient, which depicts a serious flaw in this hypothesis (Rohde, 1997; Hawkins & Porter, 2001).

3.6.5 Energy/Climate Hypotheses

Climate itself explains the latitudinal species diversity gradient. As there is a tilt in the earth on its axis, the tropics benefit from this and get more solar light energy during most of the year when compared to higher latitudes, unlike temperate and polar zones. Accordingly, higher altitudes have comparatively cooler temperatures and more variations in the seasons of the year than lower latitudes.

3.6.6 Species-energy Hypothesis

The species energy hypothesis says that limits to the richness of the system depend on energy available to the system. Hence, net primary productivity (photosynthesis) will be higher with the increase in solar energy at lower latitudes with an abundance of water. This hypothesis suggests that more primary productivity means more animals in the particular area with lower extinction rates. Some contradict this phenomenon and do not agree that there is no relationship between productivity and the number of animal species (Cardillo et al., 2005). They therefore say that the changes in the number of individuals which we attribute to latitude or productivity or too small or not in the right direction hence make it impossible to account for during the estimation of species richness (Cardillo et al., 2005). Currie et al. (2004) have critically reviewed this hypothesis by exploring the underlying potential mechanisms with the empirical support needed for its justification and what it can predict.

3.6.7 Climate Harshness Hypothesis

This hypothesis states that the main reasons of latitudinal diversity gradient are because of poor physiological tolerance of species at higher latitudes (because they are cold and dryer than tropical latitudes) than at low latitudes. Cardillo et al. (2005) further say that this hypothesis is faulty because it is well established that poor climatic tolerance of the species limits their distributions, and hence animal species will not be absent in areas with harsh climatic conditions.

3.6.8 Climate Stability Hypothesis

This hypothesis suggests that if the environment is fluctuating, it may increase the extinction rate and will ultimately retard speciation. Hence, a constant environment favors the species in their specialization, with narrower niches facilitating speciation. Taking for example temperate regions, they have fluctuations in terms of geological time scales and seasons of the year, hence temperate regions have less species diversity than the tropics. Therefore, stable environments such as tropical mountain tops with high species diversity do experience seasonal climates, e.g. they experience high rainfall during the rainy season (Brown & Lomolino, 1998).

3.6.9 Evolutionary (Historical) Hypotheses

This is further subdivided into four hypotheses, which are discussed in the below subsections.

3.6.9.1 The Historical Perturbation Hypothesis

The historical perturbation hypothesis suggests that at higher altitudes low species richness is due to an insufficient time period available for species to colonize or recolonize areas because of historical perturbations, e.g. glaciation (Brown & Lomolino, 1998; Gaston & Blackburn, 2000). This hypothesis suggests that diversity in the temperate regions cannot achieved equilibrium and the number of species will increase continuously until it is saturated (Clarke & Crame, 2003).

3.6.9.2 The Evolutionary Rate Hypothesis

This hypothesis says that in the tropics there is occurrence of more species due to higher evolutionary rates (Cardillo et al., 2005). Higher ambient temperatures, shorter generation time, and/or faster physiological processes and higher mutation rates promote higher evolutionary rates in the tropics. Nevertheless, further research is required for further confirmation of this phenomenon.

3.6.9.3 The Hypothesis of Effective Evolutionary Time

This hypothesis assumes that evolutionary time under which the ecosystem is working has been uniform. Accordingly, evolutionary speed is dependent on temperature dependent mutations, speed of selection, and generation rates, which also determine the biodiversity of that particular area (Rode et al., 2016). It is a non-equilibrium hypothesis that assumes a largely a non-saturated niche space. This hypothesis also accepts that there are several other factors which can cause latitudinal gradients in species richness. Nevertheless, Wright et al. (2006) support this hypothesis.

3.6.9.4 Biotic Hypotheses

Biotic hypotheses claim in the tropics species interact with each other which promote co-existence and facilitate species to specialize within that given area. Species interact with each other during competition, predation, mutualism, and parasitism. These phenomena are stronger in the tropics and lead to the greater speciation. Pianka (1966) further strengthened this phenomenon and said that the presence of more specialized predators increased the intensity of predation, which ultimately resulted in increased diversity. Some suggest that the changes between latitude and ecological interactions are not persistent (Lambers et al., 2002; Hillebrand, 2004).

3.6.10 The Generality of the Latitudinal Diversity Gradient

In the light of the previous studies, Hillebrand (2004) tried to generalize the latitudinal diversity gradient across habitat, organismal, and regional characteristics. Hillebrand found that marine, freshwater, and terrestrial ecosystems are present in both hemispheres and both have latitudinal gradients. He is further of the opinion that this gradient is more prominent and sharp in richer taxa. The same is observed at regional versus local scales, in larger organisms, and in marine and terrestrial versus freshwater ecosystems.

No study could directly nullify or support these hypotheses but results do suggest that a combination of energy/climate and area processes possibly contribute to the latitudinal gradient of species. Exceptions to this rule are shorebirds, penguins, freshwater zooplankton, and ichneumonidae.

REFERENCES

Ahumada, J.A., Silva, C.E., Gajapersad, K., Hallam, C., Hurtado, J., Martin, E., McWilliam, A., Mugerwa, B., O'Brien, T., Rovero, F. & Sheil, D. (2011). Community Structure and Diversity of Tropical Forest Mammals: Data from a Global Camera Trap Network. Philosophical Transactions of the Royal Society, B: Biological Sciences 366(1578), 2703–2711.

Andreasen, J.K., O'Neill, R.V., Noss, R. & Slosser, N.C. (2001). Considerations for the Development of a Terrestrial Index of Ecological Integrity. Ecological Indicators 1, 21–35.

Blackburn, T.M. & Gaston, K.J. (1997). The Relationship between Geographic Area and the Latitudinal Gradient in Species Richness in New World Birds. Evolutionary Ecology 11, 195–204.

Bokma, F.J. & Monkkonen, M. (2001). Random Processes and Geographic Species Richness Patterns: Why so Few Species in the North? Ecography 24, 43–49.

Brown, J.H. & Lomolino, M.V. (1998). Biogeography. Sinauer Associates, Sunderland.

Cardillo, M., Orme, C.D.L. & Owens, I.P.F. (2005). Testing for Latitudinal Bias in Diversification Rates: An Example Using New World Birds. Ecology 86, 2278–2287.

Carignan, V. & Villard, M.-A. (2002). Selecting Indicator Species to Monitor Ecological Integrity: A Review. Environmental Monitoring and Assessment 78, 45–61.

Carnaval, A.C., Hickerson, M.J., Haddad, C.F.B., Rodrigues, M.T. & Moritz, C. (2009). Stability Predicts Genetic Diversity in the Brazilian Atlantic Forest Hotspot. Science 323, 785–789.

Caro, T. (2010) Conservation by Proxy: Indicator, Umbrella, Keystone, Flagship, and Other Surrogate Species. Washington, DC, Island Press.

Clarke, A. & Crame, J.A. (2003). The Importance of Historical Processes in Global Patterns of Diversity. In: Macroecology Concepts and Consequences, eds. Blackburn, T.M. & Gaston, K.J., pp. 130–151. Blackwell Scientific, Oxford.

Colwell, R.K. & Hurtt, G.C. (1994). Nonbiological Gradients in Species Richness and a Spurious Rapoport Effect. American Naturalist 144, 570–595.

Colwell, R.K. & Lees, D.C. (2000). The Mid-Domain Effect: Geometric Constraints on the Geography of Species Richness. Trends in Ecology & Evolution 15, 70–76.

Colwell, R.K., Rahbek, C. & Gotelli, N. (2004). The Mid-Domain Effect and Species Richness Patterns: What Have We Learned so Far? American Naturalist 163, E1–E23.

Colwell, R.K., Rahbek, C. & Gotelli, N. (2005). The Mid-Domain Effect: There's a Baby in the Bathwater. American Naturalist 166, E149–E154.

Crutzen, P.J. (2006). The "Anthropocene". In: Earth System Science in the Anthropocene. Berlin/Heidelberg, Springer. Ehlers, E. & Krafft, T. Editors, pp. 13–18.

Currie, D.J. & Kerr, J.T. (2007). Testing, as Opposed to Supporting, the Mid-Domain Hypothesis: a Response to Lees and Colwell (2007). Ecology Letters 10, E9–E10.

Currie, D.J., Mittelbach, G.G., Cornell, H.V., Kaufman, D.M., Kerr, J.T., Oberdorff, T. & Gugan, J.-F. (2004). A Critical Review of Species-Energy Theory. Ecology Letters 7, 1121–1134.

Darwin, C. (1896). The Origin of Species by Means of Natural Selection, or the Preservation of Favored Races in the Struggle for Life, vol. 1. International Science Library.

Darwin, C. & Wallace, A. (1858). On the Tendency of Species to Form Varieties; and on the Perpetuation of Varieties and Species by Natural Means of Selection. Journal of the Proceedings of the Linnean Society of London: Zoology 3(9), 45–62.

Dirzo, R., Young, H.S., Galetti, M., Ceballos, G., Isaac, N.J. & Collen, B. (2014). Defaunation in the Anthropocene. Science 345(6195), 401–406.

Dunn, R.R., McCain, C.M. & Sanders, N. (2007). When Does Diversity Fit Null Model Predictions? Scale and Range Size Mediate the Mid-Domain Effect. Global Ecology & Biogeography 3, 305–312.

Fegraus, E.H., Lin, K., Ahumada, J.A., Baru, C., Chandra, S. & Youn, C. (2011). Data Acquisition and Management Software for Camera Trap Data: A Case Study from the TEAM Network. Ecological Informatics 6(6), 345–353.

Fordham, D., Akçakaya, H. & Alroy, J. et al. (2016). Predicting and Mitigating Future Biodiversity Loss Using Long-Term Ecological Proxies. Nature Climate Change **6**, 909–916. https://doi.org/10.1038/nclimate3086

Gaston, K.J. & Blackburn, T.M.. 2000. Pattern and Processes in Macro-Ecology. Blackwell Scientific, Oxford

Hawkins, B.A. & Diniz-Filho, J.A.F. (2002). The Mid-Domain Effect Cannot Explain the Diversity Gradient of Nearctic Birds. Global Ecology and Biogeography 11, 419–426.

Hawkins, B.A., Diniz-Filho, J.A.F. & Weis, A.E. (2005). The Mid-Domain Effect and Diversity Gradients: Is There Anything to Learn. American Naturalist 166, E140–E143.

Hawkins, B.A. & Porter, E.E. (2001). Area and the Latitudinal Diversity Gradient for Terrestrial Birds. Ecology Letters 4, 595–601.

Hillebrand, H. (2004). On the Generality of the Latitudinal Diversity Gradient. The American Naturalist 163, 192–211.

Hunt, L., Fisher, A., Kutt, A. & Mazzer, T. 2006. Biodiversity monitoring in the rangelands: A way forward, Vol. 1, report to Environment Australia, CSIRO Sustainable Ecosystems, Alice Springs.

James, C.D. (2004). Biodiversity Monitoring in Rangelands. In: Conference Papers, Australian Rangeland Society 13th Biennial Conference (eds. Bastin, G., Walsh, D. & Alice, N.S.). Springs, NT. p. 176182.

Karanth, K.U., Nichols, J.D., Kumar, N. & Hines, J.E. (2006). Assessing Tiger Population Dynamics Using Photographic Capture–Recapture Sampling. Ecology 87(11), 2925–2937.

Karanth, K.U., Gopalaswamy, A.M., Kumar, N.S., Vaidyanathan, S., Nichols, J.D. & MacKenzie, D.I. (2011). Monitoring carnivore populations at the landscapescale: Occupancy Modelling of Tigers from Sign Surveys. Journal of Applied Ecology 48, 1048–1056.

Kerr, J.T., Perring, M. & Currie, D.J. (2006). The Missing Madagascan Mid-Domain Effect. Ecology Letters 9, 149–159.

Lambers, J.H.R., Clark, J.S. & Beckage, B. (2002). Density-Dependent Mortality and the Latitudinal Gradient in Species Diversity. Nature 417, 732–735.

Lees, D.C. & Colwell, R.K. (2007). A Strong Madagascan Rainforest MDE and No Equatorward Increase in Species Richness: Re-analysis of 'The Missing Madagascan Mid-Domain Effect', by Kerr J.T., Perring M. & Currie D.J. (*Ecology Letters* 9:149–159, 2006). Ecology Letters 10, E4–E8.

Lindenmayer, D.B. & Likens, G.E. (2010) Effective Ecological Monitoring. Melbourne, Australia: CSIRO Publishing.

Lindenmayer, D.B., Piggott, M.P. & Wintle, B.A. (2013). Counting the Books While the Library Burns: Why Conservation Monitoring Programs Need a Plan for Action. Frontiers in Ecology and the Environment 11(10), 549–555.

Majer, J.D. & Beeston, G. (1996). The Biodiversity Integrity Index: An Illustration Using Ants in Western Australia. Conservation Biology 10, 65–73.

Niemeijer, D. & de Groot, R.S. (2007). A Conceptual Framework for Selecting Environmental Indicators Sets. Ecological Indicator 8, 14–25.

Oliver, I. & Parkes, D. (2003). A Prototype Toolkit for Scoring the Biodiversity Benefits (and Disbenefits) of Landuse Change, Version 5. Department of Land and Water Conservation, Sydney.

Panigrahi, S., Maski, P. & Thondiyath, A. (2023). Real-Time Biodiversity Analysis Using Deep-Learning Algorithms on Mobile Robotic Platforms. PeerJ Comput. Sci. 9, e1502 http://doi.org/10.7717/peerj-cs.1502

Piacenza, S.E., Richards, P.M. & Heppell, S.S. (2019). Fathoming Sea Turtles: Monitoring Strategy Evaluation to Improve Conservation Status Assessments. Ecological Applications 00(00), e01942. https://doi.org/10.1002/eap.1942

Pianka, E.R. (1966). Latitudinal Gradients in Species Diversity: A Review of Concepts. American Naturalist 100, 33–46.

Pimm, S.L. & Brown, J.H. (2004). Domains of Diversity. Science 304, 831–833.

Rahbek, C., Gotelli, N., Colwell, R.K., Entsminger, G.L., Rangel, T.F.L.V.B. & Graves, G.R. (2007). Predicting Continental-Scale Patterns of Bird Species Richness with Spatially Explicit Models. Proceedings of the Royal Society of London Series B 274, 165–174.

Read, J.L., Kovac, K.-J. & Fatchen, T.J. (2005). 'Biohyets': A Holistic Method for Demonstrating the Extent and Severity of Environmental Impacts. Journal of Environmental Management 77, 157–164.

Rode, J., Wittmer, H., Emerton, L. & Schröter-Schlaack, C. (2016). A Practice-Oriented Framework for Identifying Economic Instruments to Enhance Biodiversity and Human Livelihoods. Journal for Nature Conservation 33, 35–47.

Rohde, K. (1997). The Larger Area of the Tropics Does Not Explain Latitudinal Gradients in Species Diversity. Oikos 79, 169–172.

Romdal, T.S., Colwell, R.K. & Rahbek, C. (2005). The Influence of Band Sum Area, Domain Extent and Range Sizes on the Latitudinal Mid-Domain Effect. Ecology 86, 235–244.

Rosenzweig, M.L. (1992). Species Diversity Gradients: We Know More and Less than We Thought. Journal of Mammalogy 73, 715–730.

Rosenzweig, M.L. (1995) Species Diversity in Space and Time. Cambridge: Cambridge University Press.

Rosenzweig, M.L. (2003). How to Reject the Area Hypothesis of Latitudinal Gradients. In: Macroecology: Concepts and Consequences, eds. Blackburn, T.M. & Gaston, K.J., pp. 87–106. Blackwell Publishing, Oxford.

Rovero, F. & Ahumada, J. (2017). The Tropical Ecology, Assessment and Monitoring (TEAM) Network: an Early Warning System for Tropical Rain Forests. Science of the Total Environment 574, 914–923.

Rovero, F. & Marshall, A.R. (2009). Camera Trapping Photographic Rate as an Index of Density in Forest Ungulates. Journal of Applied Ecology 46, 1011–1017.

Rovero, F., Martin, E., Rosa, M., Ahumada, J.A. & Spitale, D. (2014). Estimating Species Richness and Modelling Habitat Preferences of Tropical Forest Mammalsfrom Camera Trap Data. PLoS One 9(7), e103300.

Rovero, F., Zimmermann, F., Berzi, D. & Meek, P. (2013). "Which Camera Trap Type and How Many Do I Need?" A Review of Camera Features and Study Designs for a Range of Wildlife Research Applications. Hystrix, the Italian Journal of Mammalogy 24(2), 148–156.

Sakar, S. & Margules, C. (2002). Operationalising Biodiversity for Conservation Planning. Bioscience 27, 229–308.

Schipper, J., Chanson, J.S., Chiozza, F., Cox, N.A., Hoffmann, M., Katariya, V., Lamoreux, J., Rodrigues, A.S., Stuart, S.N., Temple, H.J. & Baillie, J., et al. (2008) The Status of the world's Land and Marine Mammals: Diversity, Threat, and Knowledge. Science 322(5899), 225–230.

Schipper, J. & Rovero, F. (2018). Effective Biodiversity Assessment and Monitoring. In Encyclopedia of the Anthropocene 1(5), pp. 297–304. Elsevier. https://doi.org/10.1016/B978-0-12-809665-9.09844-X

Sergio, F., Caro, T., Brown, D., Clucas, B., Hunter, J., Ketchum, J., McHugh, K. & Hiraldo, F. (2008). Top Predators as Conservation Tools: Ecological Rationale, Assumptions, and Efficacy. Annual Review of Ecology, Evolution, and Systematics 39, 1–19.

Simcharoen, S., Pattanavibool, A., Karanth, K.U., Nichols, J.D. & Kumar, N.S. (2007). How Many Tigers *Panthera tigris* Are There in Huai Kha Khaeng Wildlife Sanctuary, Thailand? An Estimate Using Photographic Capture-Recapture Sampling. Oryx 41(4), 447–453.

Smyth, A., James, C. & Whiteman, G. (2003). Biodiversity Monitoring in the Rangelands: A way forward, Vol. 1, report to Environment Australia, CSIRO Sustainable Ecosystems, Alice Springs.

Smyth, A.K. & James, C. (2004). Characteristics of Australia's Rangelands and Key Design Issues for Monitoring Biodiversity. Austral Ecology 29, 3–15.

Storch, D., Davies, R.G., Zajicek, S., Orme, C.D.L., Olson, V., Thomas, G.H., Ding, T.S., Rasmussen, P.C., Ridgely, R.S., Bennett, P.M., Blackburn, T.M., Owens, I.P.F. & Gaston, K.J. (2006). Energy, Range Dynamics and Global Species Richness Patterns: Reconciling Mid-Domain Effects and Environmental Determinants of Avian Diversity. Ecology Letters 9, 1308–1320.

Suter, G.W. (2001). Applicability of Indicator Monitoring to Ecological Risk Assessment. Ecological Indicators 1, 101–112.

Terborgh, J. (1973). On the Notion of Favorableness in Plant Ecology. American Naturalist 107:481–501.

Wallace, A.R. (1869). The Malay Archipelago: The Land of the Orang-utan and the Bird of Paradise. A Narrative of Travel, with Studies of Man and Nature. Courier Corporation.

WildCam (2022). Are camera trap surveys a viable method for inventory of BC wildlife? Posted on January 30, 2022. info@wildcams.ca.

Willig, M.R., Kaufmann, D.M. & Stevens, R.D. (2003). Latitudinal Gradients of Biodiversity: Pattern, Process, Scale and Synthesis. Annual Review of Ecology, Evolution, and Systematics 34, 273–309.

Willig, M.R. & Lyons, S.K. (1998). An Analytical Model of Latitudinal Gradients of Species Richness with an Empirical Test for Marsupials and Bats in the New World. Oikos 81, 93–98.

Wright, S., Keeling, J. & Gillman, L. (2006). The Road from Santa Rosalia: a Faster Tempo of Evolution in Tropical Climates. Proceedings of the National Academy of Science 103, 7718–7722.

Zapata, F.A., Gaston, K.J. & Chown, S.L. (2003). Mid-Domain Models of Species Richness Gradients: Assumptions, Methods and Evidence. Journal of Animal Ecology 72, 677–690.

Zapata, F.A., Gaston, K.J. & Chown, S.L. (2005). The Mid-Domain Effect Revisited. American Naturalist 166, E144–E148.

4 Importance of Biodiversity

4.1 INTRODUCTION

All living organisms vary, and this variation may be within species or between species. When diverse organisms live in one place they interact with each other, as well as with the biota (Patel, 2014) and constitute a complex system serving as an important tool of biodiversity. Species is a group of individuals which can interbreed and produce fertile offspring. When talking about diversity in an ecosystem, this can encompass ecosystem or habitat diversity or diversity within the same species, among different species, or in a population. The structure and function of ecosystems is based on climatic changes, vegetation structure and composition, pollution levels, etc. Adaptability is an important quality of the species which help them to live in specific climatic conditions successfully. The effect of gene activity determines the variation in species and environmental conditions regulate this variation. The occurrence of species, richness, and their rate of dispersal always vary in an ecosystem. Each species has a unique biotic potential which is actually a life pattern and growth power of that particular species in their natural habitat. A hostile environment becomes dangerous for growth and development of certain species, while better climatic conditions support their biotic potential. Climatic changes change population structure. This variation helps and promotes successful regulation of the ecosystem. Biodiversity is important because it provides fuel, fodder, food, and medicines for our health. It manages day-to-day air and aquatic pollution, improves quality of soil, prevents erosion of soil, and recycles nutrients and water, resulting in a healthy ecosystem. Biodiversity protects natural resources, acts as a source of recreation, and promotes cultural and ethical development, ultimately improving and enhancing economic growth.

Generally speaking, the value of biodiversity can be categorized into anthropocentric and ethical value. An anthropocentric value is characterized by direct and indirect benefits to human beings. Ethical value, on the other hand, is considered as the concept of divine creation and speaks to respect for life with a sense of intrinsic value in nature. Biodiversity has numerous benefits for human beings. It provides a range of goods, from agricultural crops to medicines and fibers which are direct in nature and saves considerable costs if obtained the other way round. Direct biodiversity can be in the form of production and then its consumption. Consumptive use value is usually assigned to goods with no sale and purchase, hence not contributing to the economy of the country. People having no land obtain all their daily requirements for the surrounding environment. Environmental decline means non-availability of these items to these people, ending up with deterioration of their living standards. Consumption uses can also be seen in the fuel used for heating and cooking and up to about 2.4 billion people consume fuel for these activities. The importance of this energy can only be assessed when we compare by calculating the total cost of other energy sources. People have to travel long distances if they cannot find fuel or timber from nearby sources, which can lead to deforestation where population pressure is a causative agent. When we discuss the value of biodiversity in terms of production, it comes down to those goods harvested from the environment, which are bought from the producers and then sold in the local market or at an international level. Major products are construction timber, wood for fuel, fish and shellfish, fruits and vegetables, seaweed, etc. Considering their economic production, it is always estimated whether a product will be better produced naturally or synthesized. For example, after discovery of its structure, it is manufactured synthetically at far less cost than its natural production. There are several drugs whose availability is heavily dependent on the health of biodiversity and its proper functioning (Fig. 4.1). The existence and multiplication of the biological species in nature depends on the success and health of biodiversity in a given area.

DOI: 10.1201/9781003472001-4

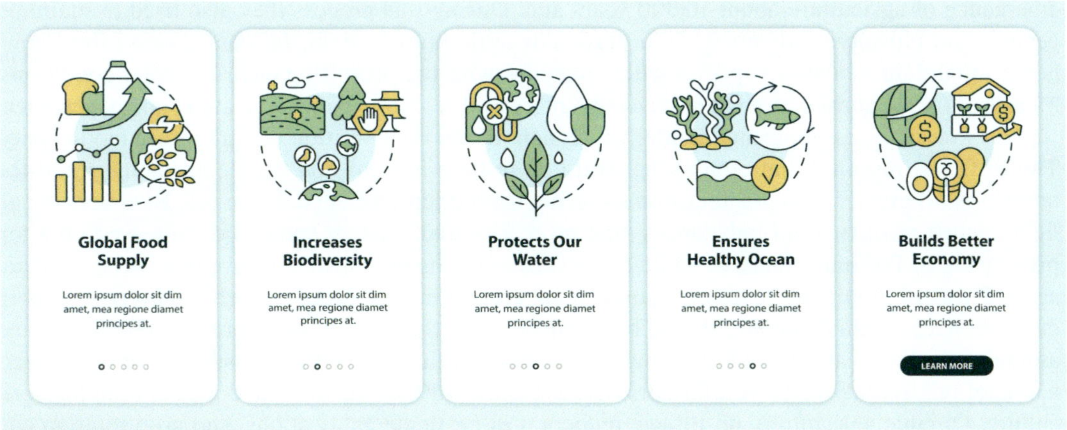

FIGURE 4.1 Importance of biodiversity. (*Source*: www.shutterstock.com.)

4.2 FOOD SECURITY, RECREATION, AND EDUCATION

4.2.1 Source of Food Production

Agriculture and food production are two fundamental outcomes of biodiversity. For their livelihoods people depend on shelter, a variety of goods, and different foods. Thus, human beings are mounting pressure on different species of organisms, as well as on their environment (FAO, 2004). Due to this building pressure many animals, plants, pollinating insects, and microorganisms which improve soil fertility are at risk.

This growing human population demands more food production from agriculture. Agriculture, therefore, needs to preserve both animals and plants with unique traits like drought-resistant plants and animals which can reproduce under harsh conditions. Sustainable use of agriculture ensures sustained food supply to human beings and, at the same time, can protect the natural resources including forests, prairies, oceans, and other ecosystems which host biodiversity. Sustainable utilization of biodiversity also helps in the conservation of genetic diversity. Agricultural biodiversity is based on a rich variety of cultivated plants and domesticated animals. Presently, 90% of human and animal food comes from just 14 mammals and some bird species.

Wheat, rice, potatoes, and maize, and potato plant species provide half of our total energy requirements. Taking the benefits of modern agriculture, farmers keep uniform and high producing crops and animals for consistent and healthy production. When farmers disturb biodiversity they kill important crops and breeds with their specialized traits. This activity worries scientists because it diminishes the quality of the gene pool. Animals and plants with unique characteristics can meet the emerging needs of the people with the provision of raw materials for the scientists to improve and produce more hardy and productive plant and animal species. The Food and Agriculture Organization (FAO) claims that out of 6,300 animal breeds, 1,350 species are endangered or have already become extinct, with concomitant loss of approximately three-quarters of the genetic diversity of agricultural crops over the last century.

At a global level, successful efforts have been employed to conserve the gene bank of plants and animals in botanical gardens and zoos, respectively. To maintain biodiversity in nature and on farms is an equally important task too. On these sites, biodiversity can evolve by competing with other species and adapting to changing climatic conditions. Farmers can develop and maintain local plants and animals of their interest and benefits with the desired traits, and hence sustenance of agriculture biodiversity totally rests with the farmers. Fishermen, farmers, forest dwellers, and pastoralists tried to select plants and animals for food and environmental conditions from the emergence of agriculture about 10,000 years ago. During this process they also tried to maintain genetic and phenotypic diversity. In a particular agriculture system, farmers possess the know-how to maintain a breed or species which matches with the prevailing environmental conditions. Measures for the conservation of biodiversity, increased research, better education, and government support are vital and important factors which help in the conservation of agriculture biodiversity. The FAO has reported that around the world more than 840 million people are hungry and the number is far greater than this for those suffering from micronutrient deficiencies. Biodiversity can fight against poverty, food imbalances, and insecurity and hence it should be conserved on a top priority basis. The importance of biodiversity further intensifies when we move from urban to rural areas. Farmers should not focus on a single plant crop or animal breed; rather, the horizon of crop cultivation and animal keeping should be expanded with the inclusion of more resistant plants and animals which can survive in harsh environmental conditions. Major responsibility for the conservation of biodiversity rests with farmers because about 40% of the total land area is used for agriculture. Organic agriculture, no-tillage, reduction in pesticide applications, and crop rotation can maintain this delicate balance with the surrounding ecosystems. The integration of plants-animals-environments can preserve the range of necessary natural processes. In the environment every community must play its pivotal role. Livestock, insects, fungi, and micro-organisms decompose organic matter and transfer nutrients to the soil. Bats, birds, bees, and butterflies facilitate pollination of fruiting plants and trees to maximize productivity. Swamps and marshes help in filtration and settlement of pollutants and provide clean water to the animal fauna around. Natural predators control unwanted and undesirable growth of any one species. Forests control flooding and protect land from erosion.

4.2.2 Ecosystem Services to Agriculture

Agriculture depends on the biodiversity of crops. Agriculture farming is not possible without the prevalence of a healthy ecosystem which is provided by the soil microorganisms. Poor biodiversity means minimum agricultural output. In intensive agriculture, the addition of excessive use of fertilizers has substituted the role of soil micro-fauna for nutrient recycling. However, the application of inorganic fertilizers at a higher rate can undermine other ecosystem services. Additional input of fertilizers can add to pollution because of their poor retention in soil and their washing out with rain. Artificial inputs can deliver 50% of the total nitrogen required in the system (Robertson & Swinton, 2005). Amongst soil fauna, the most familiar and understood contribution is from earthworms. Earthworms are very common on pastures, in mixed farming systems, and in broad-leaf woodland. The pastures can host 390 individuals per square meter, with a total of between 10 and 15 species. Considering their numbers, it might be assumed that earthworms share in soil fertility; however, their share in nitrogen contribution could be less than 1%. Nevertheless, earthworms are highly important because they initially remove the litter, break it down, and make it ready for other soil organisms. When they burrow in the soil and form casts, they maintain a good soil structure and help the water to infiltrate, ultimately facilitating aeration of the soil. Excessive ploughing—which is normally practiced in monoculture crop cultivation—significantly reduces the population of earthworms (Sastorum & Breen, 2003). Minimum mechanical applications help to increase the quantity of earthworms present in the soil (Bailey et al., 1999; Cole et al., 2006). It is unanimously agreed that the application of the monoculture agriculture system adversely impacts soil biota. In the absence of earthworms, organic acids may increase in the soil and increase soil acidification. Mere earthworms cannot be related to the whole agriculture productivity because earthworms are one part of the web of interrelated ecosystem services and have a partial contribution in agriculture productivity. Different earthworm species perform different functions at different soil depths. If the soil is not ploughed, earthworms take up a significant portion of nitrogen but their contribution becomes indistinct when wheat is grown with nitrogen-fixing crops (Schmidt & Curry, 1999). In fact it is difficult to ascertain the exact contribution of earthworms and clover in the fixing of nitrogen.

FIGURE 4.2 Eco-tourism. (*Source*: www.shutterstock.com.)

However, earthworms do help clover in germination and increasing phosphates for its growth. Nevertheless, experiments performed in field plots always displays variable results. Improvements in soil fertility induced by earthworms vary from country to country. In New Zealand, a long-term contribution from earthworms has been observed, which is in the order of 25–30%, while it is only 10% on the Dutch polders (Hoogerskamp et al., 1982). Elimination of earthworms has resulted in a drastic reduction in the porosity of soil, with an ultimate reduction in water infiltration (Lee, 1985). Westeringh (1972) observed that the absence of earthworms resulted in the buildup of undecomposed surface matter on Dutch farms.

4.2.3 Recreation and Ecotourism

Hiking, bird watching, nature walks, and scuba-diving are some of the recreational activities which are practiced in the natural environment. Ecotourism is a source of income and means of enjoyment. Eco-tourists frequently visit different countries to observe the biodiversity of that country. They view flagship species and sometimes hunt them to add to the economy of that country (Fig. 4.2).

4.2.4 Education and Scientific Value

Many books have been written on the natural environment. Similarly, many television programs and movies have been developed and produced for educational purposes as well as for entertainment for the young and for adults. Additionally, the natural environment is a source of several scientific ecological observations and publications. Although all these activities have no consumptive values, they are nonetheless an important source of knowledge about environmental knowhow and awareness and can help in imparting education to the people at the national and international levels.

4.3 HEALTHY ECOSYSTEM

A healthy ecosystem means healthy biodiversity. We will now discuss three components of ecosystem services and how they play a pivotal role in the health of the ecosystem. These services are:

1. Pollination.
2. Soil formation and protection,
3. Parasitism and predation by pests.

In recent years, concepts have changed and now the role of keystone species in ecosystem services is no longer given due importance. Now what is stressed is the functional interrelationship among different species living in a particular ecosystem, and it is further emphasized whether high biodiversity means there is a stable ecosystem which, in turn, ensures sustainability of the system. Then the concept of redundancy emerges, which means that several species play a similar role in the ecosystem (Bolger, 2000). However, the intensity of redundancy of species nevertheless depends on prevailing conditions of the environment at any given time. Adhering to this hypothesis, it is normally accepted that less redundant species are more susceptible to environmental changes and are lost first (Larsen et al., 2005). Although it has a low descriptive value, the concept of keystone species is still valid to some extent.

4.3.1 Pollination

A healthy ecosystem performs pollination, a well-known service of the ecosystem. Bees are keystone species among different pollinators. Bees collect nectar for the individual bee as well as for the colony. Compared to other insects, bees visit flowers more frequently. It is presumed that honey bees (*Apis mellifera*) pollinate up to 80% of the total flowers. However, despite this estimation bumble bees (*Bombus terrestres*) are still considered superior in pollination with better efficiency.

Bumble bees always try to move to the larger fields; nevertheless, they are more active in cool and wet climates (Santorum & Breen, 2004). Their longer tongue length facilitates them to pollinate with wider distribution ranges. Recently it has been observed that the population of bumble bees is on the decline, which is matter of concern for the healthy ecosystem. Thirteen true breed species have been counted and are distributed in the various parts of the world currently facilitating pollination of many horticultural and fruit crops. A low number of bees means low pollination, and consequently low agricultural production. For example, in Ireland people depend heavily on clover that can work without proper pollination but which is not possible without bees. Being a forage crop it contributes both to growth and gain in animal weight (Nolan & Grennan, 1998). Clover fixes nitrogen and can replace nitrogen fertilizers in extensive and organic farming. Up to 15 honey bee colonies per hectare are required for proper seeding of plants. Both the bumble bee and the honey bee can pollinate agriculturally important crops but the applicability of bumble bees is more widespread because they are more willing to cover larger distances and hence larger fields and pollination of clover flowers. These bees help in the pollination of a wide variety of crops, including strawberries, tomatoes, mustard seed, apples, strawberries, staple food crops, etc. Presently, organically grown crops are becoming more popular and hence are important too. Apples, for example, are very much dependent on bees for pollination and approximately 0.6 and 5 colonies per hectare are required for its successful pollination. In strawberries, for example, pollination increases the number of seeds present. In addition, the bee pollination improves the quality and size of the fruit. However, this is not the case for tomatoes, where wind pollination plays a greater role than honey bees. Physical shaking of the plants can also play a role in pollination. Although bees do not directly contribute in the pollination of these flowers, they do have a role in vibration of flowers, thus increasing the rate of pollination. Similarly, bees do not have much of a role in the pollination of rape seed flowers, which can be described as a future biofuel crop. Nevertheless, bee pollination is important in enhancing the timelines and the maturity of crops (Williams et al., 1989) (Fig. 4.3).

4.3.2 Domesticated Bees

Other than the services of wild bees, domesticated bees have been used for pollination purposes where hives and other amenities have been provided to bees for their successful survival and useful functioning. For the purposes of agriculture intensification, bees have been domesticated for the purpose of pollination and have been provided with required hives and other amenities. Beekeeping has been practiced in orchards for a long time and the development and establishment of

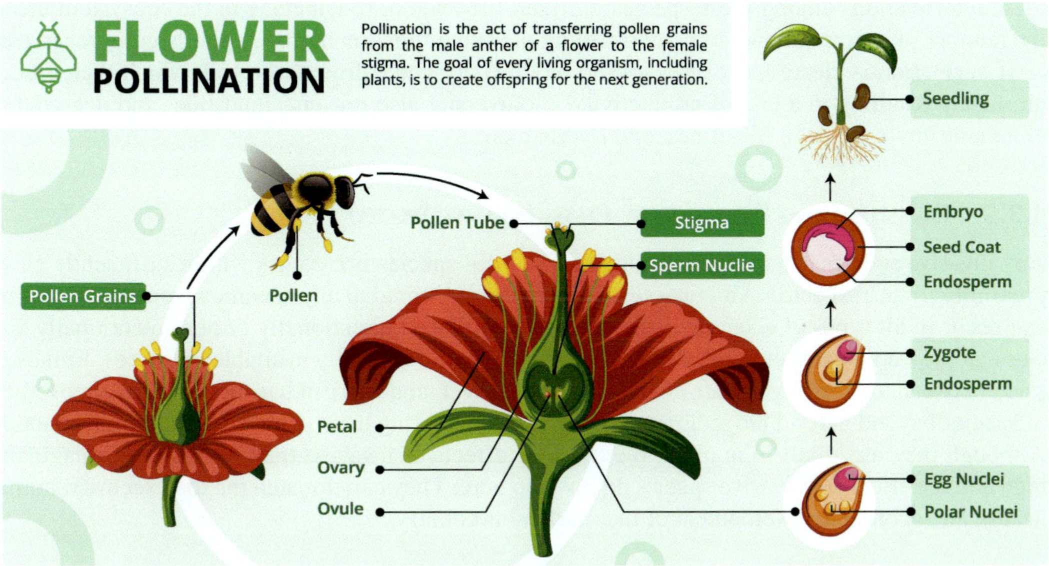

FIGURE 4.3 Pollination of flowers. (*Source*: www.Shutterstock.com.)

relationships between bee-keepers and farmers is an important topic of economic research With the decline and subsequent fluctuations in the wild honey bee population, pollination with domesticated bees has become more widespread especially in glass houses and poly tunnels. Pollination with honey bees started in around the 1980s and then bumble bees became more common; nevertheless, bumble bees are hard to manage and especially in glass houses and tunnels they always try to escape whenever they get a chance.

4.3.3 Honey Bee Pollination and Wild Plants and Gardens

Several successful attempts have been made to domesticate wild bees for pollination of various crops. However, the importance of wild bees for pollination of wild plants as well as crops grown on larger fields still cannot be ignored. These wild plants provide food to many wild as well as domestic animal species and even human being too. Other than crops, bees are also responsible for the production of gardens due to their vital role in the pollination of flowers.

4.3.4 Soil Formation and Protection

Biodiversity forms the soil, maintains its structure, and retains the desired moisture and required nutrient level for plants. Millipedes and worms perform this function by breaking down the organic matter and providing nutrients to plants. In the terrestrial ecosystem, invertebrates and fungi are vital in helping the formation of fertile soils (Bolger et al., 2000). Quantitatively, several thousand species of bacteria and other microorganisms may be contained in one gram of soil (Torsvik et al., 1994). Earthworms—which are considered a macro fauna of soil—help in the fertility of soil by physically breaking down plant leaves and dead grasses, releasing some nitrogen to plants. Mycorrhizal fungi, microbes, and smaller invertebrates further enhance the fertility of soil by decomposing this litter with the ultimate release of nitrogen for plant life. Microbial life is the biggest source of carbon after oceans. Nevertheless, information regarding smaller species and microbes (e.g., springtails (*Collembola*)), mites, and nematodes is very scanty. Similarly, there is poor knowledge about the positive and/or negative interrelationships as well as level of sensitivity of these interrelations among these species. Further, the level of redundancy in the ecosystem means the number of species known to be performing the same function has been poorly comprehended.

If vegetation is cleared, it causes loss to biodiversity and salinizes the soil, which then erodes, ultimately resulting in a loss of productivity. Biodiversity also prevents landslides and degradation of aquatic environments, coastlines, and riverbanks.

4.3.5 Parasitism and Predation by Pests (Invasive Species)

Any invasive species can transfer its parasites to local species if it comes within sufficiently closer proximity to local species. An invasive organism is not limited to any specific taxonomic group but can occur in all types of ecosystems. Externally introduced intentionally or non-intentionally species can prey on local species, and/or compete with them for locally available resources. Enhanced trade between and among countries has increased travel, and tourism has facilitated the movement of species beyond natural biogeographical barriers by creating new pathways for their introduction. Although they are small in number, the negative effects of invasive species can be unimaginably high and the effects of invasive species do not stop here: They can threaten the food security, human health, and economic development of that particular country.

4.4 IMPORTANCE OF BIODIVERSITY FOR ADAPTATIONS TO CLIMATE

During their evolutionary history human beings have faced different climatic changes and have tried to cope with this in one way or another for their successful survival. Climatic changes have occurred for a long time and now climate conditions are worse than ever before. Therefore, human beings need to take additional measures to reduce the impact of current climatic changes and those which are projected to come in the future (Leigh, 2009). Some factors like loss of biodiversity, land degradation, and damage to ecosystems further exacerbate the stressful conditions making it more complicated to live. If human beings are to live successfully in such an environment, adaptation will be highly important to address these challenges. Sustainable use of natural resources, their protection, and management can play an important role in the development of strategies which can help animals to adapt to climatic changes. Adaptation of organisms to the climatic conditions positively helps biodiversity in several activities. It restores and maintains native ecosystems, keeping all its services intact with the provision of necessary shelter for existing animal species and controlling invasive animal species. At the same time, this activity develops agroforestry in the transition zones and provides habitats for rare, threatened, and endangered species. If there is an increase in habitat conversion, overharvesting, pollution, and alien species invasions, these factors develop negative pressures on biodiversity, which ultimately reduces the adaptive capability of organisms to adapt themselves to the climatic conditions. The following example explains this phenomenon. Reduction in pollution pressure and lawful fishing (without explosives) will maintain a healthy coral reef community and help the coral reef to resist higher water temperatures, leading to a reduction in bleaching. Generally speaking, different interconnected marine reserves, protected areas, and freshwater and terrestrial ecosystems are well adapted to the changing climatic conditions, thus improving and facilitating the conservation requirements of the biodiversity.

4.4.1 Maintaining and Restoring Native Ecosystems

Conservation of biodiversity in an open area or in a controlled environment play a pivotal role in the cost-free protection of biological resources with the reduction of vulnerability of organisms to climatic changes.

Although wetlands provide many vital ecosystems on earth, at the same time they are threatened with extinction in one way or the other. They serve as vital water recharge areas, an important source for irrigation and drinking water for communities found downstream. Coastal wetlands act as a natural barrier and control storms and other hazards with ultimate reduction in the risk of settlements of disaster. A healthy mangrove ecosystem protects communities from extreme weather conditions so that they are not fully exposed to the environment; for example, the 2004 tsunami that hit Southeast Asia and the 2008 Cyclone Nargis that hit southern Myanmar. Where they can resist drastic ocean storms, wetlands host breeding and rearing of marine invertebrates, prawns, and fish. Some wetlands like freshwater lakes and rivers produce plenty of fish, which serve as balanced and nutritious food to people both rich or poor.

4.4.2 Landscape Connectivity

If a proper link is maintained between natural habitats and along altitudinal gradients, this strategy can help to develop adaptation of organisms to climate change. Further if the linkage is established between protected areas and developed and transformed productive landscapes, it can provide opportunities for the movement of species helping them in their maintenance of viable populations.

4.4.3 Adaptation in Agricultural Landscapes

In arid and semi-arid regions, agricultural productivity is very much dependent on climatic changes and subsequent rainfall patterns. It is estimated that by 2030 there will be 505 reductions in total production for rain fed agricultural crops purely due to climatic changes, which shows the vulnerability of these areas to climatic changes. Different climatic models have been developed for climate change sensitive areas like the drylands of West Asia and North Africa, which suggest that these areas will suffer from more severe temperatures following droughts. Some countries have already experienced intermittent droughts and flash floods which have devastated their economies.

4.4.4 Adaptation in Marine and Coastal Areas

Like terrestrial environments, coastal environments and fisheries do suffer from the significant impacts of climate change. The effects of rising sea levels and climate change are more prominent on mangroves and coastal wetlands. Climatic factors cause habitat fragmentation, destroy breeding habits of the animals, reduce food availability to the existing animals, which ultimately force them to move away and so affect their distribution. This factor also disturbs the established predator/prey relationships too. Perturbations in coastal areas and fluctuating climatic conditions abandon the nurseries and nursery grounds for a variety of animals, consequently threatening their biodiversity and unbalancing the ecosystem population. Structures developed near and around coastal areas primarily for the protection of terrestrial resources, including dikes and sea walls, further stress the need for marine resources and devastate their population structure.

4.4.5 Reduction in Resistance Capability

If the natural ecosystems are well restored, established, and protected, this can ameliorate the adaptive capability of organisms living in that ecosystem. Likewise, restoration, protection, and proper maintenance of mangroves can help in safeguarding coastal areas from the severe effects of rising sea levels and occurrence of other extreme weather conditions. Similarly, proper care and conservation of wetlands can help in controlling floods, thereby maintaining natural forests and improving water quality for living organisms in and around these wetlands.

At a global level, many countries are investing more and more on infrastructure to save coastal installations and to obviate the vulnerability of human settlements from flood control. Although a continuous decline in water availability for drinking and irrigation builds pressure to generate more sources for supply like construction dams, sea walls, and flood channels, this in turn will destroy natural habitats and so threaten biodiversity. All over the world many countries use forests and water reservoirs to control flooding, which could equally be applied in Pakistan if such structures are built and are operationally viable.

4.4.6 Adopting Indigenous Knowledge to Adapt to Climate Change

Indigenous people have adapted themselves to climatic changes by developing their own strategies to mitigate the effects of climate change. In some countries, land rights have been given to many territorial groups. These groups have shown much better land management and conservation strategies than those deprived of this facility. The harmful effects of climate change can be addressed much better if efforts of both government and the private sector are combined instead of individual efforts. Indigenous people are actually the power which can more conveniently and successfully adapt to the changing environment.

4.4.7 Protection of Natural Forests for Flood Control

Due to climatic changes, irregular rainfall patterns have resulted in unexpected droughts and floods. These intermittent cycles may either peak or decrease under all climatic conditions. In some countries, in the wet season flood is very common, as noted in Pakistan, when floods hit the southern areas each year and sometimes low lying plains suffer the same problem.

4.4.8 Mitigation/conservation Actions

Creation of compensatory protected areas, in-situ and ex-situ conservation of different animal species to minimize flood habitats, and control of pollution are some important factors which can mitigate climatic effects and can help in the conservation of biodiversity. In addition to these removal of vegetation, which can help water release management with moderate flow of stream water, and construction of fish passages for movement of fish and provision of fish hatchery facilities can also help. Proper implementation of fishing rights, removal of contaminants from fishing grounds, control of free-flowing pesticides, control of unnecessary sedimentation, and landscape management are some other factors which can help in the conservation of biodiversity.

4.4.8.1 Generation of Wildlife Corridors to Connect Habitats

For conservation and the successful survival of wildlife, its free movement and selection of breeding grounds of choice are both important. Some steps have been taken or could be taken to implement this strategy of conservation. The creation of compensatory protected areas, control of invasive species, revegetation of movement routes, improvement of land use management, and awareness and education about the environment and wildlife can pave the ways towards fully implementing of these strategies.

4.4.8.2 Land Use Management

Land use management covers the implementation of pollution control strategies, monitoring of water quality, management of existing wetlands, conservation of life living in and around them, and, finally, developing educational awareness programs about justifiable and fruitful utilization of land. Other than those mentioned above, if proper wildlife corridors are maintained which help free movement of wildlife, aquatic vegetation is controlled with mechanical methods, channels are built for the control of floods, diseases and their vectors are under control, encroachment of land which is actually used for living organisms is discouraged, then all these activities can help to mitigate the effects of climate change. Over the millennia, local populations have developed adaptation models to climate change. These people developed genetically different useful plants for medicinal and consumptive purposes, along with different animal breeds having wider resistance capabilities to hostile environmental conditions.

4.5 PROTECTION OF WATER RESOURCES AND POLLUTION MANAGEMENT

4.5.1 Protection of Water Resources

Biodiversity acts as a buffer and helps to control extreme drought and floods. Additionally, it protects watersheds and maintains water quality, ultimately playing a part in the protection of natural water resources using biological communities. Natural vegetation present on the land protects ecosystems from extreme events such as flood and drought and their severe effects. This vegetation cover regulates and stabilizes unnecessary water runoff and maintains water quality, as well maintaining the continuity of water cycles. Removal of vegetation and deforestation silts up the reservoirs, which reduces total water availability, spoils water quality, and ultimately deteriorates aquatic habitats. Wetlands and forests are important in water purification and mangroves trap and hold silt in place with the reduction of its impacts on marine ecosystems. Due to loss of biodiversity, some reservoirs have lost up to 19% of their water-holding capacity.

4.5.2 Breakdown of Different Pollutants and Their Absorption

Although biodiversity controls pollution, the opposite can also happen when pollution can play a destructive role and can damage biodiversity. Hence, all countries have special concerns about pollution and always try to control and mitigate its affects if it happens. Speaking generally, pollution refers to chemical contaminants that are released into the environment intentionally and/or unintentionally that harm the environment and breach its stability. Pollution can be of several types and produces several chemical compounds, whose intensity of damage always depends on their properties and concentrations. Nitrogen and phosphorus loading are of primary concern and are primary causes of biodiversity loss. They severely disturb the functioning of ecosystem on dry land, in wetlands, and coastal regions. Although both these nutrients are limiting in the natural environment, their abundance in any system results in rapid plant growth and alters the composition and function of an ecosystem. Excessive outflow of sewage and agricultural runoff are the main causes of their abundance in the natural environment and ecosystems. Now, the next question is how biodiversity manipulates this excessive and dangerous infiltration of pollutants into the environment. As mentioned above, most pollutants are from human sewage and others from oil spills. Humans create lot of pollutants like sewage and oil spills. Ecosystems and ecological processes break down these pollutants and absorb them, making water suitable for use for human beings and other organisms. It is biodiversity that processes the water and makes it suitable for use in agriculture, e.g., for staple crops, vegetables, and horticulture.

4.5.3 Regulation of Climate

Stabilized forestation contributes to the hydrological cycle and maintains rainfall in the immediate surrounding environment. This forestation then washes away pollutants and absorbs different pollutants from the air and processes them, thus purifying the air and making it fit for utilization by different organisms and human beings. In addition to that, forestation also maintains microclimates required for some organisms for their successful survival and further proliferation.

4.6 SOCIAL, CULTURAL, AND SPIRITUAL IMPORTANCE

Many people highly value biodiversity and its components aesthetically. The importance of biodiversity is well appreciated by human beings because it reveals that appreciation has several dimensions, including an appreciation of how biodiversity reveals the complex and entwined history of life on earth and how it characterizes personal experiences and various landscapes. In addition to moral, ethical, and religious values, there also are deeply logical reasons to conserve biodiversity. People watch birds and butterflies for their recreation and keep reptiles and fish as pets for their entertainment. People also raise a variety of plants, participate in native-plant societies, view natural photographs, watch television programs, and read books about nature and biodiversity for their enjoyment. Kiester (1997) has suggested that such experiences and observations help and support experts to define and elucidate biodiversity. As artists ask us to appreciate their artistic skills, likewise nature lovers help us to appreciate biodiversity (Kim et al., 2017).

Due to limited knowledge about biodiversity, self-awareness about it has recently become more common. Biodiversity develops for the history of life on earth, which inherited variation from generation to generation vertically. Hence it can be said that every species present on earth has unique history behind it. Although evolution can be defined in variety of ways, two definitions in particular are well known. The first is Darwin's "descent with modification" and the other one is Salthe's "accumulated history." Lineage of each species contains its own history and this lineage concept is important to display the images of evolution. This concept has expanded further and the vast display of life through time has become part of our culture. Taking the concept of lineage a little further, some pairs of lineages are closer to each other than others. Their close relationship depicts that these lineages have a more recent common ancestor. Methods have been developed to assess the degree of phylogenetic relationship and assist in the reconstruction of the history of life on earth. These developments have made it possible to express values of biodiversity in new ways.

4.7 ECONOMIC GROWTH

As discussed above, the conservation and maintenance of biodiversity is essential for the disposal of organic wastes, for the formation and protection of soil,, biological nitrogen fixation, crop and livestock genetics, for production of medicines, biological control of pests, and pollination of plants. Plants and microbes facilitate the degradation of chemical pollutants and organic wastes with ultimate recycling of nutrients through the ecosystem. Honey bees, bumble bees, and butterflies pollinate plants and trees, providing significant benefits to the environment and economic benefits to agriculture and natural ecosystems. These activities enhance the productivity of crops and add diversity to the ecosystem. The importance of insect pollination is well established and it is reported that one-third of the world's total food production directly or indirectly depends on crops that are insect pollinated. According to one estimate, about 130 of the crops grown in the United States are insect pollinated. Habitat fragmentation and loss have adverse effects on pollinators because these affect the food sources of pollinators, their nesting, and breeding sites, significantly decreasing the population of pollinators. Millions of tons of food worth billions of dollars is produced annually, which is attributed to different pollinators.

Water and soil biodiversity is important for constancy and effectiveness of the processes of bioremediation and bio-treatment. Humans depend on sustainable functioning of agricultural, forest, and natural ecosystems. Human activities, however, like development of natural lands are causing species extinction at an accelerated rate which is 1,000 to 10,000 times that of the natural rate of extinction of species. It is assumed that at a global level biodiversity annually provides benefits of US$2,928 billion to human beings out of which US$319 billion goes only to the USA. Conservation of biodiversity and alleviation of poverty are two of the global challenges that are indistinguishably linked together. But it is unfortunate that biodiversity is always undervalued or never valued at all and its contribution to national economies is never recognized. Rural populations specifically and urban populations in general are dependent on biodiversity for their survival and daily needs. Biodiversity conservation is one way to root out this menace under some circumstances. Goods produced from biodiversity are often of a low value, which is important for poorer people but of little value to the rich. Nevertheless, there are several well-established mechanisms which can successfully conserve biodiversity, ultimately reducing poverty. These are protection of timber forest products, paid environmental services, tourism purely based on nature, protection of fish spillovers, restoration of mangroves, promotion of agro-forestry, management of grasslands, and, finally, conservation of agrobiodiversity. Where different mechanisms for conservation of biodiversity are devised and regularly applied it is also imperative to avoid practices of unsustainable use of biodiversity and overexploitation of resources which undermines the overall integrity of biodiversity, ultimately hindering the services it is providing to human beings and other organisms.

4.8 NUTRIENT RECYCLING AND LAND IMPROVEMENT

Biogeochemical cycles transport and transform matter within the biosphere, hydrosphere, lithosphere, and atmosphere on our planet. These four categorical areas make up our planet. The biosphere hosts all living organisms, including fungi, plants, and animals. The other three spheres are the lithosphere (which includes soil, land, and the earth's crust), the atmosphere (which is composed of air and space), and the hydrosphere (which includes water such as rivers, lakes, and oceans). All the three later areas are devoid of life. However, the place where the biosphere overlaps with the lithosphere, the atmosphere, and the hydrosphere, is inhabited by living organisms.

The broader cycle that governs and maintains the functioning of our planet is made up of biogeochemical cycles. The Earth itself makes up a system which closes as well as opens. As an open system, it receives electromagnetic radiations from the sun and outer space but, when we talk about matter, it is in fact earth which is a closed system. Planet earth receives a minimum amount of matter except that which it receives from collisions of meteorites and some minor amounts of intergalactic particle trapping which is actually lost in the upper atmosphere. Therefore, the only matter that is transformed and circulated geographically is that which it contained since its birth. This proves the law of conservation of matter that matter is neither created nor destroyed but it only transforms between matter and energy. This transformation of matter takes place through biological, geological, and chemical processes which are collectively called cycles of nature or biogeochemical cycles. These cycles link all organisms and abiotic features on earth. Transformation of matter between biotic and abiotic elements on the earth is a continuous activity. Biogeochemical cycles facilitate this transfer of matter from one form to another and from one location to another. Sometimes these biogeochemical cycles are also called nutrient cycles, because these cycles also take part in the transfer of compounds that provide nutritional support to living organisms.

4.8.1 Type and Form of Elements Transported and Recycled in Biogeochemical Cycles

After death, living organisms die and decay. Their body structures collapse and disintegrate into individual molecules. Component molecules released from animal dead bodies then join the biogeochemical cycle. Elements which are transported in the biogeochemical cycles are: (1) micro elements which include boron, copper, and molybdenum Green plants, some enzymes, and nitrogen fixing bacteria use them while living animals require them in very small amounts, hence they are named as micro elements; (2) macro elements which include carbon, hydrogen, oxygen, nitrogen, phosphorous, and sulfur are the major constituents of the biogeochemical cycles. They may be further reduced to ionized or hydrated ions in aqueous solutions or to ions in the atmosphere. New organisms then recycle them and after metabolism these animals pass them on through excreta, or death/decay for their further recycling. Living organisms require them in larger amounts; hence, they are called macro elements.

4.8.2 The Importance of Biogeochemical Cycles

At the ecosystem level, biogeochemical cycles perform several functions and ensure survival of all the living organisms, including human ones. These cycles transform matter from one form to another for utilization by particular organisms. The hydrological cycle condenses water vapors to liquid and then to ice. When ice melts it is converted to liquid water which is used by human beings and other animals. Although it is in abundance in an atmosphere, nitrogen is still a limiting nutrient for plant growth. This limitation is due to the reason that most plants take up nitrogen in two solid forms, ammonium ion (NH4+) and nitrate ($NO3^{-}$) but are not capable of absorbing nitrogen in the gas form. Biogeochemical cycles, therefore, provide elements to organisms in suitable form to use which fertilize and improve the land simultaneously. These cycles also accomplish aerial transfer of molecules from one place to another. Nitrogen has the highest concentration in air, some of which is transferred to soil through the nitrogen cycle. Through biogeochemical cycles, elements are stored in their natural reservoirs. These reservoirs release them from time to time in consumable forms. When required, plants and animals use them gradually. These elements maintain the function of ecosystem. If there is any imbalance, then the ecosystem equilibrates it through biogeochemical cycles which may take few days or even several years. In the context of the functioning of biogeochemical cycles, they link living organisms with living, living organisms with non-living, and non-living organisms with non-living organisms; this link is primarily based on the flow of nutrients. The biogeochemical cycles pass through all the four different spheres. Each sphere has a particular medium with specific viscosity and density. The elemental flow in these spheres depends on the viscosity and density of that medium, and hence it differs in each sphere which regulates the flow of elements in those cycles.

4.8.2.1 Phosphorus Cycle

Phosphorus is liquid at standard temperatures and pressures. Unlike nitrogen, it is not found in gaseous state in air; rather, it is very common in soil, sediments, and water. Where present it is also cycled through these media like water, soil, and sediments. Small amounts of elemental phosphorus or its compounds may be present in very small particles present in the atmosphere. Movement of phosphorus from deposits and sediments to the living organisms is very slow. This movement of phosphorus is further slowed down when it moves back into the soil and water sediment. Among all the sedimentary cycles, the phosphorus cycle is the slowest one among all the sedimentary cycles.

4.8.2.2 Nitrogen Cycle

Atmosphere is a very rich source of nitrogen (about 78% of air) and hence all the nitrogen found in terrestrial ecosystems originates from this atmosphere. For plant life, the nitrogen cycle is by far the most important nutrient cycle found in the atmosphere (Fig. 4.4).

4.8.2.3 Oxygen Cycle

Oxygen has three main reservoirs called the atmosphere, the biosphere, and the lithosphere. Oxygen continuously moves within and between these spheres, which is known as the oxygen cycle. During the process of photosynthesis, oxygen is produced and carbon dioxide is absorbed by plants. Due to this simultaneous release and absorption, oxygen and carbon cycles are usually linked and are collectively called the oxygen-carbon cycle.

4.8.2.4 Sulfur Cycle

Natural distribution of sulfur is restricted to the sedimentary cycles where it is present in solid form. Sulfur is transported from one place to another by wind, erosion by water, and geological events like volcanic eruptions. It is normally found in compounds of sulfur dioxide, sulfuric acid, salts of sulfate, or organic sulfur. Sulfur is recycled from the ocean to the atmosphere, to land, and then to the ocean through rainfall and rivers.

4.8.2.5 Carbon Cycle

Carbon is one of the most important elements for the sustenance of life on earth. Carbon dioxide and methane gases (compounds of carbon) have substantial effects on earth's heat balance in the earth's atmosphere. Carbon absorbs infrared radiations. In this way, it can be a major contributor to global warming following climatic changes.

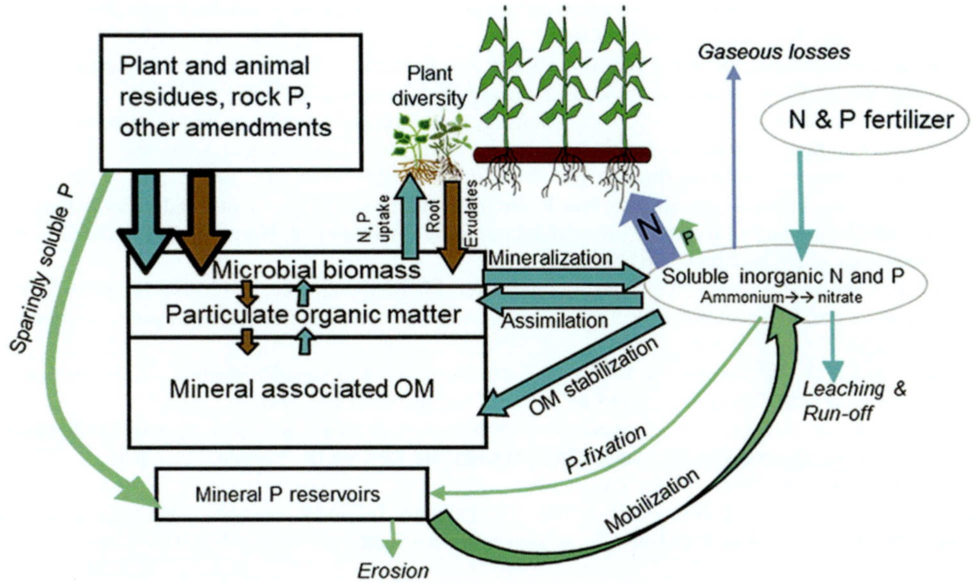

FIGURE 4.4 Ecological Nutrient Management (ENM) expands the focus of nutrient management to soil organic matter (SOM). (Courtesy of Drinkwater and Snapp, 2022; School of integrated Plant Science Cornell University Ithaca NY, USA; Open access.)

REFERENCES

Bailey, A.P., Rehman, T., Park, J., Keatinge, J.D.H. & Tranter, R.B. (1999). Towards a Method for the Economic Evaluation of Environmental Indicators for UK Integrated Arable Farming Systems. Agriculture, Ecosystems and Environment 72, 145–158.

Bolger, T. (2000). The Functional Value of Species Biodiversity – A Review. Biology and Environment 101B(3), 199–224.

Bolger, T., Schmidt, O., Purvis, G. & Curry, J.P. (2000). The Biodiversity, Function and Management of Soil Invertebrate Populations – An Irish Perspective. In: Achievement and Challenge: Rio 10 and Ireland, eds. Convery, F.C. & Feehan, J., Environmental Institute, University College Dublin.

Cole, L., Bradford, M.A., Shaw, P.J.A. & Bardgett, R.D. (2006). The Abundance, Richness and Functional Role of Soil Meso- and Macrofauna in Temperate Grassland – A Case Study. Applied Soil Ecology 33, 186–198.

Drinkwater, L.E. & Snapp, S.S. (2022). Advancing the Science and Practice of Ecological Nutrient Management for Smallholder Farmers. Frontiers in Sustainable Food Systems 6, 921216. https://doi.org/10.3389/fsufs.2022.921216.

FAO (2004). Biodiversity for Food Security, FAO 2004/26. Pages 9.

Hoogerskamp, M., Rogaar, H. & Eijsackers, H.J.P. (1982). Effect of Earthworms on Grassland on Recently Reclaimed Polder Soils in the Netherlands. In: Earthworm Ecology, ed. Satchell, J.E., Chapman & Hall, London.

Kiester, R. (1997). Aesthetics of Biological Diversity. Human Ecology Review 3, 151–157.

Larsen, T.H., Williams, N.M. & Kremen, C. (2005). Extinction Order and Altered Community Structure Rapidly Disrupts Ecosystem Functioning. Ecology Letters 7, 1109–1119.

Lee, J.-H., Kim, S.-H. & Kwon, H.-S. (2017). Mapping Interests by Stakeholders' Subjectivities Toward Ecotourism Resources: The Case of Seocheon-Gun, Korea. Sustainability 9, 93. https://doi.org/10.3390/su9010093

Lee, K.E. (1985). Earthworms: Their Ecology and Relationships With Soil and Land Use. Academic Press, London.

Leigh, A(2009). What We have Learned in 2008? Nature Climate Change 1, 4–6.

Nolan, T. & Grennan, E.J. (1998). Effect of Grazing Management on the Maintenance of White Clover, Teagasc Report 3158 Sheep Production Department, Teagasc Research Centre.

Patel, D.K. (2014). Biodiversity and Its Importance. Journal of Biodiversity and Endanger Species 2(4), 1–1. https://doi.org/10.4172/2332-2543.1000e117.

Quadros, G. (2013). Impact of Pollution on the Flora and Fauna – A Loss of Ecosystem services. National Conference on Biodiversity: Status and Challenges in Conservation "FAVEO" 2013.

Robertson, G.P. & Swinton, S.M. (2005). Reconciling Agricultural Productivity and Environment Integrity: A Grand Challenge for Agriculture. Frontiers in Ecology and Environment 3, 38–46.

Sastorum, V. & Breen, J. (2003). Bumble Bee Diversity on Irish Farmland, Tearmann – Irish. Journal of Agri-Environmental Research. In: The Economic and Social Aspects of Biodiversity: Benefits and Costs of Biodiversity in Ireland. A Report Prepared by Environment, Heritage, and Local Government, Governemnt of Ireland 2008, pp. 199.

Schmidt, M., Lauer, A., Purtauf, T., Thies, C., Schaefer, M. & Tscharntke, T. (2003). Relative Importance of Predators and Parasitoids of Cereal Aphid Control. Proceedings of Royal Society, London, B 270, 10905–190.

Schmidt, O. & Curry, J.P. (1999),The Effect of Earthworms on Biomass Production, N. Allocation and N Transfer in Wheat-Clover Intercropping Model Systems. Plant and Soil, 214 (1/2).

Torsvik, V., Goksoyr, J., Daae, F.L., Sorheim, R., Michalsen, J. & Salte, K. (1994). Use of DNA Analysis to Determine the Diversity of Microbial Communities. In: Beyond the Biomass, eds. Ritz, K., Dighton, J. & Giller, K.E., pp. 39–48. Wiley, Chichester.

Williams, I.H., Martin, A.P. & White, R.P. (1989). The Effect of Insect Pollination on Plant Development and Seed Production in Winter Oilseed Rape. Journal of Agricultural Science 109, 135–139.

5 Classifications and Phylogeny of Animals

5.1 LINNAEUS AND THE DEVELOPMENT OF CLASSIFICATION

The Greek philosopher and biologist Aristotle first classified organisms on the basis of their structural similarities. Later on in Europe, the English naturalist John Ray (1627–1705) introduced comparatively a comprehensive and more advanced system of classification, along with an introduction to advanced concepts of species. The gradual development of systematics in the 18th century was completed when Carolus Linnaeus (1707–1778) offered a scheme of classification of organisms currently practiced.

Linnaeus was a botanist by profession and was working in the University of Uppsala in Sweden. He was a very talented person and was fond of collecting different objects and then always tried to classify them. His main focus was on flowers. During his studies, Linnaeus planned and introduced an extensive system of classification both for plants and animals. After formulation of his scheme of classification, he published his great work, globally well-known and recognized as *Systema Naturae*. His scheme of classification was based on morphology, which means a comparative study of forms of organisms, and arranged specimens he collected during the course of his study. His division of animal kingdom into species giving a distinct name to each species was a very important task. In this context, he grouped different species into genera, put genera under different orders, and finally grouped morphologically similar orders under a relevant class. In this way, he came up with several classes of animals which have similarities in their forms. As he was a botanist with limited knowledge about animals, the animal genera which he designed included those animals which were very different in morphological features and were not closely related to each other. To date, many changes have been made to his scheme of study. Despite these extreme changes in his scheme, the fundamental principles are still strictly adhered to. Linnaeus's scheme of classification was designated as a hierarchical system of classification in which he arranged organisms into an ascending series of groups. He arranged animals in taxa (or taxon in the singular) and then to each particular taxon he allotted several standard taxonomic ranks to indicate the number of animals included in each taxonomic rank. Since Linnaeus's time, there has been considerable expansion in this hierarchy of taxonomic ranks. According to his scheme of study, the animal kingdom now contains seven mandatory ranks, which have been further arranged into a descending order of division and classification. These groups are called kingdom, phylum, class, order, family, genus, and species. Following Linnaeus's scheme of classification, animals are then split into at least seven taxa, each with one mandatory rank. Taxonomists have further subdivided Linnaeus's seven ranks into more than seven taxa, called superclass, subclass, infraclass, superorder, suborder, etc. (Margulius & Schwartz 1987) (Fig. 5.1). These subdivisions can be for any particular group of organisms. In Linnaeus's original classification, more than 30 taxonomic ranks have been documented to date. Indeed, taxonomists are always searching for additional ranks for fish and insects (very large and complex groups). In continuation of this classification, we now need additional ranks for the expression of different degrees of evolutionary divergence which occurred at various steps throughout the history of life.

In taxonomy. the system of classification developed by Linnaeus is called binomial nomenclature, because each species has a Latin name which is composed of two names (written in italics). In binomial nomenclature, the first word represents the genus with first letter being a capital letter and second word represents the species, which starts with a small letter. The following example

DOI: 10.1201/9781003472001-5

will further explain this. *Turdus migratorius* is the scientific name for the common robin. In these two Latin names, the first one is *Turdus*, which means thrush, and the second name is *migratorius*, which means "of migratory habit." Genera always stand only for single groups of organisms, which means that two different genera cannot have the same name. The following example further explains this. *Sitta carolinensis* is the scientific name of the white-breasted nuthatch. This species name *carolinensis* can be used with another genera too, for example, *Parus carolinensis* (Carolina chickadee) and *Anolis carolinensis* (green anole), a lizard to mean "of Carolina." Genera, however, cannot be used with different groups of animals. When we work on the "diversity of animal life," taxonomists use uninominal nouns which always start with an initial capital letter. Sometimes, species of an organism is further sub-divided into subspecies. In this case, taxonomists use trinomial nomenclature. In trinomial nomenclature, the last two words differentiate between two species. Thus, when we want to differentiate the southern form of the robin from the eastern robin, the scientific term *Turdus migratorius achrustera* (duller color) is employed for the southern type. The genus, species, and sub-species names are always written in italics. Due to small differences between two species, taxonomists do not care much about sub-species. Differences among subspecies are usually due to a few superficial characters and do not prove them to be an evolutionary distinct unit.

5.2 CHARACTERS OF A TAXON, AND PHYLOGENETIC RECONSTRUCTION

Systematics mainly focus on the reconstruction of an evolutionary tree or phylogeny. This phylogeny relates to all extant and extinct species. For reconstruction of an evolutionary tree, we study those features and characters of an organism which differ among species. In taxonomy, character is any feature which taxonomists uses to study variation within and among species. Potential useful taxonomic characters in the form of morphological, chromosomal, and molecular features exist in each organism. First of all, taxonomists observe patterns of similarity among organisms and then search for unique characters. Two organisms having similar features tell that both these organisms have received these characters from a common ancestor. Similarity in characters and their inheritance from common ancestors expresses a phenomenon commonly called homology. Nevertheless, it is not always true; sometimes similarity in characters does not reflect that these organisms have common ancestry.

Evolutionarily, when life originated independently, it possessed similar features but development followed different lineages. This phenomenon of origin of life produced patterns of similarity among organisms but it does not reflect that these organisms have the common ancestor too. Origin of life on this pattern and occurrence of different incidences during the course of evolution creates complications for the taxonomists to work with. If organisms have similar characters but do not prove their common ancestor, this phenomenon can be called homoplasy, or, more simply, non-homologous similarity. If the "variation of characters" is the basis of the reconstruction of phylogeny, then the group of characters is considered on the basis of which these organisms vary among themselves (Fig. 5.1). During a reconstruction of phylogeny, it is also considered which variant form of that character was present in the common ancestor of the entire group. Such types of characters that are considered for this purpose are called evolutionarily derived character states (Joyoti 2023).

This phenomenon can be further elaborated as follows. For example, if dentition is considered as a character in amniotic vertebrates like reptiles, birds, and mammals, then it can be said that the absence of teeth in the jaws and the presence of dentine express two different states of a character. Taking it at class level, class Aves do not have teeth but other amniotes have teeth. On the evaluation of this character state, on the basis of polarity this character demands us to determine the condition which character state such as presence or absence of teeth characterizes the most recent common ancestor of amniotes and which character state is a derivation which happened later on within the amniotes. The method which we use to examine the polarity of a variable character is called out-group comparison.

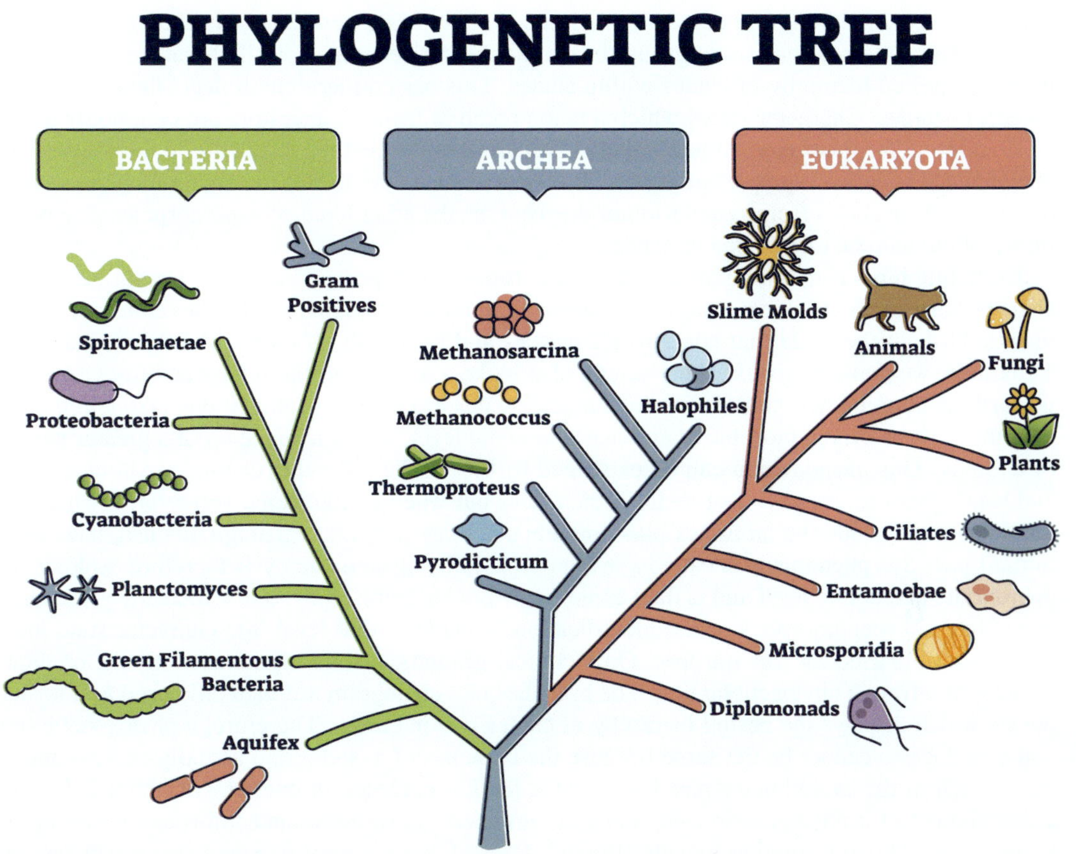

FIGURE 5.1 Phylogenetic Tree. (*Source:* www.shutterstock.com.)

Taking this method further, we consult additional groups of organisms known as an outgroup. This outgroup is phylogenetically close to the group under study but actually it does not exist in that particular group. Therefore, it can be comfortably stated that any character state that is present both in the outgroup (group used for comparison) as well as in group under study is an ancestral for the study group.

If we want to study polarization of variation in the amniotic dentition, then different groups of bony fish and amphibians make up an appropriate outgroup to these amniotes. The presence of teeth in bony fish and amphibians expresses ancestral character and the absence of teeth represents a derived feature in these group of organisms. If we go to the polarity of this feature, it informs us that modern birds lost their teeth in their ancestral pedigree. When we use different outgroups then an evaluation of the polarity of characters become most effective. All the different forms and states of characters are considered to be derived if they are present in the study group and are absent in appropriate outgroups. If organisms or species share derived character states, they form subsets within their group which are called clades, from the German word *klados*, meaning branch (Wiley et al. 1991).

Now, any derived character which members of clade share among themselves is formally called a synapomorphy (which comes from the German *synapsis*, meaning joined together and *mophf*, meaning form of state) of that clade. The phenomenon of synapomorphies indicates homology among organisms and, from the above statement, taxonomists infer that it is a particular group of organisms that form a clade. Within the amniotes, the presence of feathers and the absence of teeth are synapomorphies that identify the birds as a clade. A clade consists of a unit of animals which evolved from common descent, which means that all descendants have the same ancestral lineage. The pattern of our study group that forms the derived states of all characters take up the form of a nested hierarchy of clades within clades. This nested hierarchy reflects the goal of phylogenetic studies. Character states which a taxon receives from its ancestors are commonly called plesiomorphic for that taxon. When organisms share these ancestral states among themselves, the phenomenon is called "symplesiomorphy." It is impossible to get useful information on the nesting of clades within clades from symplesiomorphies but, on the other hand, synapomorphies give better information than the former phenomenon.

Excluding birds if mammalian and reptilian groups which possess teeth, we cannot get a valid clade because birds also have common ancestors with reptiles and mammals. Therefore, we should include birds in the clade that contains all reptiles and mammals. If during the determination of a character we make a mistake then we can also make a mistake in the inference of phylogeny. It is worth mentioning here that if certain character states are plesiomorphic at one taxonomic level, then they can be synapomorphic at another taxonomic level, which is made up of a greater variety of animals. This phenomenon can be explained further in the following example. Although birds and some gnathstomes have lost teeth in their jaws but when gnathostome vertebrates, which are amniotes and include the amniotes plus amphibians, bony fish, and cartilaginous fish, have teeth in their jaws this phenomena is called synapomorphy. Phylogenetic analysis therefore explores the appropriate taxonomic level and is then explored by phylogenetic analysis at which any given character state is synapomorph. For the identification of clade at that level, the character state mentioned above is used for this purpose. On technical grounds, taxonomists frequently differentiate a cladogram from a phylogenetic tree. The branches of a cladogram are then only used as a formal device which displays the nested hierarchy of clades within clades. Therefore, a phylogenetic tree and a cladogram cannot be the same because the branches of a cladogram actually express ancestries which in the evolutionary past have shown their occurrence in one way or another. For the establishment of a phylogenetic tree, a cladogram needs some additional information pertaining to ancestors. This information includes the durations of evolutionary lineages, or we can say just how much evolutionary change occurred in each ancestry. Despite consistency of the cladogram structure with phylogenetic trees, the cladogram is still the first approximation of the phylogenetic tree (Hull 1988).

5.2.1 Sources of Phylogenetic Information

Different characters of an organism are used to construct cladograms in comparative biochemistry, cytology, embryology, and morphology. In particular, the variable structures of integuments (which include scales, hair, and feathers), limb bones, and skull bones are fundamental to the reconstruction of the phylogeny of vertebrates. Comparative morphology uses specimens collected from living organisms and those from fossilized forms to recognize variable characters and ultimately the reconstruction of cladograms, while amino acid sequences in proteins and nucleotide sequences in nucleic acids are used in comparative biochemistry for the same purpose. The application of direct sequencing of DNA is an important tool for phylogenetic studies. Comparisons of protein sequences are usually used indirectly, however, which apply allozymic and/or immunological methods. In addition to the above methods, sometimes inferences from DNA sequences of protein-coding genes are also used for phylogenetic studies (Hills et al. 1996).

Contrary to the previous findings, it has been noted very recently that in addition to living organisms biochemistry can also be applied to fossils for phylogenetic studies. Comparative cytology uses shapes, numbers, and sizes of chromosomes and their parts for the collection of variable characters for the ultimate construction of cladograms. Unlike other subject areas, comparative cytology is almost exclusively used for living organisms but not for fossils. Consultation of the fossil record is essential for the addition of an evolutionary timescale for development and production of a phylogenetic tree. The method of radioactive dating is used to determine the age of the fossil, which shows the derived characters of a particular clade. Nevertheless, ages of different lineages on the phylogenetic tree can be estimated by comparative biochemistry.

Through evolutionary time, approximately linear rates of divergence have been observed in some protein and DNA sequences. In the comparative biochemistry, differences are measured between proteins and DNA sequences of two of the most recent common ancestors of two species, which tells us the age of these common ancestors of these two species. The evolution of proteins and DNA sequences is calibrated by the measurement of their divergence between species, the age of whose most recent common ancestor has been determined using fossils. Then molecular evolutionary calibration estimates the ages of other branches on the phylogenetic tree of animals.

5.3 THEORIES OF TAXONOMY

Taxonomy theory provides the doctrines which can be used for recognition of taxonomic groups and their further ranking. Currently, there are two tried and tested theories of taxonomy:

1. Traditional evolutionary taxonomy.
2. Phylogenetic systematics (cladistics).

Both these theories are based on evolutionary principles. The differences between these theories of taxonomy have important extrapolations and convey a message about how we apply taxonomy for the study of the evolutionary processes which have been completed in the past and/or are ongoing now. For both of these theories, it is important that the taxonomic group and the phylogenetic tree or cladogram should relate to each other. The three forms of this relationship are designated as monophyly, paraphyly, or polyphyly. A taxon is monophyletic if it includes the most recent common ancestors of the group and all its descendants. A taxon is called paraphyletic if it includes the most recent common ancestor of all members of a group and includes some but not all of the descendants of that ancestor. A taxon is polyphyletic if it does not include the most recent common ancestor of all members of a group. This condition demands that in the past the group of organisms should have had at least two separate evolutionary origins. During the evolutionary period, they acquired similar features but these features remained independent from each other. When classification of animals is based on both evolutionary and cladistics taxonomy, monophyletic groups are accepted while polyphyletic groups are rejected. Both evolutionary and cladistics taxonomy differ on the acceptance of paraphyletic groups, and from the evolution point of view this difference has important inferences.

5.3.1 Traditional Evolutionary Taxonomy

For the recognition of higher taxa and their subsequent rankings, traditional evolutionary taxonomy uses two different evolutionary principles: (1) the Principle of common ancestry; and (2) an amount of adaptive evolutionary change, as can be observed on a phylogenetic tree. It is essentially important that evolutionary taxa should use a single evolutionary origin with the reflection of distinct adaptive features. Effective development and formulation of formal procedures of evolutionary taxonomy are hereby attributed to George Gaylord Simpson, a mammalian paleontologist.

He suggests that on the evolutionary tree if a particular branch or group of animals represents a distinct adaptive zone then it is ranked at a status of higher taxon. Simpson opines that when an organism lives in a particular environment it develops a typical interaction and relationship which is called its adaptive zone. If an organism changes its structure and behavior and enters into a new zone, the population which emerges from that organism in that particular zone develops new ways and means for utilization of the resources present in that environment. If a taxon is living in a unique adaptive zone different from other organisms then it is called its grade.

For example, Simpson says that within birds, penguins have a distinct adaptive zone. The lineage which was the immediate ancestors of all penguins virtually modified their body and wing structure. This structural modification switched their aerial (flying) mode to aquatic (wading) mode. Considering the habitat, morphology, and behavior of those aquatic birds which can fly and swim on water or can wade under water, these fall in between aerial and aquatic adaptive zones. Alterations in the body and wings for swimming of penguins represent a new grade of organization. Accordingly, within birds penguins have adapted a distinct taxon and have been placed under the family *Spheniscidae.* As mentioned above, whenever animals fully occupy any area or zone with comparatively broader adaptations, then these animals are ranked higher in their corresponding taxon.

Evolutionary taxa may be either monophyletic or paraphyletic. During the recognition of paraphyletic taxa, our taxonomies distort patterns of common origin. An anthropoid primate's evolutionary taxonomy provides a good example of paraphyletic taxa.

According to the taxonomical organization family, Hominidae covers humans (genus Homo) and their immediate ancestors which have been fossilized. Accordingly, chimpanzees (genus Pan), gorillas (genus Gorilla), and orang-utans () come under the family Pongidae. Nevertheless, it has been observed that the genus Pongo pongid genera (Panand gorillas) share more common recent ancestry with family Hominidae, hence they are more closely related than the other members (Pongids) genus Pongo. Now, as the family Pongidae excludes humans—who also originated from the most common ancestors of all Pongids—this exclusion process makes Pongids a paraphyletic group of animals. Evolutionary taxonomists are of the view that animals belonging to Pongid genera are a single, family-level grade of primates with limited mental capacity bearing arboreal and herbivorous habits. Nevertheless, their family adaptive zone is the same. When we talk about humans, unlike pongids they are terrestrial in habitat and are omnivorous in feeding habits. They have considerably developed mental abilities and cultural qualities and hence at the taxonomic level of the family they encompass a distinct adaptive zone. Unfortunately, if we declare that our taxa has a distinct adaptive zone then we need to go back and need to accept a common descent in the most straightforward taxonomic manner.

Traditional evolutionary taxonomy is challenged from the two opposite directions. One challenge states that, as the phylogenetic trees can be very difficult to obtain, it is practically impossible to base our taxonomic system on common origin and adaptive evolution. It is generally said that taxonomy should represent more easily measured features displaying overall similarities among organisms under evaluation without any involvement of the phylogeny. If our taxonomy reflects those features which are easy to measure, then the principle is called phonetic taxonomy. Phonetic taxonomy does not impact animal classification to a great extent and hence scientists are losing interest in this approach. Although there are several difficulties in the reconstruction of phylogeny, zoologists still declare these efforts play a pivotal role in their systematic work. Further, they do not want to quit this goal due to methodological simplicities this goal possesses in it.

A second challenge to evolutionary taxonomy is from cladistics or phylogenetic systematics, the latter being comparatively stronger than the former. Here, cladistics says that an evolutionary taxonomy is based on the cladogram of the group which is under classification. However, criteria of common ancestry are the basis of phylogenetic systematics. Willi Hennig (a German entomologist) (1950) first proposed this approach to taxonomy, and hence it is sometimes called "Hennigian systematics." It is worth mentioning here that all the taxa which have been recognized under Hennig's cladistic system must be monophyletic.

Family Pongidae is paraphyletic and its most recent common ancestor shares common ancestry with the Hominidae. The Pongidae family is, therefore, not compatible with cladistics taxonomy and hence its recognition is not only difficult but, rather, is impossible under cladistics. Cladistic taxonomists, to avoid paraphyly, therefore currently do not use the conventional family Pongidae. Instead, chimpanzees, gorillas, and orang-utans are placed with humans in the family Hominidae. Paraphyletic groups, nevertheless, are unimportant to some scientists and even they do not always agree on their validity. Nonetheless, its important consequences become obvious when evolution is discussed. Evolutionary taxonomists claim the origin of amphibians from bony fish, birds from reptiles, and, finally, the evolution of humans from apes. These claims, however, do not carry any worth or value to the cladist. Hence, it can be inferred from these statements that amphibians, birds, or humans (descendant groups)) evolved from bony fish, reptiles, and apes (part of an ancestral group), respectively, to which their offspring do not belong. This scheme of evolution makes the ancestral group automatically paraphyletic, which is true too because bony fish, reptiles, and apes are traditionally recognized as paraphyletic groups.

Now, it must be asked how such paraphyletic groups can be recognized? This question demands further explanation and justification as to whether either of these paraphyletic groups shares some unique features among themselves which their descendants do not share? Further observations doubt positive definition and description of paraphyletic groups. The basis of their recognition are those features of an organism which are not present in its successors or offspring, because any trait which they shared from their common ancestors are also present in those descendants who have even been excluded from these analyses. For example, although apes are higher primates, they are not humans. Similarly, fish belong to those vertebrates that lack the unique characteristics of tetrapods such as amphibians and amniotes. Considering the features of these animals (apes), as well as human beings, it is very difficult to speak decisively about the evolution of humans from apes. Evolutionary taxonomists assign different adaptive zones or grades of organization to apes and human beings. When we say that humans originated from apes we mean that animals which were arboreal in habitat, with tails and smaller brain capacities are the ancestors of human beings. All three of these features are absent in human beings. The cladists, however, believe that the evolution of human beings is from something that they are not. According to the cladists, they say it is an inconsequential statement carrying no useful information. Extinct ancestral groups are always paraphyletic because these ancestral groups do not include those offspring that share their most recent common ancestor. The majority of evolutionary taxonomists, however, recognize such groups but cladists do not agree with their decision. Because of the interest of zoologists in terminal monophyletic groups (such as humans), they often construct paraphyletic groups, and then inquire about thei ancestry. Most of the time it is comfortable for the taxonomist to combine together those organisms which have approximately equally distant related features from the group we are interested in. In this combination of groups, the unique features which these groups possess are not considered and are hence ignored. In this regard, it is important to mention that those humans never became a part of a paraphyletic group, and unlike human beings most other organisms have been placed in a paraphyletic group. Invertebrates, fish, reptiles, and apes all traditionally designate paraphyletic groups which came into being by the combination of various side branches that are found when human ancestry is traced backward through the tree of life. This type of taxonomy gives a mistaken impression that all the evolution is a gradual movement toward humanity or within other groups. Pre-Darwinian views say that there is a linear and gradual development of organisms in such a way that primitive creatures are at the bottom and humans stand at the top just below the angels. But now we can say this thought is fictional. Darwin's theory of common descents states evolution as a branching process and none of the branches has an independent scale which can tell that animals evolved linearly with gradual movement toward perfection. Considering evolution as a branching process means that every branch will possess independent features in the form of a combination of ancestral and derived individuals. Contrary to the above in cladistics, we recognize the taxa based on their own unique features. Fortunately, common ancestry of groups can be depicted by following some easy ways without the construction of paraphyletic taxa. One way to accomplish this is to figure out the sister group of the taxon of our interest. Two different monophyletic taxa—if they have more recent common ancestors unrelated to other groups— are called sister groups. In this context, we can say that chimpanzees are the sister groups of humans. Gorillas in combination with chimpanzees form a sister group to humans. One of the clade that constitutes gorillas, chimpanzees, and humans—orang-utans—are the sister group of that clade. Another clade that constitutes orang-utans, chimpanzees, gorillas, and humans—gibbons—are the sister group of that clade.

5.3.2 Current State of Animal Taxonomy

The principles of evolutionary systematics are the basis of animal taxonomy, and the principles of cladistics are the basis of its revision. The principles of cladistics do not make any change in taxonomy except for the initial replacement of paraphyletic groups with monophyletic sub-groups. If taxonomy is revised in the light of cladistsic principles, there will be profound changes, of which the most major one will certainly be abandonment of the Linnaean ranks. When animal taxonomy is studied, monophyletic taxa should be used. Nevertheless, they fulfill the criteria of both evolutionary and cladistic taxonomy with the involvement of Linnaean ranks in one way or another. If commonly recognized taxa are clearly of paraphyletic grades, then this fact is well considered and alternative taxonomic schemes are studied which contain only monophyletic taxa. If taxonomy is revised according to cladistic principles only, it can create misunderstanding. Moreover, new taxonomic names are established and old ones are usually used in unfamiliar ways. For example, when we use the term bony fish in cladistic scheme of systematics, this is not restricted to finned aquatic animals but it also includes amphibians and amniotes, which also includes reptilian groups, birds, and mammals. When in the cladsistics scheme of study we use the term reptiles, then reptiles as well as snakes, lizards, turtles, and crocodilians will also include birds. Nevertheless, under this term fossil forms (synapsids), which were traditionally placed in the Reptilia, will not be included. It is mandatory for taxonomists to specify the time and place of usage of these well-known terms carefully when newer cladistics taxa or evolutionary taxa are discussed.

5.4 SPECIES

In 1859, when Thomas Henry Huxley was discussing Darwin's book *The Origin of Species*, he questioned "what is a species?" At first sight, this question looks very simple but it is difficult to answer it correctly even for those with adequate knowledge about it. The term species is in very common usage because of its simplicity and so far the term species is very commonly used because of its simplicity and explicit meaning (Fig. 5.2). Huxley's comments are as valid today as they were about 135 years ago. More and more sophistication is developing in our concept of species but diversity in different concepts of species and disagreements which exist in their usage still exists as they were present in Darwin's time (Gittleman 2024).

5.4.1 Criteria for Recognition of Species

Although biologists differ greatly about the nature of nature of species, they still lay down certain criteria which are important to consider for the identification of species. In almost all modern concepts of species, the criteria of common descent is the first one which is considered to be important and also applicable. Scientists are agreed that members of a species must have common ancestor without any restriction of having single pair of ancestors. In this context, species are historical units. Further, if the species are the smallest as well as a distinct group of organisms and share both patterns of ancestry and descent, then this will be the second criterion for the recognition of species. In the absence this criteria, separation of species from higher taxa, whose members also share common parentage, would be difficult. Morphological characters are traditionally used for the identification of such species groups but the applications of chromosomal and molecular characters for this purpose are persistently on the rise. The concept of reproductive community is a third criterion which is used for the identification of species which is applicable only to those organisms which reproduce sexually. These criteria state that the members of a species must establish their reproductive circle and zone forming a reproductive community without the interference of members of other species. There are several different concepts of species which will be discussed one by one in the following paragraphs.

FIGURE 5.2 Species and its types. (*Source:* www.shutterstock.com.)

5.4.2 Typological Species Concept

Prior to Darwin, a species was non-interconvertible and non-challengeable, which means it is a distinct entity. Each species possesses specific morphological features understandable from the pattern and/or model which have been created by nature. This species concept was designated as the typological or morphological species concept. For early recognition of species, scientists formally developed a specimen to represent the ideal form or morphology for the species, labelled it, and then stored it in the museum.

For the naming of additional specimens, scientists always compared and consulted earlier described specimens. The new specimens were assigned the same name if they possessed similar essential features of those of the previous ones. For identification and recognition of species, small differences were not accounted for and they were considered accidental deficiencies and/or limitations. If the scientists find large differences among different organisms, then they describe it as an independent species with its own name. Following these criteria, the living animal world was categorized into different species.

Although the concept of typological species is not well accepted by the evolutionists, its traditions still exist and are sometimes applied for species identification. Scientists still describe and name of the specimens stored, labeled, and displayed in museums. Hence it can confidently be stated that recognition of species morphology of an organism is still equally important. Evolution says that even though species are always changing they are still important historical units, despite minor variations During current observation, the type of specimen serves only as a guide to the general morphological features that we may expect to be present in the species.

5.4.3 Biological Species Concept

The biological species concept is very influential and was inspired by Darwinian evolutionary theory and formulated by Theodosius Dobzhansky and Ernst Mayr. This concept is very influential and was solidified during the evolutionary synthesis of the 1930s and 1940s. Although it was solidified from the earlier theories and ideas, since then it has been refined and reworded several times. Mayr (1982) said that each species is reproductively isolated from other communities and populations, establishes its own reproductive community, and occupies a unique and specific niche in nature. Hence, identification of species is based on the reproductive properties of species instead of other characteristics. Therefore, a species is a population which can interbreed, has common descent, and shares integrating characteristics. The study of morphological variations of an organism within a population, its chromosomal structure, and finally the study of its molecular genetic features will be a great help when we evaluate the geographical boundaries of those organisms which can interbreed and inhabit one place as a population or group (Panchen 1992).

When we talk about niche of particular organisms, it is expected this means that members that relate to a particular reproductive community occupy the same niche and share common ecological properties. It is well understood and well-known that reproductive communities with interbreeding qualities should maintain genetic cohesiveness among themselves, and hence it is expected that any variation among the same community group will be relatively even and will persist within species but will be uneven when compared with other organisms. As the reproductive properties of populations instead of morphology is the basis of the biological species concept, hence only organismal morphology—but not the morphology of the species—can help to diagnose biological species. Sometimes, variation in the characters can help our decisions regarding species membership. Variation in molecular characters is a very useful technique for the identification of geographical boundaries of reproductive communities. Molecular studies have identified that cryptic or sibling species can be present in a population which is too much alike morphologically and if we consider their morphological characters alone they cannot be differentiated from each other.

5.4.4 Biological Species Concept and Its Alternatives

There was a great deal of criticism of the biological species concept. Before addressing the reason for this criticism, we need to know and remember the important facts and figures about the species. First, because a species has dimensions in space and time it is difficult to locate its discrete boundaries. Second, as the species is considered as a rank in taxonomic hierarchy, as well as an evolutionary unit, this dual position of species sometimes conflicts. A third problem with the biological species concept is that the reproductive community of species is not isolated but exists rather in the same niche population which reproduce asexually. The systematics nevertheless describe species in all the groups of organisms which reproduce sexually or asexually.

5.4.5 The Species in Space and Time

The distribution of species in space is its geographic range, and when the distribution of species is through time, it is called its evolutionary duration. In both these dimensions of space and time, there are many differences between species. Species are cosmopolitan and have worldwide distributions with very large geographic ranges; however, endemic species have very restricted geographic distributions. If a species is located at a single point in space and time, then it can be recognized without any difficulty; for example, species located in our local park or woods. Nevertheless, if we compare local species with distantly located species then it will be difficult to decide whether these populations are part of a single species or whether they are a part of a different species.

The geographic range of a species may change many times during the course of its evolutionary period due to its continuity at one time and discontinuity at another time. If the geographic range of a species is not continuous, there can be points in time where the species is absent. For example, consider two populations which are living 300 miles away from each other; although they are similar they are not identical, having no related population between them. Now think for a moment whether we are studying a single species which is not evenly distributed with several breaks in it, or whether we are observing two different species that have a close relationship with each other. Further suppose that these populations had historically remained separated from each other for 50,000 years. Now the question is whether this is enough time for these two species to have evolved into an independent species and with a separate reproductive community or whether when we look at them we can see that they are still closely related and are a part of the same reproductive community. Although answers to such questions are very hard to figure out, nevertheless incongruity among the concepts of species laid down can solve such complications in species segregation and identification.

5.4.6 Evolutionary Species Concept

The time dimension regarding species mentioned above creates apparent problems for understanding the biological concept of species. The first problem is how we can assign fossil specimens to the currently recognized biological species. The second question is that if we go back into the history of life to trace the lineage of species and to cross the boundary of particular species, how far back do we have to go? By going back in the history of time, if we could follow the continuous chain of heredity of populations and achieve a point of convergence of two species, then during this lengthy journey we might have to cross the boundary of at least one species. However, it will be a difficult decision to draw a sharp line between the two species. To address this problem, Simpson (1940) proposed the evolutionary species concept and also added an evolutionary time dimension to the biological species concept. In its modified form, this concept is still valid today.

An evolutionary species therefore has its own propensities and historical fate. It maintains its uniqueness from other closely or distantly related lineages and maintains a method of reproductive cohesion. For the identification of evolutionary species, the same kinds of diagnostic features will be relevant as those discussed above. Nevertheless, in most of the cases only morphological features from fossils will be available for identification of evolutionary species. Notwithstanding that biological species both sexually and asexually producing populations/life forms are covered under the evolutionary species concept, any evolving lineage will be recognized as a species as long as it maintains its diagnostic features. Throughout the evolutionary time, if abrupt changes are visible in diagnostic features, then these changes will mark the boundaries of different species.

5.4.7 Phylogenetic Species Concept

This concept states that a species is a complex group of organisms which are diagnostically distinct. Further, they have appropriate patterns of offspring and parentage within them. This concept is the advocate of common descent covering both sexual and asexual populations. Any phylogenetic species is a strictly monophyletic unit. Practically speaking, the phylogenetic species concept differs from the evolutionary species concept in that the former recognizes the smallest groups of organisms as a separate species, which have experienced an independent evolutionary change. Contrary to the phylogenetic species concept, the evolutionary species concept recognizes any group of organisms which belongs to a geographically discontinuous population. Such populations may have some genetic divergences but still possess similar evolutionary tendencies. Applications of the phylogenetic species concept demands a description of a greater number of individuals for identification and the recognition of species, which taxonomists say is not practical at all. The phylogenetic species concept is ideal if we adhere strictly to the cladistic systematics. Only the phylogenetic species concept guarantees strictly that species are monophyletic units. This concept does not go into the detail of evolutionary processes, although supporters of the phylogenetic species concept accept the importance of the study of evolutionary processes. They are of the view that when we want to study evolutionary processes we must be very clear about the history of life. Detailed patterns of common descent need to be reconstructed to accomplish this task. These patterns should start with the smallest taxonomic units which have a history of common descent.

5.4.8 Dynamism of Species Concepts

Species concept is still controversial but it should not be taken as a negative or a discouraging exercise. Whenever any scientific field grows lively again, old concepts are reevaluated, and, if deemed suitable, they are refined and reincorporated; if not, they are replaced with newer, more progressive, ones. It is very actively debated that the field of systematics has acquired an extraordinary role and importance in biology. Current advancements in biology are parallel to those in Thomas Henry Huxley's time but are considerably larger now than any that were present in his day. Both these periods have given a great deal of consideration to understanding the meanings of species and finalizing its most acceptable definition. It is hard to predict 10 years from now which concept of species will dominate, and nothing further can said as to whether the concept of specie which is currently strongly advocated will survive at that time or not. Any future developments in the concept of species rest with the conflicting arguments in the concept of species. It is important for the zoologists to continue this debatable issue, instead of studying only the single species concept.

5.5 MAJOR DIVISIONS OF LIFE

If we start from Aristotle and move gradually to the late 1800s, every living organism was conventionally placed under either the plant or the animal kingdom. However, this kingdom system of classification had serious problems, which the following examples will explain. Under this system of division of life, plants, trees, mosses, and ferns (rooted or without roots, with the capability of photosynthesis) have been comfortably placed under the plant kingdom. However, animals including insects, fish, and mammals (which ingest their food and can move around) have been placed under the animal kingdom. Problems were faced in the classification of unicellular organisms because some life forms were debatable. Botanists classify them as plants, while zoologists consider them to be animals. Euglena, for example, moves like an animal but also has chlorophyll for photosynthesis.

Other groups, like bacteria, have been randomly assigned to the plant kingdom. To solve the problem of classification of unicellular organisms, alternative classification systems have been proposed. In 1866, Haeckel proposed a new kingdom called Protista, and included all unicellular organisms in it. First of all, he took life forms without nuclei and bounded by a membrane, e.g., bacteria and cyanobacteria (blue-green algae), and combined them with those unicellular organisms which have a nucleus in them. Finally, he recognized the important differences present between the bacteria which were devoid of a nucleus and those organisms which have a nucleus surrounded by a membrane (Aguinaldo et al. 1997).

In 1969, Whittaker proposed a five-kingdom system and reported the basic differences between prokaryotes and eukaryotes and incorporated them in his system of classification. According to his classification kingdom, Monera now contains the prokaryotes and the kingdom Protista contains the protozoa (unicellular eukaryotic organisms like protozoa) and algae (which are unicellular and eukaryotic). Nevertheless, on the basis of some peculiar differences in organization and mode of their feeding and nutrition, multicellular organisms were divided into three kingdoms. The first kingdom is Plantae, which includes algae and higher plants with the ability of photosynthesis. Kingdom

Fungi includes that group of organisms which obtain their food by absorptions, e.g., molds, yeasts, and fungi. Third kingdom Animalia contains all the vertebrates and invertebrates except protozoa. Except for some parasitic forms which are absorptive in nature, most of animal forms present under this kingdom practice the processes of food ingestion following their own internal digestion.

During the proposal of these different systems of classification, phylogenetic relationships, which are actually required for the construction of evolutionary or cladistics taxonomies, were not considered. The oldest phylogenetic events which occurred during the history of life are not clear; rather, they are very much ambiguous. For the construction of phylogeny, different organisms do not share many characters which can be compared for this purpose. Recently, a cladistic classification of all life forms has been proposed which is based purely on the phylogenetic information obtained from molecular data (the nucleotide base sequence of ribosomal RNA).

Woese et al. (1990) proposed this classification, in which above the kingdom level they proposed three monophyletic domains. These domains are eucarya (all eukaryotes), bacteria (the true bacteria), and archaea, which differed from bacteria in membrane structure and ribosomal RNA sequences. Now, if we adhere to Whittaker's three kingdoms known as Plantae, Animalia, and Fungi, the Protista become a paraphyletic group.

Cladistic classification says that the Protista must be recognized as a separate kingdom and should be further divided into separate kingdoms, namely the Ciliata, Flagellata, and Microsporidia. Cladistic classification further demands collection of phylogenetic information, e.g., amoeba, known as additional protistan groups.

Zoologists have recently studied Protistans like animals traditionally under Protozoa (animal phylum). A combination of our current knowledge and the principles of phylogenetic systematics lead to two major errors in taxonomy: (1) In the light of Linnaean hierarchy Protozoa are neither animals nor at any taxonomic level they are any valid monophyletic taxon; and (2) similarly, because of its non-monophyletic nature, the kingdom Protista is not valid.

5.6 MAJOR SUBDIVISIONS OF THE ANIMAL KINGDOM

In the classification system of the animal kingdom as proposed by Linnaeus, the phylum is the largest formal category of animals and plants alike in taxonomy. Informal, additional taxa which lie in between the phylum and the animal kingdom are produced when different animal phyla are grouped together. The embryological and anatomical characters are the basis of development and establishment of these taxa. In the past, zoologists recognized two subkingdoms, namely Protozoa, which constituted those phyla which contained unicellular organisms, and Metazoa, which comprised those phyla which contained multicellular organisms. Nevertheless, Protozoa is an invalid taxonomic group and, unlike Metazoa, it does not belong to and/or fall within the animal kingdom.

Traditionally, true animal phyla at a higher level are grouped somewhat like the details provided below. Bilateral animals, on the basis of embryological developments, are usually divided into protostomes and deuterostomes. Some phyla have peculiar characteristics and do not allow organisms present in them in these two categories, which is purely embryology-based. Phylogenetic studies is giving this traditional embryology-based classification a tough time and hence this traditional classification of the Bilateria is currently facing challenges from phylogenetic studies. Results from molecular phylogenetics have their own method of classification. They place four phyla classified as deuterostomes (Brachiopoda, Chaetognatha, Ecto-procta, and Phoronida) into Protostomia. Molecular phylogenetic studies further opine that the traditional major groupings of protostome phyla, namely acoelomates, pseudocoelomates, and eucoelomates, do not show any signs of being monophyletic. Instead, of four groups only two groups (Lophotrochozoa and Ecdysozoa (monophyletic)) constitute protostomes.

Four phyla, namely Entoprocta, Gnathostomulida, Loricifera, and Pen-tastomida, evolved when bilateria was reclassified in the recent past. Further studies are required to confirm these new groupings and to determine their relationships among themselves. Their inclusion into the classification system also needs to be worked on. Using the traditional classification, surveys are frequently arranged for animal diversity and discussions are organized to examine the implications of new innovations on the old systems of classifications.

REFERENCES

Aguinaldo, A.M.A., Turbeville, J.M., Linford, L.S., Rivera, M.C., Garey, J.R., Raff, R.A. & Lake, J.A. (1997). Evidence for a Clade of Nematodes, Arthropods and Other Moulting Animals. Nature 387, 489–493. This molecular phylogenetic study challenges traditional classification of the Bilateria.

Gittleman, J.L. (2024, February 27). Species. Encyclopedia Britannica. https://www.britannica.com/science/species-taxon

Hull, D.L. (1988). Science as a Process. Chicago, University of Chicago Press. A study of the working methods and interactions of systematists, containing a thorough review of the principles of evolutionary, phenetic, and cladistic taxonomy.

Joyoti, A. (2023). The Three Domains of Life, In: Classification of Life: The Three Major Domains. Powered by Astra Worldpress Theme.

Mayr, E. (1982). Speciation and Macroevolution. Evolution, 36(6):1119–1132.

Panchen, A.L. (1992). Classification, Evolution, and the Nature of Biology. New York, Cam-bridge University Press. Excellent explanations of the methods and philosophical foundations of biological classification.

Wiley, E.O., Siegel-Causey, D., Brooks, D.R. & Funk, V.A., (1991). The Complete Cladist: a Primer of Phylogenetic Procedures. Lawrence, University of Kansas. Printing Service. A workbook presenting detailed instruction in cladistic concepts and methods.

Woese, C.R., Kandler, O. & Wheelis, M.L. (1990). Towards a Natural System of Organ-Isms: Proposal for the Domains Archaea, Bacteria, and Eucarya. Proceedings of the National Academy of Sciences, USA 87, 4576–4579. Proposed cladistic classification for the major taxonomic divisions of life.

6 Patterns of Animal Diversity and Distribution

6.1 INTRODUCTION

Biodiversity is heterogeneously distributed across the earth. For example, there is no life in deserts and in the Polar region. However, life is abundant in coral reefs and some moist tropical forests. Other areas fall in between these two extremes. Ecologists and bio-geographers have been working for a long time to determine why these differences occur, which remains a constant and enthralling challenge to many workers in this field.

There is significant growth in investigation of variations in four areas of broad-scale spatial biodiversity. Out of these four areas, latitudinal gradients in species richness is the first one. Taxonomic covariance is another area which tells us the level of species richness. The level of relationship between energy and species is another indicator which tells us the level of abundance of any species. Sometimes the relationship of species between local and regional population tells us the status of richness of the particular species. There is no single example which can explain and substantiate the progress made in each of these areas or which can inform what the patterns of biodiversity at the global level are. Invariably, variations can be observed and exceptions exist in any given pattern of biodiversity.

6.2 SPECIES RICHNESS VARIES WITH LATITUDINAL GRADIENTS

Both terrestrial and freshwater species exist in the Tropics. A sampling of species from high to low latitudes shows an increase in species richness. Collections of species across a range of spatial locations and sites display a wide range of taxonomic groups of animals. All the available fauna follow the uniform pattern and phenomenon. Common groups of prevalent animals are protists, trees, ants, woodpeckers, and primates (Stevens, 1989; Gaston, 1996). During sampling, it has been observed that tropical areas are richer in species than temperate ones.

Talking more specifically, the presence of some non-calcareous taxa for shallow water systems is yet to be debated. However, let us consider species distribution in the marine environment. Here, two types of gradients develop which represent the richness level of different species. These are open-ocean pelagic where several species are present and then we have deep-sea taxa. Both are present in broad latitudinal gradients, which host species richness according to the habitats (Clarke & Crame, 1997).

There is a continuous rise in the further refinement of analysis of latitudinal gradients in species richness. There are several suggestions, from these analytical methods, to some important nuances (shades and tones) to this pattern. Nevertheless, how general these patterns are is still uncertain. From the data collected so far on species richness, species richness is declining with latitude. Nonetheless, this decline in species richness is faster in the northern hemisphere and slower in the southern hemisphere (Platnick, 1991; Blackburn & Gaston, 1996). Therefore, instead of lying at the equator, richness peaks may rather be at a short distance away from the equator (Roy et al., 1998; Lyons & Willig, 1999). This type of latitudinal non-uniformity can be predicted in historical climatic events, including latitudinal complexities of an area of land or ocean and the geometry. Nevertheless, information about these factors is scarce. We always consider the concept of latitudinal gradient in totality in terms of species richness. Positional variables like elevation, depth, and longitude and environmental ones like aridity and topography

DOI: 10.1201/9781003472001-6

markedly disrupt the underlying patterns of variations in species richness. Thus, the complex patterns of spatial variation are a reflection of the latitudinal pattern of change in general but all this depends on what and where the observer is looking. This gives a message that if we consider latitudinal gradients in isolation without relating them with other gradients, then this cannot be a useful or profitable method of consideration. Nonetheless, when biologists discuss species richness they always prefer to consider latitudinal gradients. The reason for this attraction in latitudinal gradients is diversity in flora and fauna present in tropical areas, rather than those present at higher latitudes.

If animals are immigrating and there is balance in speciation, then this species richness shows latitudinal gradient. This scenario creates a complex phenomenon in species distribution and richness. This latitudinal gradient for species richness introduces more species in the particular area, but it is also dependent on the emigration and extinction of species which removes them from that area. If the areas are large then the dominant effects will be from speciation and regional or global extinction. In this scenario, effects of immigration and emigration will not be of any great importance.

Biologists suggest that more than 25 different mechanisms generate systematic latitudinal variation in these processes, with the emphasis on the causes of more speciation on the tropics (Brown & Lomolino, 1998). There are several possibilities of species richness in the Tropics with greater productivity and interspecific interactions, historical disturbances, environmental stability, habitat heterogeneity, and, finally, chances of speciation, immigration, and emigration.

All these mechanisms are related to each other with slight variations in the levels of explanation they offer. When compared with other climatic zones, it is well established that climatically similar total surface area has been observed in the Tropics. The reasons for this larger surface are: (1) the temperature gradient is non-linear between the Equator and the poles and the mean temperature approximately falls between 207N and 207S; (2) latitudinal bands decrease in area when we move towards poles; and (3) the joining of the regions present immediately to the north and south of the Equator whose climates resemble to each other. Further, it is emphasized that in the Tropics if geographical range size is greater of any given species then, accordingly, species richness is also greater. This phenomenon means higher speciation rates with probabilities of decreased extinction (Rosenzweig, Rosenzweig, 1993; Rosenzweig, 1995a & Rosenzweig, 1995b).

Under these conditions, what happens is that extratropical regions have a lower number of species than tropical regions. In fact, patterns of biodiversity are influenced by area and hence it can be said that an area therefore contributes a lot to latitudinal gradients in species richness. Therefore, like other factors it is hard to say that in case of species richness an area alone is sufficient for determination of latitudinal gradients. From the above explanation it is very much obvious that generation of patterns of biodiversity can be attributed to the several mechanisms (Gaston & Blackburn, 1999). Pulling all of the different mechanisms in the same direction gives the most general and the strongest pattern (Lawton, 1996). In species richness for latitudinal gradients to date numerous mechanisms have been identified. However, none of them fulfills the requirements for a full explanation of various gradients and patterns.

6.2.1 Species–energy Relationships

The relationship between the number of species present in any particular area and the available and usable energy modulates the determination of species richness in the latitudinal gradients. This net productivity or energy can be assessed from different models and/or sometimes from other variables. The nature and possible reasons for this species energy relationship is hotly debated and this debate becomes more obvious when we study global patterns in biodiversity. When we talk about the global patterns in biodiversity there are many fundamental issues which still need resolution. A hump-shaped relationship has been observed between the availability of energy and richness of species. Species richness increases when there is an increase in energy from the low to moderate levels, followed by a decline in species richness where energy levels are at their highest (Rosenzweig & Abramsky, 1993).

A strong monotonic relationship has been observed between species richness and energy availability when we consider geographical scales across temperate to polar areas (Rosenzweig & Abramsky, 1993). For plants, the best correlation will be that water and heat means that the level of evapotranspiration and that of net primary productivity, while the best correlation for terrestrial and marine animals is that of heat (Currie, 1991). It has been observed that in temperate Europe, Eastern North America, and East Asia there is a direct increase in plant species richness with primary productivity (Adams & Woodward, 1989), whereas the richness of birds and butterflies in Britain is directly proportional to the temperature present during the appropriate season (Turner et al., 1988). In some areas of North America, there is a rise in the number of amphibians, reptiles, birds, and mammals, with annual potential of evapotranspiration. Evapotranspiration—independent of water availability—represents the net atmospheric energy balance (Currie, 1991).

If there are differences in available energy in different eco-climatic zones, then species richness and abundance will depend on the joint effects of energy availability and the area where different animal species inhabit (Chown & Gaston, 1999). Second, the area model affects the geographical-range or size of the species. This in turn impacts the species richness in that given area. As the species richness depends on these variables, hence any factor which increases one variable raises the possibility that the other variable will also increase. Both mechanisms—like energy and area—depend upon some factors which can affect the availability of biomass.

With the help of the processes of speciation and extinction we can explore and work out biomass availability, which is an end product of area and energy availability (; Wright, 1983; Wright et al., 1993). Therefore, it is presumed that species richness is low in smaller areas, although with comparatively higher energy inputs. However, species richness is poor in larger areas if energy availability is low in these areas.

6.2.2 Local and Regional Richness Are Related

Without understanding the importance of differences present in spatial scale for observed patterns, it is hard to explore variation in biodiversity on a global scale. Moreover, we also need to understand the mechanism that develops relationship of biodiversity at one scale to that at another scale. In addition to the above, we also need to comprehend the detailed information on the processes and patterns at different scales. If it follows a proportional-sampling model (type I), then regional richness is greater than local richness. There is also a possibility that they may be directionally proportionate to each other. Conversely, when there is an increase in regional richness, local richness may reach the maximum. Beyond this limit, despite a continuous increase in regional richness (type II) the local richness will not rise. In reality, most of the systems exhibit proportional sampling model (an underlying type I relationship). Nevertheless, it is quite common that regional richness has up to 75% its share in variation in local richness. However, local richness constitutes 50% of the regional richness. Hence, in southern and central Africa (Hawkins & Compton, 1992) between fig wasps and their parasitoids, type I relationships exist. Similar relationships have been reported in tiger beetles in India and North America (Pearson & Juliano, 1993). Lacustrine fish in North America(Griffiths, 1997) and primates in Africa and in South America show similar relationships (Eeley & Lawes, 1999).

6.2.3 Taxonomic Covariance in Species Richness

Most of the major animal groups present in terrestrial and freshwater environment show less species richness in temperate regions than in the Tropics. Accordingly, they are less abundant at high elevations and deserts than when compared with low elevations and forest areas, where they are in abundance. Therefore, there is then the expectation that regional richness of different groups of organisms will show positive co-variation with each other. If there is a positive relationship between local and regional richness, local richness will follow a similar trend. Nonetheless, in space these positive relationships do not guarantee uniformity in the occurrence of species richness peaks. However, mismatches in the occurrences of peaks in richness of different groups of animals are quite evident. For example, some pairs of taxa of trees, tiger beetles, amphibians, reptiles, birds, and mammals overlap at 5% of the land area across Canada and the United States. Nonetheless, this is not a universal pattern. Similarly, numbers of species in two groups present in large and small areas often correlate significantly with each other but practically these relationships are most often weak with very limited predictive value. This value is explained by latitudinal gradients in diversity in some cases (Flather et al., 1997).

It is a common observation that animal groups with similar spatial gradients in biodiversity vary significantly around their trends. These trends have been expressed in living environment and in the habitat they occupy. For example, richness peaks are observed at high or mid-latitudes in taxonomic groups like aphids, bees, sawflies, braconids, ichneumonids, and various groups of marine amphipods, freshwater invertebrates, and procellariforms. However, exceptions exist in patterns, although these are less at higher levels and frequent at lower taxonomic levels.

6.2.4 Biodiversity of Pakistan

The Republic of Pakistan is located at the convergence point of central Asia, South Asia, and Southwest Asia. Islamabad is the capital of Pakistan and Karachi is the country's largest city. Due to its geographical conditions, the country experiences extreme temperature variations. The southern slopes of the country are arid. The northern slopes of the Himalayas and the sub-mountainous tract receive annual rainfall between 760 and 1,270 mm. This region has a humid sub-tropical climate. Most of the country's area is arid. The former area has a humid sub-Tropical climate. The climate is usually cool or moderate in the extreme north due to great height. Wildlife is variable in the extreme south and the extreme north of the country, which occupies a small area. Nonetheless, the north has the most important climatic and vegetation zones or biomes. These biomes resemble those present all over the world. If we move from permanent snow and cold desert habitats and pass through alpine to mountain temperate forests and tropical deciduous forests up to the alpine dry steppes, these areas show quite rich wildlife.

This is not an end of wildlife zones. Our further movement to the arid subtropical/desert habitats, thorn forest, the refrain plains of the Indus and its tributaries finally end in rich mangrove and associated systems. Then along the coast wildlife enriches even further. In these areas, we can find a wide variety of animals with extraordinary habitats.

6.2.5 Biodiversity in Pakistani Ocean—Present Situation, Status, Threats, and Conservation Requirements

Currently, more information is available about the moon than about Pakistan's oceans. Comparatively far less money is spent on explorations of the deep sea than is spent on space research. Oceans are still poorly known about in a unique habitat on earth where it is still possible to make thrilling discoveries. Sufficient biological diversity is prevalent in coastal ecosystems. Human beings avail a wide range of goods and services from this resource. Major habitats of coastal zone and ocean life are coastal or barrier islands. In addition, there are rocky coasts, mud, or salt flats. Rock pools, sandy, pebble, or rocky beaches, dune systems, saline, brackish, and freshwater lagoons are some other biodiversity sources. These habitats do not end here; rather; they extend to estuaries and coastal river flood plains, cliffs, intertidal rocky coral reefs, seagrass beds/meadows, plains, salt marshes, and mangrove forests. Over time, human activities have highly modified these habitats.

Terrestrial and marine activities affect coasts. Both these factors are external to the coastal zones. The consequences of anthropogenic activities influence terrestrial environment. Among terrestrial influencers the major one is change in land use. Land use changes hydrological regimes. Nutrients leach out of sediments resulting from runoff. They are transported to rivers where they become scattered and river flow reduces their concentration. Marine factors are mainly natural ups and downs. Major among them are weather events like storms and cyclones. Wave patterns and coastal and ocean currents and last but not the least tsunamis are some other coastal factors that affect the spread and distribution of biodiversity. These factors affect transfer heat, nutrients, and other materials. These factors also arbitrate and monitor different geomorphological changes occurring from time to time in the oceans.

6.2.6 Climate Changes Impacts the Coastal Ecosystems in the Indian Ocean Region

Several interacting factors like direct and indirect drivers affect the loss and degradation of ecosystems. Land use creates fragmentation. This process degrades habitats. Land use primarily converts land into agriculture. This process overexploits the resources for livelihoods and commercial purposes. Sometimes farmers use chemical fertilizers. Accumulation of sewage water, excessive use of toxins, such as pesticides, and hazardous chemical all go to the sea. They enrich sea water with a variety of nutrients and other chemicals. Such compounds pollute ocean water and directly affect the loss and degradation of ocean habitats. The entry of alien invasive species which spread in the ocean waters swiftly, with their population rapidly becoming out of control has resulted in biological pollution. The interaction of anthropogenic climate change with these previous factors has reinforced their degradative effects. Indirect main drivers which are also considered ultimate root causes of degradative and habitat loss processes are as follows.

6.2.7 Maritime Zone of Pakistan

The Arabian Sea (the Sindhu Sagar) located in the south of Pakistan is one of the most unique seas. It has a tropical climate and a large marine ecosystem (LME). At a global level, it is considered as Class I, which means it is a highly productive zone. In the Indian Ocean, it covers three sub-regions, its eastern part borders Pakistan. The sub-region is unique due to physical characteristics, physiochemical qualities, and current patterns. It is famous for its upwelling phenomenon and current direction. Its current pattern is clockwise during the southwest monsoon season. Nevertheless, it shows an anti-clockwise pattern during the northeast monsoon season. It has highly diversified life with numbers of dominant species present in it.

6.2.8 Pakistan Coastline

Pakistan's coastline is 1,050 km (650 miles), of which 800 km are in Baluchistan and 250 km in Sindh. The exclusive economic zone (EEZ) covers between 150 and 200 nautical miles. Currently, the salinity of ocean water is 36 ppt. Tides in the ocean are intermediate with maximum height of about 10 to 11 feet.

6.2.9 Marine Biodiversity

It represents a variety of life present in the ocean, which includes plants, animals, and microorganisms. This definition further extends to the genes these life forms contain and the ecosystems all these life forms constitute. Bacteria, archaea, and eukaya (Eukayota) are the three major domains of life present in the marine environment. Viruses are in addition to these life forms. Although described as species of marine animals, plants, bacteria, and archaea are in the thousands but they are still considered a fraction of the existing number of species.

At global level, between 1.4 and 1.7 million species have been described (Stork, 1988), but the conservative estimate of Global Diversity Assessment opine that there are 1.75 million species

(Watson et al., 1995). More species are known on land than in the sea. It is a common presumption that the deep oceans may host 10 million species. Nonetheless, these species still need to be described (Grassle & Maciolek, 1992). Some, however, say that the number may only be 500,000. It has been observed that most of the marine species described in the benthic rather than the pelagic and oceanic areas have less diversity than coastal areas (Angel, 1993). More animal phyla are present in the marine domain than those on land. There are a total of 35 marine phyla, 11 out of 35 are present in pelagic realm (Angel, 1993), and most of the others are found in the benthic zone.

6.2.10 Importance of Marine Biodiversity

There are several reasons for complete comprehension of marine biodiversity (Norse, 1995). The importance of biodiversity can be categorized as follows:

Adaptation: Since the creation of life, marine biodiversity has always facilitated environment to adapt to the changing conditions.

Food: Fishes, invertebrates, and algae are important food commodity for living organisms, including human beings. All this comes from oceans. Conservation of biodiversity will help the utilization of unexplored resources in the future.

Raw Materials: Oceans provide raw materials for livestock feed and for building materials. In addition to the former, oceans also provide medicinal materials and polysaccharides. Oceans have an enormous potential for acquisition of future raw materials.

Climate: Existing biota on earth—along with those present in the marine environment—are important for the control of biogeochemical cycling of gases. Phytoplankton removes carbon dioxide from the surface waters and provides oxygen to the living biota. Death of phytoplankton and respiration of living animals results in the production of carbon dioxide.

Knowledge: Oceans are a great source of knowledge. Several marine ecosystems have been discovered and many are in progress. In 1976, hydrothermal vents were discovered with seral endemic animals species, e.g., tubeworms, giant clams, mussels, crabs, polychaetas, and gastropods. Furthermore, in the marine environment it has also been discovered that the animals which were considered single species are actually several species, e.g., mud crab *Scylla serrata* or, for example, *Thenus orientalis*, which currently have been described as three distinct species (Keenan et al., 1992; Burton & Davie, 2007). Second, *Capitella* (the marine worm) previously used as a pollution indicator actually comprises 15 or more sibling species. The flora and fauna of the ocean is summarized below:

Flora

Sea-Weeds: More than 300 spp ARE known to be present in the ocean (Shameel & Tanaka, 1992).

Halophytes: *Avicennia marina*. Very small patches of *Ceriops roxburghiana* and *Aegicerias corniculata*, *Rhizophora* f*mucronata*, and *Ceriops tagal* have been described. It is reported that the Indus Delta has eight plant species, most of which are now extinct in Pakistan. Among all of them salt marsh grass, *P. coarctata* is dominant. 72 species of phytoplankton (*Arthrocnemum macrostachym*) have been described to date (Chaghtai & Saifullah, 2001). Phytoplankton support planetary life and they are a critical component of that system. They regulate temperatures, draw down CO_2, and support fisheries, as well as other organisms. Decline in their population affects every step and component of the food chain.

Fauna

Protozoa-15 phyla are commonly found in the ocean (Fig. 6.1).

Mammals: 12 species of whales, dolphins, and porpoises including *Planista minor* of the Indus River are part of ocean systems (Gore et al., 2012)

FIGURE 6.1 Protozoa diversity. (*Source*: www.shutterstock.com.)

Reptilia: Snakes, turtles, including the much protected *Lepidochelys olivacea* (Khan, 2004)
Sea Birds: 106 species (Gabol & Rahim, 2006)
Fishes: About 1,000 fish species have been recorded (Moazzam & Osmany, 2024)
Invertebrata: Collection of invertebrates and publication of this information is contradictory.
Urochordata: Data on them is poorly known.
Echinodermata: 36 species of Echinodermata are known in the ocean (Murty et al., 2008).
Ectoprocta: This group of organisms is poorly known.
Chaetognatha: 16 spp of this group have been described to date (Haq & Rehman, 1973).
Mollusca: (Kazmi & Naushaba, 2004).
Annelida: This is a dominant group of organisms if we consider the total number of individuals present in the ocean and draw almost a 33% share of the total macro-faunal community. Among annelids, Archiannelids and oligochaetes have been poorly defined or not defined at all (Ishaq & Mustaquim, 1996).
Arthropoda-Insecta, **Collumbola, Picnogonida, Arachnida, Crustacea, Symphyla-Acari**: Among all of them the Crustaceans are more studied and have most diverse forms.
Branchiopoda: (Sultana et al., 2014) e.g., Cladocera and other Branchiopods
Copepoda: e.g., Ocean copepods.
Cirripedia: (Rizvi & Moazzam, 2006) e.g., Barnacles.
Ostracoda: (Siddiqui, 2006).
Decapoda
Porifera: This group has been poorly described and hence its individuals are the least known.

6.2.11 Threats to Marine Biodiversity and Their Impacts-Frequent Linkage of Existing Threats and Continuous Emergence of New Threats

6.2.11.1 Climate Change

Climatic changes have changed oceanic temperatures, patterns of water movement (including currents, eddies, and fronts), and acidity. Climate change has also altered storm events, tsunamis, and rainfall patterns. In turn, these changes have increased atmospheric carbon dioxide and damaged the ozone layer. If we look at global climate change, it is increasing sea-surface temperature. Continuous Plankton Recorder has reported unique data with northward migrations of species. Increasing carbon dioxide (CO2) levels make sea water more acidic. This acid water increases the solubility of calcium carbonate. Higher calcium carbonate levels dramatically cause calcification of corals, mollusks, coccolithophorids, pteriods, and forams.

6.2.11.2 Anthropogenic Activities

Human and marine populations are the vast biomass. This biomass globally joins the islands and coastal zones with their surrounding seas. Biodiversity changes in oceans and their management has received little attention despite huge biodiversity including both flora and fauna in the oceans. Specifically in coastal area, the impact of human beings is considerable on the oceans. Human population and recent demographic trends have directly threatened biodiversity in the coastal zone (Heywood and Watson, 1995).

6.2.11.3 Overfishing

Mass extinction biodiversity and change of fish composition is attributed to recreational fishing, illegal unregulated or unreported fishing, commercial fishing, and ghost fishing. Due to this over and illegal fishing large fish, sharks, turtles, crustaceans, and plants have disappeared. Consequently there is an increase in small fish, sea urchins, etc. with ultimate rise in their phytoplankton or macrophyte food. Worldwide modern fishing activities, has further threatened the marine biodiversity (Hiddink et al., 2008; Helfman, 2007) however differed slightly from the aforementioned view point.

6.2.11.4 Tourism and Recreational Activities

Gastropod snails are sought as souvenirs in many tropical areas. Since they play key roles in controlling prey populations, their local extinction significantly induce changes in diversity. Tourists fancy for coral skeleton and other curios have depleting effects.

6.2.11.5 Habitat Damage

Fishing gears especially bottom trawling, and coastal development damage habitat of a variety of organisms. Important of them are destruction of coral reefs, and mangroves. All these factors support coastal marine ecosystems. If they disturb natural freshwater flows and passages, consequently they destabilize coastal foreshores, coastal wetlands, and sometimes entire estuaries which coastal ecosystem. These factors have already affected birds and their habitats a lot.

6.2.11.6 Pollution (In-Sea and Land-Based, Diffuse and Point Source)

Marine pollution can be attributed to oil spills, armaments, plastic litter, different nutrients, sediments, noise, hazardous and radioactive substances. Other major threats to marine environment are increasing atmospheric greenhouse gases, and microbial build-up. Release of trace chemicals such as carcinogens, endocrine-disruptors, and info-disruptors, and discarded fishing are some other factors which endanger marine environment as well as biodiversity (Koslow, 2007; Turley et al., 2006). Hammer and Hill (2000) reported often higher diversity in disturbed habitats than undisturbed ones. This phenomenon is called the intermediate disturbance hypothesis. CFCs emission-ozone depletion leads to increase of UV-B radiation-indirect effect on marine organism including Pakistan (Hader et al., 2007).

6.2.11.7 Alien Organisms

Vessel ballast water and hull fouling transport and introduce different organism form their original living environment to the new one. They alter the ecosystem and sometimes reduce and sometimes totally eliminate the native fauna.

6.2.11.8 Construction of Dams and Barrages

On Indus River System barrages have been constructed for irrigation and hydroelectricity generation. They have dramatically reduced sediment load in water which in turn has severely affected coastal ecosystems. Occurrence of salinity changes has changed species composition.

6.2.11.9 Deforestation

Often many hundreds of kilometers inland may receive and deposit sediments in or on it. These sediments smother coral reefs and other coastal habitats. This sedimentation process impairs or destroys river passages which are essential for anadromous and diadromous species.

Threats to mangroves: Threatened by ruthless logging, damming-shrinkage of Indus Delta caused decrease in freshwater flowing in the Indus downstream from Kotri-600,000 ha in 1950s to around 86,000 presently, for reclamation, salt industry, and grazing.

Threats to corals: Corals face wide and intense threats due to agriculture runoff, shipping, overfishing, and coastal development.

Warming: Seas with high temperatures have damaged reefs. High temperature cause stress (coral bleaching). When corals lose their colorful symbiotic algae and expose their white skeletons, the process is called coral bleaching. In addition, high temperatures emit increased carbon dioxide (CO_2). This slowly decreases the pH of the ocean waters. Ocean acidification reduces growth rates of corals. It is estimated that if immediate drastic action is not initiated most of the reefs will be eradicated in the next 40 years. In terms of biodiversity loss, loss of corals and mangrove habitats are the most significant. We should also not forget the disappearance of other critical coastal habitats. Our oysters in the prepartition period used to be as oyster reefs in the Indus Estuary System. These beds deteriorated as no mitigation measures were taken against the causes of siltation, pollution, and overexploitation.

6.2.12 Importance of Conservation

Marine ecosystems are quite complex. Consideration of developing science of restoration ecology should be an integral part for formulation of any strategy for conservation of coastal biodiversity. The best way to conserve marine diversity in the coastal area is conservation of habitat and landscape flora and fauna. Rehabilitation and replanting of mangrove plants is fruitful, and rewarding. Currently there is sufficient human attention on the decline of terrestrial biodiversity as well as its conservation. In this context oceans are neglected. This practice should be avoided and oceans should also be given equal importance in terms of restoration of biodiversity. Maintenance of stable relationships between marine biodiversity and humans is also essential. Any nation or international organization alone cannot maintain health of oceans. Consequently pressure is continuously building up on the oceans. Human being is destroying the coastline. Damage this protection oceans let them flood frequently. Human beings are not polluting the coastal waters, but also over and irregular fishing the existing fish population. This activity critically changes the oceanic food webs. In many cases, the current degradations are probably irreversible within the timescale of a human life.

Unlike other parts of the country, an abundant variety of flora and fauna is prevalent in the mountainous areas like the Himalaya, Karakorum, and Hindukush Ranges. Alpine grazing lands, sub-alpine scrub, and temperate forests are the most faunal population in these mountainous ranges.

These areas offer efficient grazing and feeding areas for wildlife. They also provide hiding and breeding places for wildlife. Shapu, musk deer, marmots, tragopan, monal pheasants, snow leopard, the black and the brown bears, otter, wolf, lynx, Himalayan ibex, markhor, bharal, Marco Polo's sheep are the most common wildlife species found in these areas. At higher elevations, the snow partridge and snowcock are common. At lower elevations, the dominant species are goral, several species of flying squirrels, chakor, partridge, and pheasants (koklass, kaleej, and cheer), as well as Rhesus monkey, common langur, red fox, black bear, common leopard, a variety of cats, and musk deer (over a limited area). Among musk deer, Marco Polo's sheep, the snow leopard, and the brown bear are endangered. The Tibetan wild ass and the blue sheep populations have been significantly reduced. Within Pakistan's boundaries, the cheer pheasant is reported extinct. Hence it is included in the IUCN Red Data Book. The western horned tragopan disappeared from within Pakistani territory in the past and was presumed extinct. Nonetheless, it has been relocated in Indus Kohistan. Although it was relocated, its existing individuals are nonetheless low in numbers (Clements, 2000; Lepage, 2007).

Current data show that Pakistan is rich in wild populations. Parallel to this claim is a host of numbers of endangered animal species. Common endangered species are several species of birds, whales, turtles, woolly flying squirrel, markhor, blue whale, Pakistan sand cat, and the ibex wild goat. Certain mammals like Snow leopard, the Marco Polo sheep, and the Indus river dolphin were reported abundant in the past. Unfortunately now they have become the rarest. Such changes and reduction in population has forced Pakistan authorities to build more national parks. However, it will not work for those species which are at the verge of extinction. It is too little as well as too late for restoration of those species. Some other factors which are posing threats to wildlife are deforestation (estimated at 1% annually), overgrazing, soil erosion, rampant hunting and fishing, and agricultural practices. Due to these, it is estimated that several animal species and at least 12% of the flora is threatened. However, the exploration of the real status of most of the wildlife species is still pending and not fully known.

Keeping in view the current global climatic changes, we should understand and fully comprehend that it can have serious consequences on wildlife too. The initiation of the salvation of Biodiversity Ecosystem Functioning problem is still at its very early stage. Scientists, however, are pondering this issue and working on its resolution. Although scientists still cannot always comprehend the intricate biological dynamics, we should take it seriously before it is too late.

6.2.13 Three Main Strategies Can Conserve Biodiversity

These are:

National Park: This is a reserved area for wildlife. It is built on natural or semi-natural land and is restricted for most of development. It is owned by the Pakistani Government. It is set aside for human recreation and enjoyment and for animal and environmental protection.

The Northeast Greenland National Park is the largest national park in the world which fulfills the IUCN criteria. It was established in 1974. Pakistan has the following National Parks (Pakistan Wildlife Foundation):

01. Margallah Hills National Park: Islamabad, established 1980, area 15,883 ha.
02. Hazarganji Chiltan National Park: Baluchistan, established 1980, area 15,555 ha.
03. Hingol National Park: Baluchistan, established 1997, area 6, 19,043 ha.
04. Kirthar National Park: Sindh, established 1974, area 3, 08,733 ha.
05. Ayubia National Park: Khyber Pakhtunkhwa, established 1984, area 3,122 ha.
06. Chitral Gol National Park: Khyber Pakhtunkhwa, established 1984, area 7,750 ha.
07. Sheikh Badin National Park: Khyber Pakhtunkhwa, established 1999, area 15,540 ha.
08. Saiful Maluk National Park: Khyber Pakhtunkhwa, established 2003, area 12,026 ha.

09. Lulosar Dodipat National Park: Khyber Pakhtunkhwa, established 2003, area 75,058 ha.
10. Broghil Valley National Park: Khyber Pakhtunkhwa, established August 25, 2010, Area 134,744 ha.
11. Khunjerab National Park: Gilgit-Baltistan, established 1975, area 2, 27,143 ha.
12. Hunderab Shandoor National Park: Gilgit-Baltistan, established 1993, area 1, 65,000 ha.
13. Deosai National Park: Gilgit-Baltistan, established 1993, area 3, 63,600 ha.
14. Central Karakoram National Park: Gilgit-Baltistan, established 1993, area 9, 73,845 ha.
15. Qurumber National Park: Gilgit-Baltistan, established August 2, 2011, area 74,000 ha.
16. Machhiara National Park: AJ & K, established 1996, area 13,532 ha.
17. Ghamot National Park: AJ & K, established 2004, area 27,271 ha.
18. Pir Lasoorha National Park: AJ & K, established 2005, area 1,580 ha.
19. Toli Pir National Park: AJ & K, established 2005, area 1,000 ha.
20. Musk Deer Gorez National Park: AJ & K, established 2007, area 52,815 ha.
21. Deva Vatala National Park: AJ & K, established 2007, area 2,993 ha.
22. Poonch River Mahaseer National Park: AJ & K, established 2011
23. Lal Suhanra National Park: Punjab, established 1972, area 51,368 ha.
24. Chinji National Park: Punjab, established 1987, area 6, 097 ha.
25. Kala Chitta National Park: Punjab, established Dec. 2008, area 36,965 ha.
26. Murree-Kotli Sattian-Kahuta National Park: Punjab, established Sep. 2009

A wildlife refuge: Alternatively it is also called it is also called a wildlife sanctuary. Sanctuary may occur naturally such as island. It may also be geographic territory, or a declared protected area. Nonetheless, sanctuary protects animal species from hunting, predation, or competition. Government by law officially designates these territories. Its land may be publicly or privately owned.

Game reserves: For game reserve government sets aside part of land for the maintenance of wildlife for tourism and/or for hunting purposes. Africa as well as several other countries has many game reserves. It is a natural habitat and ideal situation for indigenous wildlife. These game reserves are used for protection of wildlife and its natural propagation and growth. Take an example of rhinoceros, elephant, buffalo, leopard, and lion, living altogether at one place in Africa. Tourists and general public can visit these areas freely. Most are open to the public, and tourists. People visit them for sightseeing and hunting of wild animals (Mazhar, 2020).

6.2.14 National Conservation Strategy (NCS)

National conservation strategy (NCS) recognizes that the existing protected area system is incomplete. Further, this protected area system does not represent animal communities, plants, and the ecosystems. In most of the cases, the ecological criteria is not well taken in during demarcation of the boundaries of existing protected areas. Many protected areas are too small and isolated from other communities. Law enforcement programs are very weak and management planning capabilities are limited. Plans are not available which can identify hot spots of the existing species. Moreover, there is a serious deficiency of the specific management practices required for rare and endangered species. There is no mention of the participation of local people and if they participate what financial benefits and incentives they will get. At national level, this practice is injurious for biodiversity conservation. Due to these factors, Pakistan's biodiversity needs critical attention because most of it is severely threatened. For conservation of biodiversity, the NCS suggest and recommends programs which incur Rs 1.08 billion in investment for their accomplishment. These programs intend to involve community management of game reserves, development of new wetland reserves, development of new national parks, and propagation of medicinal plants and germ plasm preservation programs.

To initiate a biodiversity conservation program, we first need to recognize the areas which host endangered species or those which are near extinction. After having sufficient knowledge, we can transform these areas into protected areas.

Denis Gabor says that "till now man has been up against nature, from now on he will be up against his own nature."

It is a fact that biodiversity is declining quite swiftly. In this world, we were given an opportunity since our birth to observe and enjoy nature in its natural beauty. Unfortunately, we did the reverse. We destroyed biodiversity instead of protecting it. Do we want to forbid our children the opportunity to make amends? Care for the future of our children demands that we save our earth and its inhabitants. Ultimately, it demands that we care about biodiversity.

6.2.15 Threatened Species in Pakistan

Marco Polo Sheep (*Ovis ammon polii*), Bharal or Blue Sheep (*Pseudois nayaur*), Goral Sheep (*Naemorhedus baileyi*), snow leopard (*Panthera uncia or Uncia uncia*), Houbura Bustard (*Chlamydotis undulata*), the Indus Dolphin (*Platanista minor*) (Braulik et al., 2012), Markhor (*Capra falconeri* f*Falconeri*) Himalayan or Siberian Ibex (*Capra sibirica*), Urial (*Ovis orientalis vignei*), Sind Wild Goat (*Capra aegagrus blythi*), and Chiltan Wild Goat (*Capra aegagrus chialtanensis*) are threatened animal species of Pakistan.

The Marine Turtle Project in Pakistan was launched for the protection of Green Turtles (*Chelonia mydas*), which is thought to be the last hope for the survival of this turtle. Present estimates show that total Snow leopard population on the globe is 7,000 to 10,000 number. Out of this total, Pakistan hosts approximately 300 in number. Similarly, Indus Dolphin has very limited population. Some 400–500 individuals are found in the river Indus between the Kotri and Guddu Barrages. Due to its limited population, it has been declared as an endangered species. Another endangered mammal listed in the category of threatened species is the woolly flying squirrel.

Considering the current status of wildlife in the country the government of Pakistan has short listed several parks for the conservation and protection of these endangered species in specific species and other wildlife in general. The most important of them are Lal Sohanra National Park (Bahawalpur/Punjab), Kirthar National Park (Sindh), Khunjerab National Park (Northern Areas), Chiltan Hazarganji, National Park (Baluchistan), Margalla Hills National Park (Islamabad), and Chitral Gol National Park (NWFP). Moreover, Pakistan has 99 wildlife sanctuaries, 96 game reserves, and 14 national parks for the preservation and conservation of biodiversity. Kirthar National Park is distinctive in the sense that it was the first park on the UN list of protected areas.

6.2.16 Issues and Challenges

Since the ratification of the CBD, Pakistan has been facing certain challenges. These are political instability, increasing population, legal impediments, poverty, lack of financial resources, and limited public and stakeholder participation. In Pakistan, there are several challenges and disasters which disturb biodiversity and hamper its conservation and management. Relevant institutes have no technical capacity to handle biodiversity issues. People are not aware of biodiversity and local knowledge present about biodiversity is not accessible to them. A lack of mainstreaming of biodiversity with other sectors, lack of transfer of technology and expertise, inadequate engagement of scientific community, climate change, and natural disasters are some other key challenges disturbing biodiversity and hampering its conservation and management. It is critical now to think about the rehabilitation of biodiversity devising long-term strategies for its sustainability. If not acted upon now, biodiversity will vanish entirely. Then the next generations will have no chance to enjoy wildlife in its natural environment and will be limited to seeing only pictures, movies, and stuffed toy animals.

6.3 TERRESTRIAL ARTHROPODS

Insects are the major group of invertebrates present in the terrestrial environment. Pakistan hosts 2,000 species of insects. In some cases insect species are in quite large number and in a given area number of invertebrates especially insects are always higher than vertebrates. Pakistan hosts 13 representatives insect orders. Orders Protura, Thysanura, and Diplura consist of small numbers of species. They live in kitchens, food stores, and in the soil. From these orders, 150 species exist in Pakistan.

Order Orthoptera contains 150 species of grasshoppers and crickets in Pakistan. The majority of the members of the family Acrididae are grasshoppers which live on vegetable matter. However, family Tettigoniidae consists of bush crickets. Bush crickets feed on other insects. Among them some are large, beautiful insects. They sing and attract the attention of those nearby. Order Dermaptera contain insects are called earwigs, many species of them live in Pakistan. Earwigs spend most of their time in stores, sewage pipes, dirty kitchens etc. They principally feed on vegetation and small insects for their survival.

Order Psocoptera comprises two recorded species of insects from Pakistan which are quite insignificant, e.g., dust lice or book lice. They feed on pollen and fungus myceli and lie beneath the bark of the trees. Some others live in buildings damaging books and other belongings of historical importance.

Order Thysanoptera consists of thrips. These are very small slender-shaped insects. They have two pairs of wings. These wings are thread-shaped wings with broad fringes. The order Hemiptera contains bugs and more than 159 bugs have been recorded under this order.

Order Nemoptera contains two species called lacewings and alderflies which have been reported in Pakistan.

Some orders of insects with species recorded from Pakistan have been given below.

Order Mecoptera consists of many species of scorpion flies:

Order Siphonoptera contains 25 species of fleas
Order Lepidoptera contains 315 species of butterfly
Order Hymenoptera contains wasps and ants
Order Diplopoda contains several species of millipedes
Order Chilopoda contains several species of centipedes
Order Arachinoidea contains 84 species of arachnids

In addition to these insects, 75 species of nematodes and 225 species of mollusks have also been recorded.

6.4 FISH

Both marine and freshwater aquatic resources are immense in Pakistan. The marine fauna has already been discussed above. Rivers, canals, lakes, and streams are our freshwater fisheries resources. Other water reservoirs are the result of construction of dams and barrages. In the aquatic environment, fish are the most successful vertebrates. Huge and expanded marine aquatic resources (three-fourths of the globe) with one-fourth freshwater resources are the main reasons behind this successful succession and expansion in aquatic environment. It is estimated that all over the globe there are more than 20,000 species of fish; in Pakistan, however, 531 species have been estimated. Of these, 233 are presumed to be freshwater fish, while the remaining number are marine in nature. From the freshwaters of Pakistan, 179 species have so far been described and recorded, although this number is still debatable. Like other animals, similarities and differences among fish are their basis of classification. Animals with similar features are grouped together. Mahseer is the national fish of Pakistan. Detail of some important fish species are given below.

Family Bagridae

Aorichthys aor (Singhari)
Batasio pakistanicus (Batasio)
Mystus bleekeri (Bleekri)
Mystus cavasius (Teengara)
Mystus horal (Horaki Keengar)
Mystus vitatus (Keengar)
Rita rita (Khagga)

Family Chandiae

Chanda baculis (Baculis shisha)
Chanda nama (Nama shisha)
Chanda ranga (Ranga shisha)
Channa gachua (Dauli)
Channa marulius (Saul)
Channa punctata (Daula)
Channa striatus (Sauli)

Family Cobitidae

Botia birdi (Botia)
Botia lohachata (Botia)

Family Cyprinidae

Amblypharyngodon mola (Amblypharyngodon mola)
Aspidoaria morar (Aam Chilwa)
Barbodes sarana (Khurni)
Barilius bendelisis (Patha Chilwa)
Barilius naseeri (Naseeri Chilwa)
Barilius pakistanicus (Pakistani Chilwa)
Barilius vagra & Barilius modestus (Lahori Chilwa)
Brachydanio rerio (Zebra Machli)
Carassius auratus (Sunehri machli)
Catla catla (Theila)
Cirrhinus mrigala (Mori)
Cirrhinus reba (Sunni)
Ctenopharyngodon idela (Grass carp)
Cyprinion watsoni (Watsoni sabzag)
Cyprinus carpio (Gulfam)
Esomus danricus (Soomara Machli)
Hypopthalmichthys molitrix (Silver carp)
Labeo boga (Bhangan)
Labeo boggut (Bhangana)
Labeo calbasu (Kalbans)
Labeo dero Pahari (rohu)
Labeo dyocheilus pakistanicus (Torki)
Labeo gonius (Sareeha)
Labeo rohita (Rohu)
Osteobrama cotio (Paliro)
Puntius chola Kola (popra)
Puntius conchonius (Gulabi barb)

Puntius gelius (Gelius popra)
Puntius punjabensis (Punjabi popra)
Puntius *sophore* (Sophore popra)
Racoma labiata (Chun)
Salmostoma bacaila (Choti Chal)
Salmtostoma punjabensis (Punjabi Chal)
Securicula gora (Bari Chal)
Tor putitora (Mahseer)

Among freshwater fish fauna *Cyprinus carpio*(common carp), three species of Chinese carps *Ctenopharyngodon idella* (grass carp), *Hypophthalmichthys molitrix* (silver carp), and *Hypophthalmicthys nobilis* (bighead carp) are exotic species. All these fish are herbivorous in nature and, due to the fast growth of some of the species, they are commercially important and are cultured on a mass scale in semi-intensive fish culture systems.

Family Mastcembelidae

Mastacembelus arm atus (Bam)
Mastacembelus pancalus (Grooj)
Nandidae Nandus nandus (Pata)

Family Neomacheilidae

Schistura curtistigma (Shistura)
Schistura prashari (Shistura)
Schistura punjabensis (Shistura)
Schistura shadiwalensis (Shistura)
Shistura nalbanti (Shistura)

Family Notopteridae

Gudusia chapra (Pali)
Notopterus chitala (Cheetal Pari)
Notopterus notopterus (But Pari)

Family Osphranomidae

Colisa fasciata (Bari kanghi)
Colisa lalia Choti (kanghi)
Oreochromis mossambica (Tilapia)

Family Salmonidae

Oncorhynchus mykiss (Rainbow Trout)

Family Schilbeidae

Eutropichthys vacha (Jhali)
Pseudeutropius atherinoides (Chaali)
Clupisoma garua (Bachwa)
Clupisoma murius (Aahi)

Family Siluridae

Hetero pneustes fossilis (Singhi)
Ompok bimaculatus (Pafta)
Walago atu (Mulee)

Family Sisoridae

Bagarius bagarius (Goonch, Fauji Khagga)
Gagata cenia (Gageeta)
Glyptothorax cavia (Kani tingara)
Glyptothorax stocki (Stocki Pahari Khagga)
Nangra nangra (Nangra)
Nangra robusta (Nangra)

Family Synbranchidae

Monopterus cuchia (Cuchia)

Family Xenentodontidae

Aplocheilus panchax (Lal jheengra)
Xenentodon cancila (Kaan)

(Reference: Punjab Fisheries Manual 2014.)

6.5 AMPHIBIANS

In Pakistan, amphibian fauna is very scarce. Presently, there are two vital orders (Caudata and Gymnophiona) of amphibians but unfortunately not a single species from here exists. Anura is the only clade that represents the amphibian fauna in Pakistan's geographical boundaries. It has a heterogeneous assemblage of 21 amphibian species (Sarwar et al., 2016). Anurans are distributed among Bufonidae, Megophryidae, Microhylidae, and Dicroglossidae families and 12 genera (Frost, 2016). Amphibian fauna of Pakistan originate from sea level. Distribution then expands up to the Himalayas and Karakoram up to 4,000 meters high. This distribution stretches across the latitude starting from the Indian boundary ending in its western borders (Khan, 2014). The two genera of family Bufonidae called Genus *Duttaphrynus* and *Bufotes*, are widely distributed toad genera in Pakistan. Genus *Duttaphrynus himalayanus* has very wide distribution. It expands throughout the Himalayas.

Distribution of this genus has also been reported in adjacent countries too, including India and China. *Duttaphrynus melanostictus* is nevertheless mainly distributed in the Hazara division in the province of Khyber Pakhtunkhwa (Khan, 2001).

Duttaphrynus stomaticus is one of the widely distributed and most well-adapted anuran species of Pakistan. It is equally prevalent in both lowlands as well as highlands.

Unlike the above-mentioned species, the olive toad, *Duttaphrynus olivaceus*) is common only in the lowland areas of Northwestern Baluchistan (Auffenberg & Rehman, 1977).

Three species of genus *Bufotes* called the *Bufotes psedudoraddei* (endemic), *Bufotes surdus*, *Bufotes zugmayeri* with several others have very wide distribution in the western areas of Pakistan inclusive of Baluchistan (Khan, 2014). *Bufotes latastii* (Laddakh toad)), however, inhabits the highlands (Skardu) of the Himalayas (Skardu) (Khan, 1997).

Scutiger nyingchiensis, commonly known as the Tibetian toad, and which belongs to the family Megophyridae, is only found in the northern highlands (highlands of Deosai) (Khan, 2014). *Uperodon systoma*, a member of family Microhylidae is common in the capital territory (North Punjab). The Murree hills host a subterranean frog naturally secretive in habit. *U. systoma* (Masroor, 2011) and *Microhyla oronata* have the same habitat (Rais et al., 2012). Two species of genus Allopaa called *Allopa hazarensis* and *Allopa barmoachensis* inhabit the western, central, and eastern Himalayas (Kashmir, Hazara). *Chrysopaa sternosignata*, which is also known as the Baluchistan Karez frog, inhabits the Kashmir valley and the whole province of Baluchistan. In the Baluchistan highlands it lives in irrigation channels.

Scutiger nyingchiensis is an altitude-adapted frog and is mostly present in the grass fields of northwestern mountains and areas of the Kashmir valley (Khan, 2006). *Euphlyctis* has two species. Both these two species have wide distribution ranges in Anura and prefer to live in plain areas, especially in Punjab.

Hoplobatrachus tigrinus, *Fejervarya syhadrensis* and members of genus *Sphaerotheca* (*Sphaerotheca breviceps* and *Sphaerotheca strachani*) live in plains with relatively high populations. Distribution of members of Genus *Sphaerotheca strats* starts and is in the channels along the river and expansion in its distribution of habitat goes up to the coastal areas in Sindh (Begum et al., 2014). The Murree frog, *Nanorana vicina*, inhabits the Kashmir valley and North-central Pakistan (Rais et al., 2014). Due to arid and semi-arid habitats of Pakistan it hosts only 22 species of amphibians. All the 22 species of frogs have been recorded and described. Out of these, nine frog species are endemic. Bufonids and ranids may have Himalayan origin. Nevertheless, Southeast Asian frogs and Palearctic toads are dominant in the plains. There are four amphibian families known as Bufonidae, Microhylidae, Megophryidae, and Ranidae, which belong to Pakistan. A summary of some common amphibian families with genus, species, and common names has been given below.

6.6 REPTILES

Pakistan hosts 174 species of reptiles. Out of this total, only 13 species are prevalent and currently thriving. All the others are declining and scarce in number. Like other living animal species they do not belong to a single group; rather, they are a mixture of Palaearctic, Indo-Malayan, and Ethiopian forms. The composition of these species of reptiles is: lizards, 88; snakes, 72; turtles, 10 (marine, 2, freshwater, 8); tortoises, 2; crocodile, 1; and gavial, 1 (Safi & Khan, 2014; Khan et al., 2015). Due to overhunting, the marsh crocodile is in danger. This species is near extinction and only a few numbers are left. These living species of marsh crocodile are found in Sindh and in some areas of Baluchistan (Groomridge & Wright, 1982). It is not clear whether the gharial has become extinct or is still surviving because a very small number has been seen between the Sukkur and Guddu barrages several years ago. Hence, the situation of this animal is very precarious. For collection of skins, heavy hunting of the monitor species has been observed, which are also not in a good condition.

In Pakistan, 72 species of snakes have been found. Out of this total, only 26 are poisonous (14 marine and 12 terrestrial snake species). Well-known poisonous species are the Indian cobra, common krait, saw-scaled viper, and Russel's viper. Their pictures have been provided below for the information of the reader.

The Chagai Desert consists of five species of lizards and one snake (Hashmi et al., 2013). In addition to these, six more species of reptiles are found here and hence it has become of special interest for researchers due to the presence of reptiles here and on the Pakistan–Iran borders. Pakistan's southern beaches host an important population of marine turtles. According to IUCN (1990) gharial, central Asian monitor, green and olive turtle, mugger, central Asian cobra, Indian python, are included in that category of reptilian species which are internationally threatened. Class Reptilia has 16 orders of which 12 are extinct and 4 are living. Order Rhynchocephalia has a single representative lizard called Sphenodon, which is restricted to New Zealand and none of its representatives is found in Pakistan. Representatives of the remaining three orders are found in Pakistan which have been summarized below.

6.6.1 Order Crocodilia

Family *Crocodylidae* (crocodiles) – 1 species

- *Crocodylus palustris* (Mugger Crocodile, Indian Crocodile) Indus Crocodile or Marsh Crocodile) –it is also called a national reptile of Pakistan.

Family *Gavialidae* (gharials) 1 species.

6.6.2 Order *Squamata*

Around 153 species of squamata are found in Pakistan.

6.6.2.1 Suborder *Lacertilia*

There are 86 species of Lacertilia (lizards and relatives) in Pakistan.

- **Family *Agamidae*** (agamas): 22 species
  - *Brachysaura minor* (Hardwicke's Bloodsucker)
  - *Calotes versicolor* (Oriental Garden Lizard, Eastern Garden Lizard or Changeable Lizard)
  - *Japalura kumaonensis* (Kumaon Mountain Lizard)
  - *Paralaudakia badakhshana* (Badakhshana Rock Agama)
  - *Paralaudakia caucasia* (Caucasian Agama)
  - *Paralaudakia himalayana* (Himalayan Agama)
  - *Laudakia agrorensis* (Agror Agama)
  - *Laudakia melanura* (Black Agama)
  - *Laudakia fusca* (Yellow-headed Rock Agama)
  - *Laudakia nupta* (Large-scaled Rock Agama): 2 subspecies
    - *Laudakia n. fusca*
    - *Laudakia n. nupta*
  - *Laudakia tuberculata* (Kashmir Rock Agama)
  - *Phrynocephalus clarkorum* (Clark's Toad-headed Agama)
  - *Phrynocephalus euptilopus* (Alcock's Toad-headed Agama)
  - *Phrynocephalus luteoguttatus* (Yellow-speckled Toad-headed Agama)
  - (Source:alchetron.com)
  - *Phrynocephalus maculatus* (Blacktail Toadhead Agama)
  - *Phrynocephalus ornatus* (Ornate Toadhead Agama)
  - *Phrynocephalus scutellatus* (Gray Toadhead Agama)
  - *Trapelus agilis pakistanensis* (Brilliant Ground Agama)
  - *Trapelus rubrigularis* (Red-throated Agama)
  - *Trapelus ruderatus* (Horn-scaled Agama or Baluch Ground Agama)
  - *Uromastyx asmussi* (Iranian Mastigure)
  - *Uromastyx hardwickii* (Hardwick's Spiny-tailed Lizard or Indian Spiny-tailed Lizard).
- **Family *Chamaeleonidae*** (chameleons): 1 species
  - *Chamaeleo zeylanicus* (Indian Chameleon or Ceylon Chameleon)
- **Family *Gekkonidae*** (eyelid and lidless geckoes): 33 species
  - *Agamura femoralis* or *Rhinogecko femoralis* (De Witte's Gecko or Sharp-tailed Spider Gecko)
  - *Agamura persica persica* (Persian Spider Gecko)
  - *Bunopus tuberculatus* (Baluch Rock Gecko)
  - *Crossobamon eversmanni* (Baluchistan Sand Gecko or Comb-toed Gecko): 2 subspecies
    - *Crossobamon e. eversmanni* or *Crossobamon maynardi*
    - *Crossobamon e. lumsdenii*
  - *Crossobamon orientalis* (Sind Gecko)
  - *Cyrtodactylus walli* (Chitral Gecko or Swat Stone Gecko)
  - *Cyrtopodium agamuroides* (Nikolsky Spider Gecko)
  - *Cyrtopodium baturensis* (Batura thin-toed Gecko or Batura Gecko)
  - *Cyrtopodium dattanensis* (Khan's Bow-fingered Gecko or Hazara Gecko)
  - *Cyrtopodium fortmunroi* (Fort Munro Gecko)
  - *Cyrtopodium indusoani* (Soan Sakaser Gecko)

    - *Cyrtopodium kachhensis* (Kutch Gecko): 2 subspecies
        - *Cyrtopodium k. ingoldbyi* (Ingoldby's Stone Gecko)
        - *Cyrtopodium k. kachhensis* (Warty Rock Gecko)
    - *Cyrtopodium mintoni* (Minton's Gecko)
    - *Cyrtopodium montiumsalsorum* (Salt Range Rock Gecko)
    - *Cyrtopodium rhodocaudus* (Red-tailed Gecko)
    - *Cyrtopodium rohtasfortai* (Rohtas Fort Gecko)
    - *Cyrtopodium scaber* (Keeled Rock Gecko)
    - *Cyrtopodium watsoni* (Quetta Rock Gecko)
    - *Eublepharis macularius* (Leopard Gecko)
    - *Hemidactylus brookii brookii* (Brook's House Gecko or Spotted Indian House Gecko)
    - *Hemidactylus flaviviridis* (Yellow-bellied House Gecko)
    - *Hemidactylus frenatus* (Pacific House Gecko or South Asian Waif Gecko)
    - *Hemidactylus imbricatus* (Carrot-tail Viper Gecko)
    - *Hemidactylus leschenaultii* (Leschenault's Leaf-toed Gecko or Bark Gecko)
    - *Hemidactylus persicus* (Persian Leaf-toed Gecko)
    - *Hemidactylus triedrus triedus* (Termite Hill Gecko or Blotched Gecko)
    - *Hemidactylus turcicus turcicus* (Mediterranean House Gecko)
    - *Microgecko depressus* (Low-lying Gecko)
    - *Microgecko helenae* (Banded Dwarf Gecko)
    - *Ptyodactylus homolepis* (Pakistan Fan-fingered Gecko): 2 subspecies
        - *Ptyodactylus h. homolepis*
        - *Ptyodactylus h. socotranus*
    - *Teratoscincus microlepis* (Small-scaled Wonder Gecko or Baluch Plate-tailed Gecko)
    - *Teratoscincus scincus* (Turkestan Plate-tailed Gecko or Common Wonder Gecko or Frog-eyed Gecko)
    - *Tropiocolotes persica* (Persian Sand Gecko or Persian Pygmy Gecko)
- **Family *Lacertidae*** (sand lizards, wall lizards and true lizards): 12 species
    - *Acanthodactylus cantoris* (Indian Fringe-fingered Lizard): 2 subspecies
        - *Acanthodactylus c. blanfordii* (Mekran Fringe-fingered Lizard)
        - *Acanthodactylus c. cantoris* (Indian Fringe-fingered Lizard)
    - *Acanthodactylus micropholis* (Yellow-tailed Sand Lizard)
    - *Eremias acutirostris* (Reticulate Desert Lacerta)
    - *Eremias fasciata* (Yellow-headed Desert Lacerta)
    - *Eremias scripta* (Caspian Desert Lacerta)
    - *Eremias velox persica* (Persian Steppe Lacerta)
    - *Mesalina brevirostris* (Sort-nosed Desert Lacerta)
    - *Mesalina guttulata* (Small-spotted Lizard or Long-tailed Desert Lacerta)
    - *Ophisops elegans (Elegant Snake-eyed Lizard)*: 2 subspecies
        - *Ophisops e. blanfordi* (Blanford's Snake-eyed Lizard)
        - *Ophisops e. elegans* (Elegant Snake-eyed Lizard)
    - *Ophisops jerdonii* (Jerdon's Snake-Eyed Lizard or Punjab Snake-eyed Lizard)
    - *Ophisops microlepis* (Indian Snake-eyed Lizard)
    - *Scapteira aporosceles* (Chagai Desert Lacerta)
- **Family *Scincidae*** (skinks): 15 species
    - *Ablepharus grayanus* (Minor Snake-eyed Skink)
    - *Ablepharus pannonicus* (Asian Snake-eyed Skink)
    - *Chalcides ocellatus* (Ocellated Skink)
    - *Eumeces schneiderii zarudnyi* (Schneider's Skink, Berber skink or Zarudny's Skink)
    - *Eurylepis indothalensis* (Thal skink)
    - *Eurylepis taeniolatus* (Alpine Punjab Skink or Yellow-bellied Mole Skink)

- *Eutropis dissimilis* (Striped Grass Mabuya or Striped Grass Skink)
- *Eutropis macularia* (Bronze Mabuya, Bronze Grass Skink)
- *Ophiomorus blanfordi* (Makran Sand Swimmer)
- *Ophiomorus brevipes* (Short-toed Sand Swimmer)
- *Ophiomorus raithmai* (Indus Sand Swimmer)
- *Ophiomorus tridactylus* (Three-toed Snake Skink or Indian Sand Swimmer)
- *Scincella ladacensis*: 2 subspecies
  - *Scincella l. ladacensis* (Ladak Ground Skink)
  - *Scincella l. himalayana* (Himalayan Ground Skink)

- **Family *Varanidae*** (monitor lizards): 2 species
  - *Varanus flavescens* (Yellow Monitor)
  - *Varanus griseus*: 2 subspecies
    - *Varanus g. caspius* (Eastern Desert Monitor)
    - *Varanus g. koniecznyi* (Thar Desert Monitor)

6.6.2.2 Suborder Ophidia/Serpentes

There are 97 species of serpents (snakes) found in Pakistan.

- **Family *Boidae*** (boas): 2 species
  - *Eryx johnii* (Indian Sand Boa or Red Sand Boa)
  - *Eryx tataricus speciosus* (Tartar Sand Boa)
  - *Gongylophis conicus* or *Eryx conicus* (Rough-scaled Sand Boa)
- **Family *Colubridae*** (colubrids): 32 species
  - *Herpetoreas platyceps*
  - *Herpetoreas sieboldii*
  - *Amphiesma stolata stolata*
  - *Boiga trigonata*: 2 subspecies
    - *Boiga t. melanocephala*
    - *Boiga t. trigonata*
  - *Coluber fasciolatus*
  - *Coluber karelini*: 2 subspecies
    - *Coluber k. karelini*
    - *Coluber k. mintonorum*
  - *Coluber ravergeiri ravergeiri*
  - *Coluber rhodorachis*: 3 subspecies
    - *Coluber r. kashmirensis*
    - *Coluber r. ladacensis*
    - *Coluber r. rhodorachis*
  - *Coluber ventromaculatus*: 3 subspecies
    - *Coluber v. bengalensis*
    - *Coluber v. indusai*
    - *Coluber v. ventromaculatus*
  - *Enhydris pakistanica*
  - *Lycodon aulicus aulicus*
  - *Lycodon striatus*: 2 subspecies
    - *Lycodon s. bicolor*
    - *Lycodon s. striatus*
  - *Lycodon travancoricus*
  - *Lytorhynchus maynardi*
  - *Lytorhynchus paradoxus*
  - *Lytorhynchus ridgewayi*

    - *Natrix tessellata tessellate*
    - *Oligodon arnensis*
    - *Oligodon taeniolatus*
    - *Psammophis condanarus condanarus*
    - *Psammophis leithii*
    - *Psammophis lineolatus*
    - *Psammophis schokari*
    - *Pseudocyclophis persica*
    - *Ptyas mucosus*
    - *Sibynophis sagittarius*
    - *Spalerosophis arenarius*
    - *Spalerosophis diadema*: 2 subspecies
      - *Spalerosophis d. diadema*
      - *Spalerosophis d. schirazianus*
    - *Telescopus rhinopoma*
    - *Xenochrophis cerasogaster*
    - *Xenochrophis piscator piscator*
    - *Xenochrophis sanctijohannis*
- **Family *Elapidae*** (kraits and cobras): 4 species
    - *Bungarus caeruleus* (Common Krait)
    - *Bungarus sindanus* (Sind Krait): 2 subspecies
    - *Bungarus s. razai* (Northern Punjab Krait)
      - *Bungarus s. sindanus* (Sind Krait)
    - *Naja naja* (Indian Cobra or Spectacled Cobra)
    - *Naja oxiana* (Central Asian Cobra or Oxus Cobra)
- **Family *Hydrophiidae*** (sea snakes): 14 species
    - *Astrotia stokesii* (Stokes' Sea Snake)
    - *Enhydrina schistosa* (Beaked Sea Snake, Hook-nosed Sea Snake, Common Sea Snake, or Valakadyn Sea Snake)
    - *Hydrophis caerulescens* (Dwarf Sea Snake or Many-toothed Sea Snake)
    - *Hydrophis cyanocinctus* (Annulated Sea Snake or Blue-banded Sea Snake)
    - *Hydrophis fasciatus* (Striped Sea Snake)
    - *Hydrophis lapemoides* (Persian Gulf Sea Snake)
    - *Hydrophis mamillaris* (Bombay Sea Snake)
    - *Hydrophis ornatus* (Ornate Reef Sea Snake)
    - *Hydrophis spiralis* (Yellow Sea Snake)
    - *Lapemis curtus* (Shaw's Sea Snake)
    - *Microcephalophis cantoris* (Cantor's Small-headed Sea Snake)
    - *Microcephalophis gracilis* (Graceful Small-headed Sea Snake or Slender Sea Snake)
    - *Pelamis platurus* (Yellow-bellied Sea Snake or Pelagic Sea Snake)
    - *Praescutata viperina* (Viperine Sea Snake)
- **Family *Leptotyphlopidae*** (thread snakes): 2 species
    - *Leptotyphlops blanfordii* (Blanford's Worm Snake or Sind Thread Snake)
    - Blanford's Worm Snake or Sind Thread Snake (*Leptotyphlops blanfordii*) (Source: revolvy.com)
    - *Leptotyphlops macrorhynchus* (Long-nosed Worm Snake or Beaked Thread Snake)
- **Family *Pythonidae*** (pythons): 1 species
    - *Python molurus molurus* (Indian Python, Black-tailed Python or Sind Python)
- **Family *Typhlopidae*** (blind snakes): 5 species
    - *Ramphotyphlops braminus* (Brahminy Blind Snake)
    - *Typhlops ahsanai* (Ahsan's Blind Snake)

    - *Typhlops porrectus* or *Typhlops ductuliformes* (Slender Worm Snake)
    - *Typhlops madgemintonai*: 2 subspecies
        - *Typhlops m. madgemintonai* (Madge's Blind Snake)
        - *Typhlops m. shermanai* (Sherman's Blind Snake)
- **Family *Viperidae*** (vipers and pit vipers): 7 species
    - *Daboia russelii russelii* (Daboia or Russell's Viper)
    - *Echis carinatus* (Indian Saw-scaled Viper): 3 subspecies
        - *Echis c. astolae* (Astola Saw-scaled Viper)
        - *Echis c. multisquamatus* (Multiscale Saw-scaled Viper)
        - *Echis c. sochureki* (Sochurek's Saw-scaled Viper)
- *Echis pyramidum* (Northeast African Carpet Viper, Egyptian Saw-scaled Viper)
- *Eristicophis macmahonii* (McMahon's Viper, Asian Sand Viper, Leaf-nosed Viper)
- *Gloydius himalayanus* (Himalayan Pit Viper)
- *Macrovipera lebetina obtusa* (West-Asian Blunt-nosed Viper, Levant Blunt-nosed Viper)
- *Pseudocerastes persicus persicus* (Persian Horned Viper, False Horned Viper)

6.6.3 Order *Testudines*

This order has 15 species that represent Pakistan.

- **Family *Cheloniidae*** (sea turtles): 4 species
    - *Chelonia mydas* (Green 'Sea' Turtle)
    - *Caretta caretta* (Loggerhead Sea Turtle)
    - *Eretmochelys imbricata bissa* (Hawksbill Turtle)
    - *Lepidochelys olivacea* (Olive Ridley Turtle)
- **Family *Dermochelyidae*** (softshell sea turtles): 1 species
    - *Dermochelys coriacea* (Leatherback 'Sea' Turtle or Leatherback)
- **Family *Geoemydidae*** (Asian river turtles, Asian leaf turtles, Asian box turtles, and roofed turtles) - 4 species
    - *Geoclemys hamiltonii* (Black Pond Turtle, Spotted Pond Turtle, or Indian Spotted Turtle)
    - *Hardella thurjii* (Brahminy River Turtle or Crowned River Turtle)
    - *Kachuga smithii smithii* (Brown-roofed Turtle)
    - *Kachuga tecta* (Indian Roofed Turtle)
- **Family *Testudinidae*** (true tortoises): 2 species
    - *Geochelone elegans* (Indian Star Tortoise)
    - *Testudo horsfieldi horsfieldii* (Afghan Tortoise, Horsfield's Tortoise or Central Asian Tortoise)
- **Family *Trionychidae*** (softshell turtles): 4 species
    - *Aspideretes gangeticus indica* (Ganges Soft-shelled Turtle)
    - *Aspideretes hurum* (Peacock Soft-shelled Turtle)
    - *Chitra indica* (Narrow-headed Soft-shelled Turtle)
    - *Lissemys punctata andersoni* (Indian Flap-shelled Turtle)

6.7 BIRDS

Pakistan hosts 786 species of birds. Out of this total, 39 are accidental (which means their presence is either rare or accidental) in Pakistan. One species included in the list given below existed in the past but is now extirpated (a species that exists in other countries of the world but not found in Pakistan). The shaheen falcon is the symbol of the Pakistan Avicultural Foundation and Pakistan Air Force. The *Alectoris chukar* (common name chukar) has been declared as the official national bird of Pakistan.

6.7.1 Order: Gaviiformes Family: Gaviidae (Loons)

In size, Loons are equal to a small goose or a large duck. The resemblance of the shape of loons with these birds is more obvious during swimming. Worldwide, there are five species of loons. Pakistan nevertheless has two species which are *Gavia arctica* (Black throated diver) and *Gavia immer*(Great northern diver).

6.7.2 Order: Podicipediformes Family: Podicipedidae (Grebes)

Grebes are diving freshwater birds. Their size is small to medium. Their lobed toes help them in their excellent swimming and diving skills. When on land they feel uncomfortable due to the far back position of their feet under the body. Worldwide, there are 20 species of grebes while Pakistan hosts five of them. These are eared grebe, *Podiceps nigricollis*, little grebe, *Tachybaptus ruficollis*, red-necked grebe, *Podiceps grisegena*, horned grebe, *Podiceps auritus* and great crested grebe, *Podiceps cristatus.*

6.7.3 Order: Procellariiformes Family: Procellariidae (Porcellariids)

The procellariids are true petrels. They are medium in size and constitute the main group of petrels. Their nostrils are united. The medium septum and a long outer functional primary play a role in the union of these nostrils. Species present in Pakistan are Flesh-footed shearwater, Jouanin's petrel, *Bulweria fallax, Ardenna carneipes*, Wedge-tailed shearwater, *Ardenna pacificus*, Persian shearwater, *Puffinus persicus*, and Short-tailed shearwater, *Ardenna tenuirostris.*

6.7.3.1 Family: Oceanitidae (Austral Storm)

The austral storm petrels are the smallest seabirds and relate to petrels. When they hover they pick up small fish from the surface and feed on them. They also feed on planktonic crustaceans. Sometimes their flight is bat-like. Usually, they flutter during flight.

6.7.4 Order: Phaethontiformes Family: Phaethontidae (Tropic birds)

Tropicbirds inhabit tropical oceans. They are white in color and slender shaped with exceptionally long central tail feathers. Their heads and long wings are marked in black. There are two species of tropic birds present in Pakistan, named Red-billed tropicbird, *Phaethon aethereus*, and Red-tailed tropicbird, *Phaethon rubricauda.*

6.7.5 Order: Suliformes Family: Sulidae (Sulids)

The sulids have two types of birds called boobies and the gannets. Both are coastal sea birds. Their size is from medium to large. These birds dive and plunge for fish. The Masked booby, *Sula dactylatra*, is commonly found in Pakistan.

6.7.5.1 Family: Phalacrocoracidae (Coastal Fish-Eating Sea Birds)

This family consists of cormorants and shags. They are large to medium in size coastal fish-eating sea birds. There are many variations in the color of the plumage. Plumage color varies from bird to bird. The majority of birds have dark plumage, some with black-and-white plumage, while a few others have colorful plumage. Common species found in Pakistan are the Indian cormorant, *Phalacrocorax fuscicollis*, little cormorant, *Microcarbo niger*, pygmy cormorant, *Microcarbo pygmeus*, and great cormorant, *Phalacrocorax carbo.*

6.7.5.2 Family: Anhingidae (Darters)

Due to their long thin necks, these birds are often called darters or snake birds. When they swim, they submerge their bodies in water and display a snake-like appearance. The females have smaller bills than the males. Male plumage is dark-brown to black in color. The nape bears an erectile crest. The neck and underparts of the females have paler plumage. The feet of the darters are completely webbed, with short legs. The legs are set far back on the body. Their plumage resembles that of cormorants and is permeable to water. After diving, they dry their wings by spreading and exposing them to the air and sunshine. Worldwide, there are four species in this family. Only one species, the Oriental darter, *Anhinga melanogaster*, is found in Pakistan.

6.7.6 Order: Pelecaniformes Family: Pelecanidae (Gray pelican)

Pelicans represent this family. They are large-sized water birds. They have a distinctive pouch under their beak is their characteristic feature. Like other birds under the order Pelecaniformes, the feet of gray pelicans have four toes which are webbed. The representative birds of this family which are found in Pakistan are the Great white pelican, *Pelecanus onocrotalus*, the Dalmatian pelican, *Pelecanus crispus*, and the Gray pelican, *Pelecanus philippensis*.

6.7.6.1 Family: Ardeidae

This family constitutes bitterns, herons, and egrets. Among them, egrets and herons are large-sized wading birds, although some are medium in size. Although all the birds in this family have long necks and legs, bitterns normally have shorter necks and are comparatively more suspicious. Unlike other long-necked birds like storks, ibises, and spoonbills, these birds retract their necks, while egrets and bitterns do not. Birds of this family found in Pakistan are Western reef heron, *Egretta gularis*, Goliath heron, *Ardea goliath*, Eastern great egret, *Ardea modesta*, Intermediate egret, *Ardea intermedia*, Purple heron, *Ardea purpurea*, Little egret, *Egretta garzetta*, Indian pond heron, *Ardeola grayii*, Cattle egret, *Bubulcus ibis*, Striated heron, *Butorides striata*, Black-crowned night heron, *Nycticorax nycticorax*, Yellow bittern, *Ixobrychus sinensis*, Little bittern, *Ixobrychus minutus*, Cinnamon bittern, *Ixobrychus cinnamomeus*, Black bittern, *Ixobrychus flavicollis*, and Great bittern, *Botaurus stellaris*.

6.7.6.2 Family: Threskiornithidae (Terrestrial and Wading Birds)

This includes ibises and spoonbills which are largely terrestrial and wading birds. Their wings are broad and long with 11 primary and about 20 secondary feathers. Although they are of quite good size they are good soarers and strong fliers. Worldwide, this family has 36 species but only four are found in Pakistan, which are Black-headed ibis, *Threskiornis melanocephalus*, Eurasian spoonbill, *Platalea leucorodia*, Red-naped ibis, *Pseudibis papillosa*, and Glossy ibis, *Plegadis falcinellus*.

6.7.7 Order: Ciconiiformes Family: Ciconiidae (Storks)

Storks are large wading birds with long stout bills. They have long necks, as well as long legs. Storks do not utter any sound and are thought to be mute. At nest, they clatter their bills, which is an important mode of their communication. They build big nests which can be used again and again for many years. Many species which fall under this family are migratory. Important members of this family are Painted stork, *Mycteria leucocephala*, Woolly-necked stork, *Ciconia episcopus*, Greater adjutant, *Leptoptilos dubius*, Black-necked stork, *Ephippiorhynchus asiaticus*, White stork, *Ciconia ciconia*, Black stork, *Ciconia nigra, and* Asian openbill, *Anastomus oscitans*.

6.7.8 Order: Phoenicopteriformes Family: Phoenicopteridae (Falimngos)

Flamingos are very tall birds. Their height ranges from three to five feet. Both Western and Eastern hemispheres host these birds. They are social birds. They are filter feeders. Their preferred foods are shellfish and algae. Their beak shape is very odd. Nevertheless, they use it to separate mud and silt from the food during food consumption. Their beak is very unique to them and they use it in an upside down manner. Worldwide, this family has three species. Out of six, two species are found in Pakistan. The representative species of this family are Greater flamingo, *Phoenicopterus roseus* and lesser flamingo, *Phoenicopterus minor.*

6.7.9 Order: Anseriformes Family: Anatidae (Ducks, Geese, Swans)

This family constitutes ducks and duck-like waterfowl. Common examples are geese and swans. These birds have webbed feet and flattened bills and hence they are well adapted to an aquatic habitat. Their feathers have an oily coating on them which helps to shed water when present in water. The representative species of this family are Fulvous whistling duck, *Dendrocygna bicolor,* Lesser white-fronted goose, *Anser erythropus*, Greater white-fronted goose, *Anser albifrons*, Whooper swan, *Cygnus cygnus*, Lesser whistling duck, *Dendrocygna javanica*, Mute swan, *Cygnus olor,* Tundra swan, *Cygnus columbianus*, Greylag goose, *Anser anser,* Bar-headed goose, *Anser indicus*, Ruddy shelduck, *Tadorna ferruginea*, Common shelduck, *Tadorna tadorna*, Knob-billed duck, *Sarkidiornis melanotos*, Cotton pygmy goose, *Nettapus coromandelianus*, Eurasian wigeon, *Mareca penelope*, Falcated duck, *Mareca falcata*, Gadwall, *Mareca strepera*, Baikal teal, *Sibirionetta formosa*, Eurasian teal, *Anas crecca*, Mallard, *Anas platyrhynchos*, Spot-billed duck, *Anas poecilorhyncha*, Northern pintail, *Anas acuta*, Garganey, *Spatula querquedula*, Northern shoveler, *Spatula clypeata*, Marbled teal, *Marmaronetta angustirostris*, Red-crested pochard, *Netta rufina*, Common pochard, *Aythya ferina*, Ferruginous pochard, *Aythya nyroca*, Baer's pochard, *Aythya baeri*, Tufted duck, *Aythya fuligula*, Greater scaup, *Aythya marila*, Long-tailed duck, *Clangula hyemalis*, Velvet scoter, *Melanitta fusca*, Common goldeneye, *Bucephala clangula*, Smew, *Mergellus albellus*, Red-breasted merganser, *Mergus serrator,* Common merganser, *Mergus merganser,* and White-headed duck, *Oxyura leucocephala*.

6.7.10 Order: Accipitriformes

6.7.10.1 Family: Pandionidae (Osprey)

The Osprey is the only species in this family. It has worldwide distribution. It is a medium to large size raptor and specifically feeds on fish. Osprey, *Pandion haliaetus* is an example of a member bird of this family.

6.7.10.2 Family: Accipitridae (Hawks, Eagles, Kites, Harriers, and Old World Vultures)

This family includes hawks, eagles, kites, harriers, and Old World vultures. They are also called birds of prey. These birds have keen eyesight, powerful talons, and strong legs. They tear flesh from their prey with powerful hooked beaks, the characteristic feature of these birds. Important examples of these birds have been given below.

Crested honey buzzard, *Pernis ptilorhynchus*, Black-winged kite, *Elanus caeruleus*.

Black kite, *Milvus migrans*, Brahminy kite, *Haliastur indus*, White-bellied sea eagle, *Haliaeetus leucogaster,* Pallas's fish eagle, *Haliaeetus leucoryphus*, White-tailed eagle, *Haliaeetus albicilla*, Lammergeier, *Gypaetus barbatus*, Egyptian vulture, *Neophron percnopterus*, White-rumped vulture, *Gyps bengalensis*.

Indian vulture, *Gyps indicus*, Himalayan griffon, *Gyps himalayensis*, Eurasian griffon, *Gyps fulvus*, Cinereous vulture, *Aegypius monachus*, Red-headed vulture, *Sarcogyps calvus*, Short-toed snake eagle, *Circaetus gallicus*, Western marsh harrier, *Circus aeruginosus*, Hen harrier, *Circus cyaneus*, Pallid harrier, *Circus macrourus*, Pied harrier, *Circus melanoleucos*, Montagu's harrier, *Circus pygargus*, Shikra, *Accipiter badius*, Besra, *Accipiter virgatus*, Eurasian sparrowhawk, *Accipiter nisus*, and Northern goshawk, *Accipiter gentilis*.

White-eyed buzzard, *Butastur teesa*, Himalayan buzzard, *Buteo burmanicus*, Long-legged buzzard, *Buteo rufinus*, Oriental honey buzzard, *Pernis ptilorhynchus*, Upland buzzard, *Buteo hemilasius*.

Black eagle, *Ictinaetus malaiensis*, Lesser spotted eagle, *Clanga pomarina*, Indian spotted eagle, *Clanga hastata*, Greater spotted eagle, *Clanga clanga*, Tawny eagle, *Aquila rapax*, Steppe eagle, *Aquila nipalensis*, Imperial eagle, *Aquila heliaca*, Golden eagle, *Aquila chrysaetos*, Bonelli's eagle, *Aquila fasciata*, Booted eagle, *Hieraaetus pennatus, and* Mountain hawk-eagle, *Nisaetus nipalensis.*

6.7.11 Order: Falconiformes Family: Falconidae (Falcons)

They are also birds of prey and kill their prey with their beaks instead of their talons as hawks, eagles, and kites do. Worldwide, there are 62 species falcons but in Pakistan there are only 14 species. There has been a rapid decline in their population due to their demand for the falcon hunting trade. Species commonly found in Pakistan are Lesser kestrel, *Falco naumanni*, Eurasian kestrel, *Falco tinnunculus*, Red-necked falcon, *Falco chicquera*, Amur falcon, *Falco amurensis*, Sooty falcon, *Falco concolor*, Merlin, *Falco columbarius*, Eurasian hobby, *Falco subbuteo*, Oriental hobby, *Falco severus*, Lanner falcon, *Falco biarmicus*, Laggar falcon, *Falco jugger.*

Saker falcon, *Falco cherrug*, Gyrfalcon, *Falco rusticolus*, Shaheen falcon, *Falco peregrinus peregrinator*, Barbary falcon, *Falco pelegrinoides*, Peregrine falcon, *Falco peregrinus*—the breeding subspecies, *peregrinator*, is used in the logo of the Pakistani Air Force.

6.7.12 Order: Galliformes Family: Phasianidae (Chakoor, Himalayan Snowcock, Gray Francolin, the Kaliji Pheasant, the Indian Peafowl, etc.)

Quails, partridges, snowcocks, francolins, spurfowls, tragopans, monals, pheasants, peafowls, and jungle fowls are the important members of the family Phasianidae. These birds have short but broad wings. Worldwide, there 156 species representative of this family, but only 17 species are found in Pakistan. Over the last decennia, a considerable decline has been observed in the numbers of several species. Species found in Pakistan are Snow partridge, *Lerwa lerwa*, Himalayan snowcock, *Tetraogallus himalayensis*, Chukar, *Alectoris chukar* (the National bird of Pakistan).

Sand partridge, *Ammoperdix heyi*, See-see partridge, *Ammoperdix griseogularis*, Black francolin, *Francolinus francolinus*, Gray francolin, *Francolinus pondicerianus*, Common quail, *Coturnix coturnix*, Rain quail, *Coturnix coromandelica*, Jungle bush-quail, *Perdicula asiatica*, Western tragopan, *Tragopan melanocephalus*, Koklass pheasant, *Pucrasia macrolopha*, Himalayan monal, *Lophophorus impejanus*, Red junglefowl, *Gallus gallus*, Kalij pheasant, *Lophura leucomelanos*, Cheer pheasant, *Catreus wallichii*, and the Indian peafowl, *Pavo cristatus.*

6.7.13 Order: Gruiformes Family: Gruidae (Cranes)

Cranes have long necks and long legs and are usually big birds. Although herons and cranes look similar they differ in flying behavior. During flying, cranes normally outstretch their necks, not pulling their necks back as herons do. Most crane species dance during courtship behavior and display noisy courtships. Worldwide, there are 15 species of cranes, but only four species are found in both India and Pakistan. Demoiselle and common cranes are migratory species, while the sarus crane is a resident species in Pakistan. In Pakistan, they are caught for keeping them as pet birds. Their hunting is very common particularly in the Northwest of Pakistan. Due to extensive hunting their number is on the decline. The sarus crane is in abundance in India. They have almost disappeared in Pakistan. In 2011, after 10 years of struggle its only one pair was sighted in tharparker area. Indiscriminate hunting is the major reason for this decline. The Tibetan crane is the state bird of Kashmir. Common representative species of the family Gruidae in Pakistan are the Demoiselle crane, *Anthropoides virgo.*

Siberian crane, *Grus leucogeranus*, Sarus crane, *Grus antigone*, Tibetan crane, *Grus nigricollis*, and Common crane, *Grus grus.*

6.7.13.1 Family: Rallidae (Rails, Crakes, Coots, Gallinules)

Rallidae consists of rails, crakes, coots, and gallinules. It is a really large family. Members of this family vary in size and normally range from small to medium. Naturally they live near lakes, swamps, and rivers where there is dense vegetation and the environment is damp. Generally, these birds remained hidden. They are shy and secretive in nature. They have strong legs and long toes. This is their adaptation to move on soft, bumpy, and irregular surfaces. They are weak fliers because of their short and rounded wings. Important representative members of this family are Slaty-legged crake, *Rallina eurizonoides*, Slaty-breasted rail, *Lewinia striata*, Water rail, *Rallus aquaticus*, Brown crake, *Amaurornis akool*, White-breasted waterhen, *Amaurornis phoenicurus*, little crake, *Porzana parva*, Baillon's crake, *Porzana pusilla*, Spotted crake, *Porzana porzana*, Ruddy-breasted crake, *Porzana fusca*, Watercock, *Gallicrex cinerea.*

Gray-headed swamphen, *Porphyrio poliocephalus*, Common moorhen, *Gallinula chloropus*, and Eurasian coot, *Fulica atra*.

6.7.14 Order: Otidiformes Family: Otididae (Bustards)

Various species of bustards are representative of this family. They live in dry open country in a terrestrial environment with grasslands and prairies. Their feeding habit is omnivore. For their steady movement they have strong legs and big toes. Unlike some other birds, they build their nests on the ground. During their movement they peck for food. Their wings are long and broad. Wingtips are fingered. During flight they practice striking patterns. During courtship many birds show an interesting display. The houbara bustard is the provincial bird of Baluchistan. Due to massive hunting, their number has considerably declined. Important representatives of this family are the Great bustard, *Otis tarda* (Fig. 6.2).

FIGURE 6.2 Great bustard, *Otis tarda*. (*Source:* www.shutterstock.com.) Indian bustard, *Ardeotis nigriceps*, Houbara bustard, *Chlamydotis undulata*. Macqueen's bustard, *Chlamydotis macqueenii*, Lesser florican, *Sypheotides indicus*, and Little bustard, *Tetrax tetrax*.

6.7.15 Order: Charadriiformes Family: Turnicidae (Buttonquails)

The buttonquails resemble true quails. They are small in size, dull in color, and running in habit. The Females are brighter in color than the males. The male is responsible for incubation of the eggs and caring for the young ones. This family has three species commonly found in Pakistan. These are small buttonquail, *Turnix sylvatica*, yellow-legged buttonquail, *Turnix tanki*, and barred buttonquail, *Turnix suscitator.*

6.7.15.1 Family: Jacanidae (Jacanas)

This family consists of birds called jacanas or tropical waders. They are found throughout the Tropics. They inhabit all the tropical areas. In shallow lakes, they can easily walk on the floating vegetation due to their huge feet and claws. These feet and claws are their identifying feature too. Worldwide, this family has eight species, of which only two are found in Pakistan. These are Pheasant-tailed jacana, *Hydrophasianus chirurgus* and Bronze-winged jacana, *Metopidius indicus.*

6.7.15.2 Family: Rostratulidae (Snipes)

This family contains a single species called the painted snipe. They have long bills and short legs. Their snipes are similar in shape to true snipes. Nevertheless, the color of their snipes is brighter. The Greater painted snipe, *Rostratula benghalensis*, is the only representative of this family.

6.7.15.3 Family: Dromadidae (Crab Plover)

This family has only one living representative called the crab-plover, *Dromas ardeola*. It resembles waders, but apparently looks more like a plover. When we compare its very long gray legs and strong heavy black bill it looks more like a tern. It has a long neck and its bill is designed for eating crabs. Its plumage is black and white and its feet are partially webbed.

6.7.15.4 Family: Haematopodidae (Oystercatchers)

The oystercatchers are plover-like birds which they are very noisy. These birds are large in size and have strong bills. They use these bills for smashing or prizing open mollusks. At world level, there are 11 species of this family but Pakistan hosts only one species, which is the Eurasian oystercatcher, *Haematopus ostralegus.*

6.7.15.5 Family: Ibidorhynchidae (Ibisbills)

The ibisbill is the most distinctive representative of this family. The adult ibisbill has a gray color and a white belly. They are more close to waders. Their legs are red. They have a curved bill, pointing downward. The face is black and has a band on the breast. Pakistan hosts a single species of this family called the Ibisbill, *Ibidorhyncha struthersii.*

6.7.15.6 Family: Recurvirostridae (Avocets and Stilts)

The avocets and stilts are representatives of this family. These birds are large in size and wading in habit. The legs of the avocets are long. Their bills are curved and pointed upward. However, stilts have very long legs. Their bills are straight, thin, and long. Worldwide, this family has nine species, although Pakistan hosts only two species. These are the black-winged stilt, *Himantopus himantopus* and the pied avocet, *Recurvirostra avosetta.*

6.7.15.7 Family: Burhinidae (Thick-Knees Waders)

Representative birds of this family are mainly tropical waders. Their global distribution falls in tropical zones, although some species breed in Australia and temperate Europe. Their size ranges from medium to large. Their bills are strong black or yellow/black in color. Their eyes are yellow in color with cryptic plumage. Although these birds are considered waders, most of them still prefer to inhabit arid or semi-arid habitats. Three species are found in Pakistan, including the Eurasian stone-curlew, *Burhinus oedicnemus*, the Indian stone-curlew, *Burhinus indicus*, and the Great stone-curlew, *Esacus recurvirostris.*

6.7.15.8 Family: Glareolidae (Pratincoles, Coursers)

This family contains birds called pratincoles which wade in water. The legs of these birds are short. Their wings are pointed and they have forked tails. Coursers, however, have short wings and long legs with long pointed bills curved downward. Five species of this family are found in Pakistan. The Small pratincole resides in Pakistan and breeds successfully here, hence also known as resident breeders. The representative species of this family are Cream-colored courser, *Cursorius cursor*, Indian courser, *Cursorius coromandelicus*, Collared pratincole, *Glareola pratincola*, Oriental pratincole, *Glareola maldivarum*, and small pratincole, *Glareola lactea*.

6.7.15.9 Family: Charadriidae (Plovers, Dotterels, and Lapwings)

Plovers, dotterels, and lapwings are common members of this family. Their bodies are compact and their size is size is small to medium. Their necks are short and thick with long and pointed wings. All over the world mostly they prefer to live near water. I addition they can also comfortably live in an open country. At global this family has 66 species and only 13 are found in Pakistan. They are Northern lapwing, *Vanellus vanellus*, Yellow-wattled lapwing, *Vanellus malabaricus*, Red-wattled lapwing, *Vanellus indicus*, Sociable lapwing, *Vanellus gregarius*, White-tailed lapwing, *Vanellus leucurus*, Pacific golden-plover, *Pluvialis fulva*, European golden-plover, *Pluvialis apricaria*, Black-bellied plover, *Pluvialis squatarola*, Common ringed plover, *Charadrius hiaticula*, Little ringed plover, *Charadrius dubius*, Snowy plover, *Charadrius alexandrinus*, Lesser sandplover, *Charadrius mongolus*, Greater sandplover, and *Charadrius leschenaultii.*

6.7.15.10 Family: Scolopacidae (Sandpipers, Curlews, Godwits, Shanks, Tattlers, Woodcocks, Snipes, Dowitchers, and Phalaropes)

This is very large family with diversified member species. Their size is small to medium. These birds inhabit shore areas and include sandpipers, curlews, godwits, shanks, tattlers, woodcocks, snipes, dowitchers, and phalaropes. The majority of the members of this species thrive on small invertebrates which they pick up from mud or soil. Particularly at the coast, these multiples species can feed in the same habitat. Due to variations in the length of their legs and bills, there is no feeding competition among them. Under this family, 30 species have been recorded in Pakistan. These are Eurasian woodcock, *Scolopax rusticola*, Jack snipe, *Lymnocryptes minimus*, Solitary snipe, *Gallinago solitaria*, Wood snipe, *Gallinago nemoricola*, Pintail snipe, *Gallinago stenura*, Common snipe, *Gallinago gallinago*, Black-tailed godwit, *Limosa limosa*, Bar-tailed godwit, *Limosa lapponica*, Whimbrel, *Numenius phaeopus*, Eurasian curlew, *Numenius arquata*, Spotted redshank, *Tringa erythropus*, Common redshank, *Tringa totanus*, Marsh sandpiper, *Tringa stagnatilis*, Common greenshank, *Tringa nebularia*, Green sandpiper, *Tringa ochropus*, Wood sandpiper, *Tringa glareola*, Terek sandpiper, *Xenus cinereus*, Common sandpiper, *Actitis hypoleucos*, Ruddy turnstone, *Arenaria interpres.*

Great knot, *Calidris tenuirostris*, Red knot, *Calidris canutus*, Sanderling, *Calidris alba*, Little stint, *Calidris minuta*, Temminck's stint, *Calidris temminckii*, Sharp-tailed sandpiper, *Calidris acuminata*, Curlew sandpiper, *Calidris ferruginea*, Dunlin, *Calidris alpina*, Broad-billed sandpiper, *Calidris falcinellus*, Ruff, *Calidris pugnax*, and Red-necked phalarope, *Phalaropus lobatus*.

6.7.15.11 Family: Stercorariidae (Pomarine Jaeger and Parasitic Jaeger)

The members of this family have gray-brown plumage and are medium to large in size. Their wings are marked white and they are long-distance migrants. Mostly they inhabit temperate and arctic regions and build their nests on the ground. Worldwide, there are seven species of this family, but only two species are found in Pakistan. These are the Pomarine jaeger, *Stercorarius pomarinus* and the Parasitic jaeger, *Stercorarius parasiticus*.

6.7.15.12 Family: Laridae (Large Seabirds, Gulls, Terns, and Skimmers)

Large seabirds, like gulls, terns, and skimmers represent this family. They are medium to large size birds. Among them, gulls are typically white. Sometimes they are gray with black markings on the head and wings with gray or white plumage. Their bills are long and stout and their feet are webbed. Gulls with slender bills breed on the coast of Pakistan. Terns are also included in this family but they are sea birds. Most terns are divers and hunt fish. Some are insectivores and pick up insects off the surface of freshwater. Terns generally live a long life and several species live more than 30 years. Skimmers are smaller in number, live in the Tropics, and are similar to terns. Their lower mandible is elongated. This facilitates their feeding when they skim waters for small fish. The common representative species of this family are the Common gull, *Larus canus*.

Lesser black-backed gull, *Larus fuscus*, Heuglin's gull, *Larus heuglini*, Vega gull, *Larus vegae*, Caspian gull, *Larus cachinnans*, Sooty gull, *Ichthyaetus hemprichii*, Pallas's gull, *Ichthyaetus ichthyaetus*, Brown-headed gull, *Chroicocephalus brunnicephalus*, Black-headed gull, *Chroicocephalus ridibundus*, Slender-billed gull, *Chroicocephalus genei*, Gull-billed tern, *Gelochelidon nilotica*, Caspian tern, *Hydroprogne caspia*, Lesser crested tern, *Thalasseus bengalensis*, Sandwich tern, *Thalasseus sandvicensis*, Great crested tern, *Thalasseus bergii*.

River tern (bird), *Sterna aurantia*, Common tern, *Sterna hirundo*, White-cheeked tern, *Sterna repressa*, Black-bellied tern, *Sterna acuticauda*, Little tern, *Sternula albifrons*, Saunders's tern, *Sternula saundersi*, Bridled tern, *Onychoprion anaethetus*, Whiskered tern, *Chlidonias hybrida*, White-winged tern, *Chlidonias leucopterus*, Brown noddy, *Anous stolidu*, and Indian skimmer, *Rynchops albicollis*.

6.7.16 Order: Pterocliformes Family: Pteroclidae (Sandgrouse)

Representatives of this species have sturdy and compact bodies. Like pigeons, their heads and necks are small. Their wings are long and pointed and sometimes their tail is too for fast flight. Flocks fly at dawn and dusk to watering holes. There are feathers on their legs up to the toes. Worldwide, 16 species fall under this order. Out of these, seven species occur in Pakistan. The representative species of this family are Pin-tailed sandgrouse, *Pterocles alchata*, Chestnut-bellied sandgrouse, *Pterocles exustus*, Spotted sandgrouse, *Pterocles senegallus*, Black-bellied sandgrouse, *Pterocles orientalis*, Crowned sandgrouse, *Pterocles coronatus*, Lichtenstein's sandgrouse, *Pterocles lichtensteinii*, and Painted sandgrouse, *Pterocles indicus*.

6.7.17 Order: Columbiformes Family: Columbidae (Pigeons and Doves)

Representatives of this family have stout bodies with short necks. Bills are slender shaped. At the base of the upper beak there is a waxy fleshy covering called the cere. Representative species of this family are Rock pigeon, *Columba livia*, Hill pigeon, *Columba rupestris*, Snow pigeon, *Columba leuconota*, Yellow-eyed pigeon, *Columba eversmanni*, Common wood pigeon, *Columba palumbu*.

Speckled wood pigeon, *Columba hodgsonii*, European turtle dove, *Streptopelia turtur*, Oriental turtle dove, *Streptopelia orientalis*, Eurasian collared dove, *Streptopelia decaocto*, Red collared-dove, *Streptopelia tranquebarica*, Spotted dove, *Spilopelia chinensis*, Laughing dove, *Spilopelia senegalensis*, Common emerald dove, *Chalcophaps indica*, Ashy-headed green pigeon, *Treron phayrei, and* Yellow-footed green pigeon, *Treron phoenicoptera*.

6.7.18 Order: Psittaciformes Family: Psittaculidae (Parakeets)

The representatives of this family are the Alexandrine parakeet, *Psittacula eupatria*, Rose-ringed parakeet, *Psittacula krameri*, Slaty-headed parakeet, *Psittacula himalayana*, and Plum-headed parakeet, *Psittacula cyanocephala*. The rose-ringed parakeet is the territorial bird of Islamabad.

6.7.19 Order: Cuculiformes Family: Cuculidae (Cuckoos, Roadrunners, and Anis)

Cuckoos, roadrunners, and anis are representatives of this family. These birds have different sizes. Their bodies are slender shaped with strong legs. Their tails are long. Brood parasitism has been observed in the Old World cuckoos. Representative bird species of this family are Pied cuckoo, *Clamator jacobinus*, Large hawk-cuckoo, *Hierococcyx sparverioides*, Common hawk-cuckoo, *Hierococcyx varius*, Indian cuckoo, *Cuculus micropterus*, Common cuckoo, *Cuculus canorus*, Himalayan cuckoo, *Cuculus saturatus*, Oriental cuckoo, *Cuculus optatus*, Lesser cuckoo, *Cuculus poliocephalus*, Asian koel, *Eudynamys scolopacea*, Sirkeer malkoha, *Phaenicophaeus leschenaultii*, and Greater coucal, *Centropus sinensis* (Fig. 6.3).

6.7.20 Order: Strigiformes Family: Tytonidae (Owls)

The representative birds of this family are barn owls. They are medium to large in size. Their heads are large and their faces are heart-shaped. Their legs are long with powerful talons. Worldwide, this family has 16 species and there is only one species found in Pakistan, which is called the barn owl, *Tyto alba*.

6.7.20.1 Family: Strigidae (Typical Owls)

This family comprises typically small to large size owls. They are nocturnal birds. They prey, and spend a solitary life. Their eyes and ears are large in size and face forward. Their beak is hawk-like. They have a conspicuous circle of feathers around each eye which is called a facial disk. The species of owls which fall under this family are Indian scops owl, *Otus bakkamoena*. Collared scops owl, *Otus lettia*, Pallid scops owl, *Otus brucei*, Eurasian scops owl, *Otus scops*, Oriental scops owl, *Otus sunia*, Eurasian eagle-owl, *Bubo bubo*, Rock eagle-owl, *Bubo bengalensis*, Dusky eagle-owl, *Bubo coromandus*, Brown fish owl, *Ketupa zeylonensis*, Mottled wood owl, *Strix ocellata*, Brown

FIGURE 6.3 Greater coucal, *Centropus sinensis*. (*Source*: www.shutterstock.com.)

wood owl, *Strix leptogrammica*, Tawny owl, *Strix aluco*, Himalayan owl, *Strix nivicolum*, Desert owl, *Strix hadorami*, Collared owlet, *Glaucidium brodiei*, Asian barred owlet, *Glaucidium cuculoides*, Little owl, *Athene noctua*, Brown hawk-owl, *Ninox scutulata*, Long-eared owl, *Asio otus*, and Short-eared owl, *Asio flammeus*.

6.7.21 Order: Caprimulgiformes Family: Caprimulgidae (Nightjars)

Representative birds of this order are medium-sized. They are nocturnal in habit. These birds have very short bills and legs with long wings. Small feet are characteristics of most of the birds, which are a hindrance to walking. Their wings are long and pointed. Their soft plumage resembles bark or leaves. This plumage helps birds to camouflage themselves to avoid predators. Examples of representative birds of this family are Gray nightjar, *Caprimulgus jotaka*, Eurasian nightjar, *Caprimulgus europaeus*, Egyptian nightjar, *Caprimulgus aegyptius*, Sykes's nightjar, *Caprimulgus mahrattensis*, Large-tailed nightjar, *Caprimulgus macrurus*, Indian nightjar, *Caprimulgus asiaticus*, and Savanna nightjar, *Caprimulgus affinis*.

6.7.22 Order: Apodiformes Family: Apodidae (Alpine swift)

Swifts, small sized birds, represent this family. The legs of these birds are very short and hence they barely walk on the ground, keeping instead to perching on vertical surfaces. They keep flying for the major part of their lives. Many swifts have long wings, which when swept back look similar to a crescent or a boomerang. Nine species of swifts have been recorded in Pakistan. The representative species of Apodidae are Himalayan swiftlet, *Aerodramus brevirostris*, White-throated needletail, *Hirundapus caudacutus*, Asian palm-swift, *Cypsiurus balasiensis*, Alpine swift, *Tachymarptis melba*, Common swift, *Apus apus*, Pallid swift, *Apus pallidus*, Blyth's swift, *Apus leuconyx*, Little swift, *Apus affinis*, and House swift, *Apus nipalensis*.

6.7.23 Order: Coraciiformes Family: Alcedinidae (Kingfishers)

Kingfishers with large heads are representative of this family. They are medium-sized birds with long pointed bills. Their legs are short with stubby and thick tails. Worldwide, this family has 93 species, while there are only five species in Pakistan. These are Common kingfisher, *Alcedo atthis*, White-throated kingfisher, *Halcyon smyrnensis*, Black-capped kingfisher, *Halcyon pileata*, Crested kingfisher, *Megaceryle lugubris*, and Pied kingfisher, *Ceryle rudis*.

6.7.23.1 Family: Meropidae (Bee Eaters)

This family comprises bee-eaters. These birds are similar to passerine birds. Most of their member species live in Africa. Others are found in southern Europe, Madagascar, Australia, and New Guinea. Their bodies are slender with characteristic rich colored plumage. Their central tail feathers are usually long. All members are very colorful, with long bills turned downward. Their wings are pointed and if seen from a distance they give a swallow-like appearance. Worldwide, this family has 26 species and only four species occur in Pakistan. Representative species of this family are Green bee-eater, *Merops orientalis*, Blue-cheeked bee-eater, *Merops persicus*, Blue-tailed bee-eater, *Merops philippinus*, and European bee-eater, *Merops apiaster*.

6.7.23.2 Family: Coraciidae (Rollers)

Rollers represent this family. They resemble crows in size. However, apparently they are closer to the kingfishers and bee-eaters. They are very colorful and blue and brown colors dominate on their bodies. The outer toe on the feet is free, and two inner front toes are connected. Worldwide, this family has 12 species but only two are found in Pakistan. These are the European roller, *Coracias garrulus*, and the Indian roller, *Coracias benghalensis*.

6.7.23.3 Family: Upupidae (Hoopoes)

Members of this family are black, white, and orange to pink in color. The presence of a large erectile crest on their heads is their characteristic feature. Worldwide, this family has two species and there is one species found in Pakistan, which is called the Hoopoe, *Upupa epops*.

6.7.23.4 Family: Bucerotidae (Hornbills)

As the name indicates, their bill is like the horn of a cow, but there is no twist in it. Nevertheless, sometimes there is a casque on the upper mandible. Most often the bill has a bright color. Worldwide, there are 57 species under this family but only single species is present in Pakistan which is called the Indian gray hornbill, *Ocyceros birostris*.

6.7.24 Order: Piciformes Family: Megalaimidae (Barbets)

This family contains a group of birds called barbets, also called plump birds. Their heads are large with short necks. Their name is attributed to the bristles present on the periphery of their heavy bills. Most of the species in this family have bright colors. Species present in Pakistan are Great barbet, *Psilopogon virens*, Brown-headed barbet, *Psilopogon zeylanicus*, Coppersmith barbet, *Psilopogon haemacephalus*, and Blue-throated barbet, *Psilopogon asiaticus*.

6.7.24.1 Family: Indicatoridae (Honeyguides)

These birds feed on wax and the feeding behavior of these birds is unique. These birds guide traditional honey hunters to bees' nests for harvesting honey. After this harvest, they feed on the remaining contents of the hive. Worldwide, this family has 17 species and there is only one species in Pakistan, which is called the Yellow-rumped honeyguide, *Indicator xanthonotus*.

6.7.24.2 Family: Picidae (Woodpeckers)

The size of woodpeckers is small to medium. Their tails are stiff. They have chisel-like beaks, and short legs. They capture insects with their long tongues. Several species have only three toes, while some have four toes—two pointing forward and two pointing backward. Most of the woodpecker species tap on tree trunks with their beaks, which is very noisy. There are several species found in Pakistan, which are Eurasian wryneck, *Jynx torquilla*, Speckled piculet, *Picumnus innominatus*, Brown-capped woodpecker, *Yungipicus nanus*, Gray-capped woodpecker, *Yungipicus canicapillus*, Brown-fronted woodpecker, *Dendrocoptes auriceps*, Fulvous-breasted woodpecker, *Dendrocopos macei*, Rufous-bellied woodpecker, *Dendrocopos hyperythrus*, Sind woodpecker, *Dendrocopos assimilis*, Himalayan woodpecker, *Dendrocopos himalayensis*, Yellow-crowned woodpecker, *Leiopicus mahrattensis*, Rufous woodpecker, *Micropternus brachyurus*, Scaly-bellied woodpecker, *Picus squamatus*, Gray-faced woodpecker, *Picus canus* and Black-rumped flameback, *Dinopium benghalense*.

6.7.25 Order: Passeriformes Family: Pittidae (Pittas)

Pittas are medium-sized birds. They match the standard of passerine birds and are sturdy in nature. This is very much obvious from their stout bills. Their legs are fairly long but their tails are short. Many pittas are bright in color and spend most of their time on wet forest floors. From these habitats they eat invertebrates like insects, snails, etc. Worldwide, this family has 32 species, while a single species occurs in Pakistan, which is called the Indian pitta, *Pitta brachyura*.

6.7.25.1 Family: Alaudidae (Larks)

Larks are small in size and live in a terrestrial environment. These birds display flights and sing extravagant songs. Most larks are dull in appearance. Larks feed on insects and seeds. There are several species of larks present in Pakistan which are Singing bushlark, *Mirafra cantillans*,

Indian bushlark, *Mirafra erythroptera*, Black-crowned sparrow-lark, *Eremopterix nigriceps*, Ashy-crowned sparrow-lark, *Eremopterix grisea*, Bar-tailed lark, *Ammomanes cincturus*, Rufous-tailed lark, *Ammomanes phoenicurus*, Desert lark, *Ammomanes deserti*, Greater hoopoe-lark, *Alaemon alaudipes*, Bimaculated lark, *Melanocorypha bimaculata*, Greater short-toed lark, *Calandrella brachydactyla*, Hume's short-toed lark, *Calandrella acutirostris*, Lesser short-toed lark, *Alaudala rufescens*, Asian short-toed lark, *Alaudala cheleensis*, Sand lark, *Alaudala raytal*, Crested lark, *Galerida cristata*, Eurasian skylark, *Alauda arvensis*, Oriental skylark, *Alauda gulgula*, and Horned lark, *Eremophila alpestris*.

6.7.25.2 Family: Hirundinidae (Martins)

Martins are well adapted to feed in the air. The body is cylindrical with long pointed wings. The bill is short with a wide gap. Using their feet, they prefer perching over walking. Their toes are partially joined at the base. Pakistan hosts several species of this family. These are Sand martin, *Riparia riparia*, Pale martin, *Riparia diluta*, Gray-throated martin, *Riparia chinensis*, Eurasian crag martin, *Ptyonoprogne rupestris*, Pale crag martin, *Ptyonoprogne obsoleta*, Dusky crag martin, *Ptyonoprogne concolor*, Barn swallow, *Hirundo rustica*, Wire-tailed swallow, *Hirundo smithii*, Red-rumped swallow, *Cecropis daurica*, Streak-throated swallow, *Petrochelidon fluvicola*, Common house martin, *Delichon urbica*, and Asian house martin, *Delichon dasypus*.

6.7.25.3 Family: Motacillidae (Wagtails, Longclaws, and Pipits)

Wagtails, long claws, and pipits are common representatives of this family. With medium to long tails they are small passerine birds. Their body is cylindrical and they feed in open country on ground insects. This family has several species known as White wagtail, *Motacilla alba*, Black-backed wagtail, *Motacilla lugens*, White-browed wagtail, *Motacilla maderaspatensis*, Citrine wagtail, *Motacilla citreola*, Yellow wagtail, *Motacilla flava*, Gray wagtail, *Motacilla cinerea*, Oriental pipit, *Anthus rufulus*, Richard's pipit, *Anthus richardi*, Tawny pipit, *Anthus campestris*, Blyth's pipit, *Anthus godlewskii*, Long-billed pipit, *Anthus similis*, Tree pipit, *Anthus trivialis*, Olive-backed pipit, *Anthus hodgsoni*, Meadow pipit, *Anthus pratensis*, Red-throated pipit, *Anthus cervinus*, Rosy pipit, *Anthus roseatus*, Water pipit, *Anthus spinoletta*, Upland pipit, *Anthus sylvanus*, and American pipit, *Anthus rubescens*.

6.7.25.4 Family: Campephagidae (Cuckooshrikes)

These birds are passerine in nature and range in size from small to medium. Mainly, they are grayish with white and black patches on them. Nevertheless, some species are brightly colored. At a global level, this family has 82 species, while there are eight species in Pakistan. These are Large cuckoo shrike, *Coracina macei*, Black-winged cuckoo shrike, *Coracina melaschistos*, Black-headed cuckoo shrike, *Coracina melanoptera*, Rosy minivet, *Pericrocotus roseus*, Small minivet, *Pericrocotus cinnamomeus*, White-bellied minivet, *Pericrocotus erythropygius*, Long-tailed minivet, *Pericrocotus ethologus* and Scarlet minivet, *Pericrocotus flammeus*.

6.7.25.5 Family: Pycnonotidae (Bulbuls)

These are medium-sized songbirds. Some species are very brightly colored, with colors varying from red, to yellow, to orange. Their supercilia, throats, and cheeks are also of the same color. Most of the bulbuls are drab. Their plumage is uniform in color and ranges from olive-brown to black. Some species have a distinctive crest. The following species are commonly found in Pakistan. These are Black-headed bulbul, *Brachypodius atriceps*, White-eared bulbul, *Pycnonotus leucotis*, White-cheeked bulbul, *Pycnonotus leucogenys*, Red-vented bulbul, *Pycnonotus cafer*, and Black bulbul, *Hypsipetes leucocephalus*.

6.7.25.6 Family: Regulidae (Crests)

A small group of birds called kinglets or crests represent this family. Their inclusion in the Old World warblers is quite common. Because of their resemblance to titmice, they are frequently considered to be members of that family. Worldwide, this family has seven species but there is only single species in Pakistan, which is called Goldcrest, *Regulus regulus.*

6.7.25.7 Family: Aegithinidae (Ioras)

These birds resemble bulbul. They live in scrub or open forest. They are drab in coloration. Sexually, they are dimorphic and the males have bright yellow and green plumage. Worldwide, this family has four species, but only a single species lives in Pakistan, e.g., the Common iora, *Aegithina tiphia.*

6.7.25.8 Family: Bombycillidae (Waxwings)

These birds have soft silky plumage and some of their wing feathers have unique red tips. Their tips look like sealing wax and hence they get their name from the Bohemian and cedar waxwings. These birds live in northern forests and are arboreal in nature. In summer, they thrive on insects and in winter they live on berries. Worldwide, this family has three species but there is only one species which occurs in Pakistan, which is the Bohemian waxwing, *Bombycilla garrulus.*

6.7.25.9 Family: Hypocoliidae (Gray Hypocolius)

A small bird called the gray hypocolius is its common representative. It is also called the Middle Eastern bird. It resembles the waxwing in its shape and soft plumage. Normally, they are uniformly gray in color. A black triangular mask is present around their eyes. The only species found in Pakistan is the Hypocolius, *Hypocolius ampelinus.*

6.7.25.10 Family: Cinclidae (Dippers)

This family comprises dippers or perching birds. In the Americas, Europe, and Asia they live in aquatic environments. Their bobbing and/or dipping movements are their characteristic feature. Worldwide, this family has five species. Out of these, two occur in Pakistan. These are the White-throated dipper, *Cinclus cinclus* and the Brown dipper, *Cinclus pallasii.*

6.7.25.11 Family: Troglodytidae (Wrens)

Wrens are ordinary birds. They are small in size and sing loudly. These birds have thin bills, which turn downward. They have short wings. Several species often hold their tails upright. All species of this family feed on insects. Globally, there are 80 species of this family. One of them is a new world species. Only one species occurs in Pakistan, that is the Eurasian wren, *Troglodytes troglodytes.*

6.7.25.12 Family: Prunellidae (Accentors)

These birds are common in the Palearctic region. They are small birds drab in color and artificially look like a sparrow. Worldwide, there are 13 species which fall under this family and there are six species which are found in Pakistan. These are Alpine accentor, *Prunella collaris*, Himalayan accentor, *Prunella himalayana*, Robin accentor, *Prunella rubeculoides*, Rufous-breasted accentor, *Prunella strophiata*, Brown accentor, *Prunella fulvescens* and Black-throated accentor, *Prunella atrogularis.*

6.7.25.13 Family: Turdidae (Thrushes)

Representative birds of this family mainly occur in the Old World. They are small to medium-sized birds and mainly feed on the ground. They are plump and soft plumaged in characteristics. Mostly they are insectivores, although sometimes they are insectivores or sometimes even omnivores. Most of the birds found in this family have attractive songs. Representative species of this family are Orange-headed thrush, *Geokichla citrina*, Alpine thrush, *Zoothera mollissima*, Scaly thrush, *Zoothera dauma*, Tickell's thrush, *Turdus unicolor*, Black-breasted thrush, *Turdus dissimilis*, White-collared blackbird, *Turdus albocinctus*, Gray-winged blackbird, *Turdus boulboul*, Tibetan blackbird, *Turdus maximus*, Chestnut thrush, *Turdus rubrocanus*, Dark-throated thrush, *Turdus ruficollis*, Dusky thrush, *Turdus naumanni*, Redwing, *Turdus iliacus*, Song thrush, *Turdus philomelos*, and Mistle thrush, *Turdus viscivorus.*

6.7.25.14 Family: Cisticolidae (Warblers)

Warmer southern regions of the Old World host most of the members of this family. Although these birds are drab brown in color, sometimes they have a gray appearance. They prefer to live in open grassland or scrub areas. The representative birds of this species are Zitting cisticola, *Cisticola juncidis*, Striated prinia, *Prinia criniger*, Gray-crowned prinia, *Prinia cinereocapilla*, Rufous-fronted prinia, *Prinia buchanani*, Gray-breasted prinia, *Prinia hodgsonii*, Graceful prinia, *Prinia gracilis*, Jungle prinia, *Prinia sylvatica*, Yellow-bellied prinia, *Prinia flaviventris*, and Plain prinia, *Prinia inornata.*

6.7.25.15 Family: Scotocercidae (Streaked Scrub Warbler)

6.7.25.16 Family: Cettiidae (Warbler)

This family constitutes two species of warblers called Brownish-flanked bush warbler, *Cettia fortipes and* Cetti's warbler, *Cettia cetti.*

6.7.25.17 Family: Locustellidae (Warblers)

This family constitutes the following members called Long-billed bush warbler, *Locustella major*, Grasshopper warbler, *Locustella naevia*, Striated grassbird, *Megalurus palustris*, and Bristled grassbird, *Chaetornis striatus.*

6.7.25.18 Family: Acrocephalidae (Warbler)

This family contains the following members called the Mustached warbler, *Acrocephalus melanopogon*, Paddyfield warbler, *Acrocephalus Agricola*, Blunt-winged warbler, *Acrocephalus concinens*, Eurasian reed warbler, *Acrocephalus scirpaceus*, Blyth's reed warbler, *Acrocephalus dumetorum*, Great reed warbler, *Acrocephalus arundinaceus*, Clamorous reed warbler, *Acrocephalus stentoreus*, Booted warbler, *Iduna caligata*, Sykes's warbler, *Iduna rama*, and Upcher's warbler, *Hippolais languida.*

6.7.25.19 Family: Phylloscopidae (Warbler)

This family contains the following members called Willow warbler, *Phylloscopus trochilus*, Common chiffchaff, *Phylloscopus collybita*, Mountain chiffchaff, *Phylloscopus sindianus*, Plain leaf warbler, *Phylloscopus neglectus*, Tickell's leaf warbler, *Phylloscopus affinis*, Sulphur-bellied warbler, *Phylloscopus griseolus*, Pale-rumped warbler, *Phylloscopus chloronotus*, Brooks's leaf warbler, *Phylloscopus subviridis*, Yellow-browed warbler, *Phylloscopus inornatus*, Hume's leaf warbler, *Phylloscopus humei*, Greenish warbler, *Phylloscopus trochiloides*, Large-billed leaf warbler, *Phylloscopus magnirostris*, Tytler's leaf warbler, *Phylloscopus tytleri*, Western crowned leaf warbler, *Phylloscopus occipitalis*, Blyth's leaf warbler, *Phylloscopus reguloides*, Golden-spectacled Warbler, *Phylloscopus burkii*, Gray-hooded warbler, *Phylloscopus xanthoschistos*, and Whistler's warbler, *Phylloscopus whistleri.*

6.7.25.20 Family: Sylviidae (Insectivorous Passerine Birds)

This family contains small passerine birds. They are insectivorous. They mainly breed in Asia, Europe, and, to a lesser extent, Africa. Most of them are not well distinguished but their distinctive songs are their identity. This family constitutes the following species. These are Greater whitethroat, *Sylvia communis*, Lesser whitethroat, *Sylvia curruca*, Small whitethroat, *Sylvia minula*, Hume's whitethroat, *Sylvia althaea*, Asian desert warbler, *Sylvia nana*, Barred warbler, *Sylvia nisoria*, Eastern Orphean warbler, *Sylvia crassirostris*, Menetries's warbler, *Sylvia mystacea*, Yellow-eyed babbler, *Chrysomma sinense*, and Jerdon's babbler, *Chrysomma altirostre.*

6.7.25.21 Family: Muscicapidae (Flycatchers)

Old World flycatchers are small passerine birds. Mostly they are native to the Old World. As mentioned above, they are smaller in size, thrive on insects, and are arboreal in habit. These birds appear quite differently. Mostly their calls are harsh and songs are weak. Common members of this family are Common rock thrush, *Monticola saxatilis*, Blue-capped rock thrush, *Monticola cinclorhyncha*, Chestnut-bellied rock thrush, *Monticola rufiventris*, Blue rock thrush, *Monticola solitarius*, Blue whistling thrush, *Myophonus caeruleus*, Spotted flycatcher, *Muscicapa striata*, Siberian flycatcher, *Muscicapa sibirica*, Rusty-tailed flycatcher, *Ficedula ruficauda*, Red-breasted flycatcher, *Ficedula parva*, Taiga flycatcher, *Ficedula albicilla*, Kashmir flycatcher, *Ficedula subrubra*, Ultramarine flycatcher, *Ficedula superciliaris*, Slaty-blue flycatcher, *Ficedula tricolor*, Verditer flycatcher, *Eumyias thalassina*, Rufous-bellied niltava, *Niltava sundara*, Common nightingale, *Luscinia megarhynchos*, Bluethroat, *Luscinia svecica*, Indian blue robin, *Luscinia brunnea*, White-bellied redstart, *Luscinia phaenicuroides*, Himalayan rubythroat, *Calliope pectoralis*, Chinese rubythroat, *Calliope tschebaiewi*, Himalayan bluetail, *Tarsiger rufilatus*, Golden bush robin, *Tarsiger chrysaeus*, Rufous-tailed scrub robin, *Cercotrichas galactotes*, White-rumped shama, *Copsychus malabaricus*, Indian robin, *Saxicoloides fulicata*, Rufous-backed redstart, *Phoenicurus erythronota*, Blue-capped redstart, *Phoenicurus caeruleocephalus*, Black redstart, *Phoenicurus ochruros*, Common redstart, *Phoenicurus phoenicuru*, White-winged redstart, *Phoenicurus erythrogaster*, Blue-fronted redstart, *Phoenicurus frontalis*, White-capped redstart, *Chaimarrornis leucocephalus*, Plumbeous redstart, *Rhyacornis fuliginosus*, Little forktail, *Enicurus scouleri*, Spotted forktail, *Enicurus maculatus*, Siberian stonechat, *Saxicola maurus*, White-browed bushchat, *Saxicola macrorhynchus*, White-tailed stonechat, *Saxicola leucurus*, Pied bushchat, *Saxicola caprata*, Gray bushchat, *Saxicola ferreus*, Hooded wheatear, *Oenanthe monacha*, Hume's wheatear, *Oenanthe alboniger*, Northern wheatear, *Oenanthe oenanthe*, Mourning wheatear, *Oenanthe lugens*, Finsch's wheatear, *Oenanthe finschii*, Variable wheatear, *Oenanthe picata*, Pied wheatear, *Oenanthe pleschanka*, Red-tailed wheatear, *Oenanthe xanthoprymna*, Desert wheatear, *Oenanthe deserti*, Isabelline wheatear, *Oenanthe isabellina*, and Brown rock chat, *Cercomela fusca*.

6.7.25.22 Family: Rhipiduridae (Fantail)

Birds of this family are small in size and insectivorous in feeding habit. Aerial feeding is their peculiarity. Representative members of this family are the White-throated fantail, *Rhipidura albicollis* and the White-browed fantail, *Rhipidura aureola*.

6.7.25.23 Family: Stenostiridae (Fantail and Fly Catchers)

There are two representative species of this family. They are the Yellow-bellied fantail, *Chelidorhynx hypoxanthus* and *the* Gray-headed canary flycatcher, *Culicicapa ceylonensis*.

6.7.25.24 Family: Monarchidae (Paradise Flycatcher)

The monarch flycatchers are small to medium in size. They are insectivorous passerines and hunt flies by catching. Worldwide, there are 99 species and only one exists in Pakistan which is Indian paradise flycatcher, *Terpsiphone paradisi*.

6.7.25.25 Family: Leiothrichidae (Laughing Thrushes and Babblers)

This family constitutes the following species. These are Rufous-chinned laughingthrush, *Ianthocincla rufogularis*, White-throated laughingthrush, *Ianthocincla albogularis*, Streaked laughingthrush, *Garrulax lineatus*, Variegated laughingthrush, *Garrulax variegatus*, Chestnut-crowned laughingthrush, *Garrulax erythrocephalus*, Common babbler, *Turdoides caudatus*, Afghan babbler, *Turdoides huttoni*, Striated babbler, *Turdoides earlei*, Large gray babbler, *Turdoides malcolmi*, Jungle babbler, *Turdoides striatus*, Red-billed leiothrix, *Leiothrix lutea*, and Rufous sibia, *Heterophasia capistrata*.

6.7.25.26 Family: Pellorneidae (Babblers)

This family consists of two members called Puff-throated babbler, *Pellorneum ruficeps* and Rufous-vented grass babbler, *Laticilla burnesii*.

6.7.25.27 Family: Timaliidae (Babblers and Timaliids)

Both these birds are diverse in color and size. Soft fluffy plumage is their characteristic feature. Common member species of this family are Spot-breasted scimitar babbler, *Pomatorhinus erythrocnemis*, Rusty-cheeked scimitar babbler, *Pomatorhinus erythrogenys* and Black-chinned babbler, *Stachyridopsis pyrrhops.*

6.7.25.28 Family: Vireonidae (Shrike Babbler)

This family constitutes two species, these are Himalayan shrike-babbler, *Pteruthius flaviscapis* and Green shrike-babbler, *Pteruthius xanthochlorus.*

6.7.25.29 Family: Panuridae (Bearded Reedling)

This family consists of single species called Bearded reedling, *Panurus biarmicus.*

6.7.25.30 Family: Aegithalidae (Long Tailed Tits)

Small passerine birds are the members of this family. Their tails are medium to long. They weave their nests in the trees and hence they are called weaver birds. Most of them eat a mixed diet including insects. These are White-browed tit-warbler, *Leptopoecile sophiae*, White-cheeked tit, *Aegithalos leucogenys*, Black-throated tit, *Aegithalos concinnus*, and White-throated tit, *Aegithalos niveogularis.*

6.7.25.31 Family: Paridae

These birds are small in size. They have short stout bills and are mainly stocky woodland species. Some have crests. They are very adaptable. They feed on seeds and insects. Ten species are common in Pakistan. These are Black-breasted tit, *Periparus rufonuchalis*, Rufous-vented tit, *Periparus rubidiventris*, Black-crested tit or spot-winged tit, *Periparus melanolophus*, Gray-crested tit, *Lophophanes dichrous*, Cinereous tit, *Parus Cinereus*, Green-backed tit, *Parus monticolus*, Himalayan black-lored tit, *Parus xanthogenys*, Azure tit, *Cyanistes cyanus*, Yellow-breasted tit, *Cyanistes flavipectus*, and Fire-capped tit, *Cephalopyrus flammiceps.*

6.7.25.32 Family: Sittidae (Nuthatches)

These are woodland birds and are small in size. They have the unique ability to climb down trees. Their heads are big. Their tails are short but they have powerful bills and feet. Worldwide, this family has 24 species, but only five are found in Pakistan. These are Chestnut-bellied nuthatch, *Sitta castanea*, Kashmir nuthatch, *Sitta cashmirensis*, White-cheeked nuthatch, *Sitta leucopsis*, and Persian nuthatch or eastern rock nuthatch, *Sitta tephronota.*

6.7.25.33 Family: Tichodromidae (Wallcreepers)

This bird belongs to the nuthatch family and is small in size. Its plumage is gray and black with stunning crimson. It has only single family called the Wallcreeper, *Tichodroma muraria.*

6.7.25.34 Family: Certhiidae (Treecreepers)

These birds are small in size and are called woodland birds. Their upper side is brown while underside is white. Their bills are thin and curved downward to extricate insects from bark. Their tail feathers are stiff and resemble those of woodpeckers. These stiff feathers help birds to stay on vertical trees. The representative species of this family are Hodgson's treecreeper, *Certhia hodgsoni* and Bar-tailed treecreeper, *Certhia himalayana.*

6.7.25.35 Remizidae (Penduline Tits)

They are small birds of passerine nature. They relate to true tits. They are insectivore in feeding habit. There are three representative species of this family which have been recorded in Pakistan. They are Eurasian penduline tit, *Remiz pendulinus*, Black-headed penduline tit, *Remiz macronyx* and White-crowned penduline tit, *Remiz coronatus.*

6.7.25.36 Family: Nectariniidae (Sunbirds and Spiderhunters)

Like the bird mentioned above, these birds are very small in size and feed on nectar. They can also take insects when they feed their young. They fly quite swiftly and directly on their short wings. However, most species are like the humming bird and take up nectar but usually they perch for food. Worldwide, this family has 131 species but only three are found in Pakistan. These are Purple sunbird, *Cinnyris asiaticus*, Mrs. Gould's sunbird, *Aethopyga gouldiae*, and Crimson sunbird, *Aethopyga siparaja*.

6.7.25.37 Family: Dicaeidae (Flowerpeckers)

These are very small birds. They are stout and most often have bright colors. They have short tails, tubular tongues, and short thick curved bills. This family has a total of 44 species, but only two species occur in Pakistan. One of them is the Thick-billed flowerpecker, *Dicaeum agile*.

6.7.25.38 Family: Zosteropidae (White-Eyes)

They are mostly small non-distinguishable birds. Their plumage is dull in color. Sometimes they appear greenish-olive. Nevertheless, some species have a white or bright yellow breast or lower parts. Buff flanks are common in several members of this species. In many species, white rings are common around the eyes. Worldwide, this family has 96 species. Out of these, there is only one in Pakistan, which is the Oriental white-eye, *Zosterops palpebrosus*.

6.7.25.39 Family: Oriolidae (Orioles)

They are colorful passerine birds. New World orioles are different from these birds. Worldwide, there are 29 species. Only a single species exists in Pakistan, which is the Indian golden oriole, *Oriolus kundoo*.

6.7.25.40 Family: Laniidae (Shrikes)

Members of this family are passerine birds. They catch other birds and small animals. They stab uneaten portions of their prey on thorns. A typical shrike has a hooked beak and resembles a bird of prey. Representative species of this family are Red-backed shrike, *Lanius collurio*, Isabelline shrike, *Lanius isabellinus*, Red-tailed shrike, *Lanius phoenicuroides*, Brown shrike, *Lanius cristatus*, Bay-backed shrike, *Lanius vittatus*, Long-tailed shrike, *Lanius schach*, Gray-backed shrike, *Lanius tephronotus*, Great gray shrike, *Lanius excubitor* and Lesser gray shrike, *Lanius minor*.

6.7.25.41 Family: Tephrodornithidae (Common Woodshrike)

This family has single species called the Common woodshrike, *Tephrodornis pondicerianu*.

6.7.25.42 Family: Dicruridae (Drongos)

Representative birds of this family are mostly dark gray or black in color. Sometimes their color contains metallic tints. Their tail is long and forked. In some Asian specie, tail decorations are well elaborated. Their legs are short. Like shrike, they sit very upright during perching. Normally they catch the fly in the air but sometimes they pick up their prey from the ground. Worldwide, this family has 24 species. Only two species are found in Pakistan. These are the Black drongo, *Dicrurus macrocercus* and the Ashy drongo, *Dicrurus leucophaeus*.

6.7.25.43 Family: Corvidae (Jays, Crows, Choughs, Magpies, Treepies, Nutcrackers, and Ground Jays)

This family consists of aforementioned birds. In the order Passeriformes, they are above average in size. In this family, larger species are quite intelligent. The species included in this group are Eurasian jay, *Garrulus glandarius*, Black-headed jay, *Garrulus lanceolatus*, Gold-billed magpie, *Urocissa flavirostris*, Rufous treepie, *Dendrocitta vagabunda*, Gray treepie, *Dendrocitta formosae*, Eurasian magpie, *Pica pica*, Eurasian nutcracker, *Nucifraga caryocatactes*, Large-spotted nutcracker, *Nucifraga multipunctata*, Red-billed chough, *Pyrrhocorax pyrrhocorax*, Yellow-billed chough, *Pyrrhocorax graculus*, Eurasian jackdaw, *Corvus monedula*, House crow, *Corvus splendens*, Rook, *Corvus frugilegus*, Carrion crow, *Corvus coroneand*, Large-billed crow, *Corvus macrorhynchos*, Brown-necked raven, *Corvus ruficollis*. Common raven, *Corvus corax*, and Hooded crow, *Corvus cornix*.

6.7.25.44 Sturnidae (Starlings)

They are passerine birds of medium in size. They are very gregarious. Their flight is direct and strong. They prefer to live in open country. They feed on fruits and insects. In particular, their dark plumage with a metallic sheen is characteristic feature. Representative birds in this species are Spot-winged starling, *Saroglossa spilopterus*, Jungle myna, *Acridotheres fuscus*, Bank myna, *Acridotheres ginginianus*, Common myna, *Acridotheres tristis*, Pied myna, *Gracupica contra*, Daurian starling, *Agropsar sturninus*, Chestnut-tailed starling, *Sturnia malabarica*, Brahminy starling, *Sturnia pagodarum*, Rosy starling, *Pastor roseus*, and European starling, *Sturnus vulgaris*.

6.7.25.45 Family: Ploceidae (Weavers)

These birds are small in size and have a relationship with finches. They have rounded conical bills. They feed on seeds. In this family, the males of many species have bright colors. Colors of these birds are variable. They may be red or yellow or sometimes may be black. Sometimes this color variation is common during breeding season. Worldwide, there are 116 species, but only two species occur in Pakistan. These are the Streaked weaver, *Ploceus manyar* and *the* Baya weaver, *Ploceus philippinus*.

6.7.25.46 Family: Estrildidae (Avadavats, Estrildid Finches)

Representative birds in this family are small in size and are also passerine. The passerine birds of the Old World tropics and Australia is their alternate designation. Their bills are short, thick, and pointed and hence they are also called conical seed eaters. They are gregarious and quite social in habit. Although habits and structure resemble each other, there significant variations in colors and patterns of plumage. Worldwide, this family has 141 species, but only four exist in Pakistan. These are Red avadavat, *Amandava amandava*, White-throated munia, *Euodice malabarica*, Nutmeg mannikin, *Lonchura punctulata*, and Spice finch, *Lonchura punctulata*.

6.7.25.47 Family: Emberizidae (Buntings, Sparrows, Seedeaters, and Allies)

A large group of passerine birds represent this family. They have a distinct shape of bills and eat seeds. In America, most of these species are called sparrows, while in Europe they are buntings. Nevertheless, these birds have no relations with Old World sparrows. Many species in this family have very distinct patterns of heads. The representative members of this species are Crested bunting, *Emberiza lathami*, Pine bunting, *Emberiza leucocephalos*, Rock bunting, *Emberiza cia*, Godlewski's bunting, *Emberiza godlewskii*, Gray-hooded bunting, *Emberiza buchanani*, Ortolan bunting, *Emberiza hortulana*, Chestnut-breasted bunting, *Emberiza stewarti*, Striolated bunting, *Emberiza striolata*, Chestnut-eared bunting, *Emberiza fucata*, Little bunting, *Emberiza pusilla*, Yellow-breasted bunting, *Emberiza aureola*, Chestnut bunting, *Emberiza rutila*, Black-headed bunting, *Emberiza melanocephala*, Red-headed bunting, *Emberiza bruniceps*, Reed bunting, *Emberiza schoeniclus*, and Corn bunting, *Emberiza calandra*.

6.7.25.48 Family: Fringillidae (Siskins, Crossbills, and Allies)

These are also passerine birds and seed eaters. Their size is small to moderate. They have strong beaks and in some species it is large and in some species it is conical. Representative birds of this family have 12 tail feathers with nine primaries. Generally, these birds have 12 tail feathers with nine primaries. Their flight is a sort of bouncing flight. There are alternate sessions when they flap and glide, while keeping their wings closed. Most of them can sing well. Worldwide, there are 137 species in this family and there are 31 species that are found in Pakistan. These are Common chaffinch, *Fringilla coelebs*, Brambling, *Fringilla montifringilla*, Plain mountain-finch, *Leucosticte nemoricola*, Black-headed mountain-finch, *Leucosticte brandti*, Dark-breasted rosefinch, *Carpodacus nipalensis*, Common rosefinch, *Carpodacus erythrinus*, Beautiful rosefinch, *Carpodacus pulcherrimus*, Pink-browed rosefinch, *Carpodacus rhodochrous*, White-browed rosefinch, *Carpodacus thura*, Red-mantled rosefinch, *Carpodacus rhodochlamys*, Caucasian great rosefinch, *Carpodacus rubicilla*, Red-fronted rosefinch, *Carpodacus puniceus*, Red crossbill, *Loxia curvirostra*, Yellow-breasted greenfinch, *Chloris spinoides*, European goldfinch, *Carduelis carduelis*, Twite, *Linaria flavirostris*, Common linnet, *Linaria cannabina*, Fire-fronted serin, *Serinus pusillus*, Brown bullfinch, *Pyrrhula nipalensis*, Orange bullfinch, *Pyrrhula aurantiaca*, Red-headed bullfinch, *Pyrrhula erythrocephala*, Hawfinch, *Coccothraustes coccothraustes*, Black-and-yellow grosbeak, *Mycerobas icterioides*, Collared grosbeak, *Mycerobas affinis*, Spot-winged grosbeak, *Mycerobas melanozanthos*, White-winged grosbeak, *Mycerobas carnipes*, Spectacled finch, *Callacanthis burtoni*, Crimson-winged finch, *Rhodopechys sanguinea*, Mongolian finch, *Rhodopechys mongolica*, Trumpeter finch, *Bucanetes githaginea*, and Desert finch, *Rhodospiza obsoleta*.

6.7.25.49 Family: Passeridae (Sparrows)

Sparrows are small in size and passerine in habit. Generally, sparrows have short tails. They are plum, brown, or gray in color. They eat seed with their powerful beaks. In addition to seeds, they can also consume insects. The provincial bird of Sind is also the Sindh sparrow. The representative species of this family are House sparrow, *Passer domesticus*, Spanish sparrow, *Passer hispaniolensis*, Sind sparrow, *Passer pyrrhonotus*, Russet sparrow, *Passer innamomeus*, Dead Sea sparrow, *Passer moabiticus*, Eurasian tree sparrow, *Passer montanus*, Chestnut-shouldered petronia, *Gymnornis xanthocollis*, Rock sparrow, *Petronia petronia*, Pale rockfinch, *Carpospiza brachydactyla*, White-winged snowfinch, *Montifringilla nivalis*, and Black-winged snowfinch, *Montifringilla adamsi* (Fig. 6.4).

FIGURE 6.4 Black-winged snowfinch, *Montifringilla adamsi*. (*Source*: www.shutterstock.com.)

6.8 MAMMALS OF PAKISTAN

Pakistan hosts 173 species of mammals. Out of this total, only three species are native to Pakistan. There are also some near endemic sub-species. Some species belong to the Palaearctic monarchy. These species commonly inhabit the Himalayas and uplands of Baluchistan. Some species belong to the Indo-Malayan kingdom. Primarily, they inhabit the foothills of the Himalayas, the Indus plains, and the Thar Desert. The dry southwest part of Pakistan, the Makran coast, is also part of their habitat (Roberts 1997).

Some species from the above given number relate to the Ethiopian region. Out of 18 mammalian orders, only five are found in Pakistan. These orders also include the smallest living mammals. Roberts (1977) observed and reported mammalian species which are endemic to Pakistan as Yellow desert bat (*Nyticeius pallidus*), wooly flying squirrel (*Eupetaurus cinereus*), pygmy gerboa (*Salpingotus michaelis*), Hotson's long-tailed hamster (*Calomyscus hotsoni*), Murree vole (*Hyperacrius wynnei*), and Indus river dolphin (*Platanista minor*). Other endemic and/or near endemic sub-species which have been reported from Pakistan are Chiltan markhor (*Capra falconeri chialtanensis*), Suleman markhor (*Capra falconeri jerdoni*), Pakistan sand cat (*Felis margarita scheffeli*), Balochistan bear (*Ursus thibetanus gedrosianus*), and Punjab urial (*Ovis vignei punjabiensis*) are the common examples. The status of these species, however, is disputable among taxonomists. Roberts (1977) reported extinction of the swamp deer (*Cervus duvauceli*), Lion (*Panthera leo)*, and tiger (*Panthera tigris*) during the last century. However, about 400 years ago, the Indian single horned rhinoceros (*Rhinoceros unicornis*) disappeared from this region. Several extremely rare marine species are present on the borders of Pakistan. All of these are in quite small numbers. Terrestrial border species of Pakistan are the cheetah (*Acinonyx jubatus*), Indian wild ass (*Equus hemionus khur*), and hangul (*Cervus elaphus hanglu*). However, they have already been declared extinct (NCCW, 1978). Among threatened species in Pakistan are the Indus dolphin, snow leopard, gray wolf, Suleman markhor, brown bear, lynx, Marcopolo sheep. Out of a total 173 total mammal species reported in Pakistan, 12 are critically endangered, 11 are endangered, 14 are vulnerable, and the other 10 are near threatened. The brown bear is the largest living mammal in Pakistan. Markhor (*Capra falconeri)* is the national animal of Pakistan. Before going further and showing the existence and/or extinction of various mammals in order, we need to know their different tags. After a proper assessment of wildlife status in Pakistan, the IUCN has issued tags for categorization. These tags are used universally for study of the conservation status of any particular species:

Extinct (Ex) means that without any doubt the organism no longer exists.

Extinct in the wild (EW) means that an organism no longer exists in its previous range. Nonetheless, it is present only as a naturalized population or in captivity.

Critically endangered (CR). There is a possibility that these species will be extinct in the wild in the very near future.

Endangered (EN) means that an organism is facing an extreme risk of extinction in its natural habitat in the wild.

Vulnerable (VU). The status of these species is very risky. They might become extinct soon in the wild. **Near threatened** (NT) means that this species presently has no risk of extinction but it might be at risk in the future.

Least concern (LC) means that currently there is no apparent risk to this species which has been identified.

Data deficient (DD) means that it is hard to make any assessment that there is any risk to this species.

Some species have been assessed following the above criteria but others have been assessed applying the following criteria:

Lower risk/conservation dependent (LR/cd) means that these species have been included in a conservation program; these species may move to a high risk category if conservation programs are halted.

Lower risk/near threatened (LR/nt) means that the species has not been included in conservation programs but the species is approaching the status when it may be classified as vulnerable.

Lower risk/least concern (LR/lc) means that there is no risk of extinction for this particular species.

Presently, class Mammalia (Sub-class Theria, Infra-class Eutheria) has 12 distinct orders, whose representative species found in Pakistan have been summarized below.

6.8.1 Order: Proboscidea Family: Elephantidae (elephants)

Elephants are the largest land animals and are living both in nature and in captivity. There are three living species of elephants which represent this family. These are the Asian elephant, *Elephas maximus* (EN) and the Indian elephant, *Elephas maximus indicus* (EN) (extirpated).

6.8.2 Order: Primates Family: Cercopithecoidea

The order Primates contains lemurs, lorisoids, monkeys, apes, and human beings. Rhesus macaque (*Macaca mulatta*) (LR/nt) is the territorial animal of Islamabad. The gray langur, *Semnopithecus entellus* (LR/nt) is also an important member of this family.

6.8.3 Order: Rodentia (Rodents)

This order has the largest number of mammals with over 40% of living mammalian species. Most of them, however, are small in size with the exception of Capybara, which can weigh up to 45 kg (100 lbs). Rodents have two incisors in the lower jaw, as well as in the upper jaw. Growth of these incisors is continuous throughout their life. However, these animals gnaw their food persistently. This gnawing process keeps their growth under control. Important families of this order along with their representatives have been given below.

Family: Hystricidae (Old World porcupines)

Indian porcupine, *Hystrix indica* (LR/lc).

Family: Sciuridae (squirrels)

Kashmir flying squirrel, *Eoglaucomys fimbriatus* (LC)
Woolly flying squirrel, *Eupetaurus cinereus* (EN)
Red giant flying squirrel, *Petaurista petaurista*
Northern palm squirrel, *Funambulus pennantii* (LR/lc)
Long-tailed marmot, *Marmota caudata* (LR/nt)
Himalayan marmot, *Marmota himalayana* (LC)

Family: Gliridae (Dormice)

Baluchistan forest dormouse, *Dryomys niethammeri* (VU)
Forest dormouse, *Dryomys nitedula* (LR/nt)

Family: Dipodidae (Jerboas)

Small five-toed jerboa, *Allactaga elater* (LR/lc)
Hotson's jerboa, *Allactaga hotsoni* (LR/lc)
Baluchistan pygmy jerboa, *Salpingotus michaelis* (LR/lc)
Blanford's jerboa, *Jaculus blanfordi* (LR/lc)
Chinese birch mouse, *Sicista concolor* (LR/lc)

Family: Calomyscidae

Baluchi mouse-like hamster, *Calomyscus baluchi* (LR/lc)
Hotson's mouse-like hamster, *Calomyscus hotsoni* (EN)

Family: Cricetidae

Gray dwarf hamster, *Cricetulus migratorius* (LR/nt)
White-tailed mountain vole, *Alticola albicauda* (LR/nt)
Silver mountain vole, *Alticola argentatus* (LR/lc)
Southern mole vole, *Ellobius fuscocapillus* (LR/lc)
True's vole, *Hyperacrius fertilis* (LR/lc)
Murree vole, *Hyperacrius wynnei* (LR/lc)
Juniper vole, *Microtus juldaschi* (LR/lc)

Family: Muridae (mice, rats, voles, gerbils, hamsters, etc.)

Arabian spiny mouse, *Acomys dimidiatus* (LC).
Swarthy gerbil, *Gerbillus aquilus* (LR/lc)
Indian hairy-footed gerbil, *Gerbillus gleadowi*
Baluchistan gerbil, *Gerbillus nanus* (LC)
Indian desert jird, *Meriones hurrianae* (LR/lc)
Persian jird, *Meriones persicus* (LR/lc)
Libyan jird, *Meriones libycus* LC
Sundevall's jird, *Meriones crassus* (LC)
Great gerbil, *Rhombomys opimus* (LR/lc)
Indian gerbil, *Tatera indica* (LR/lc)
Kashmir field mouse, *Apodemus rusiges* (LR/lc)
Ward's field mouse, *Apodemus wardi* (LR/lc)
Lesser bandicoot rat, *Bandicota bengalensis* (LR/lc)
Indian bush rat, *Golunda ellioti* (LR/lc)
Sand-colored soft-furred rat, *Millardia gleadowi* (LR/lc)
Soft-furred rat, *Millardia meltada* (LC)
Rock-loving mouse, *Mus saxicola* LR/lc
Earth-colored mouse, *Mus terricolor* (LR/lc)
House mouse, *Mus musculus* (LC)
Brown spiny mouse, *Mus saxicola* (LR/lc)
Short-tailed bandicoot rat, *Nesokia indica* (LC)
Chestnut white-bellied rat, *Niviventer fulvescens* (LR/lc)
Turkestan rat, *Rattus turkestanicus* (LR/lc)
House rat, *Rattus rattus* (LC)
Brown rat, *Rattus norvegicus* (LC)

6.8.4 Order: Lagomorpha (Lagomorphs)

This order has two families, namely the Leporidae, which includes hares and rabbits and, second, the Ochotonidae, which includes pikas. Lagomorphs resemble rodents. They were classified as the superfamily of rodentia until the early 20th century but now a separate order has been assigned to them. Characteristically, several features differentiate them from rodents. One of them is the presence of incisors in the upper jaw. Rodents have four incisors in the upper jaw while rodents have only two incisors in the upper jaw.

Family: Ochotonidae (Pikas)

Ladakh pika, *Ochotona ladacensis* (LR/lc)
Black-lipped pika, *Ochotona curzoniae* (LR/lc)
Large-eared pika, *Ochotona macrotis* (LR/lc)
Nubra pika, *Ochotona nubrica* (LR/lc)
Royle's pika, *Ochotona roylei* (LR/lc)
Afghan pika, *Ochotona rufescens* (LR/lc)

Family: Leporidae (rabbits, hares)

Indian hare, *Lepus nigricollis* (LR/lc)
Woolly hare, *Lepus oiostolus* (LR/lc)
Cape hare, *Lepus capensis* (LC)

6.8.5 Order: Erinaceomorpha (Hedgehogs and Gymnures)

This family has only single order. Hedgehogs and gymnures fall under this family. Gymnures resemble large rats. The presence of spines on the whole body is a characteristic feature of hedgehogs.

Family: Erinaceidae (hedgehogs, gymnures)

Long-eared hedgehog, *Hemiechinus auritus* (LR/lc)
Indian long-eared hedgehog, *Hemiechinus collaris* (LR/lc)
Brandt's hedgehog, *Hemiechinus hypomelas* (LR/lc)
Indian hedgehog, *Hemiechinus micropus* (LR/lc)

6.8.6 Order: Soricomorpha (Shrews, Moles, and Solenodons)

The different forms of shrew feed on insects. The moles have a stout body and are burrowers in their habitat. Shrews and solenodons nevertheless closely resemble mice morphologically. This order also has a single family.

Family: Soricidae (shrews)

Gmelin's white-toothed shrew, *Crocidura gmelini* (LR/lc)
Pale gray shrew, *Crocidura pergrisea* (DD)
Zarudny's shrew, *Crocidura zarudnyi* (LR/lc)
Asian gray shrew, *Crocidura suaveolens*
Etruscan shrew, *Suncus etruscus* (LC)
Asian house shrew, *Suncus murinus* (LR/lc)
Anderson's shrew, *Suncus stoliczkanus* (LR/lc)
Kashmir shrew, *Sorex planiceps* (LR/lc)

6.8.7 Order: Chiroptera (Bats)

The modification of forelimbs into wings is a distinguishing feature of bats. Hence, in class Mammalia, they are the only mammals which can fly. Bats have several families and species and are 20% of all the mammals present on earth.

Pteropodidae (flying foxes, Old World fruit bats)

Indian flying-fox, *Pteropus giganteus* (LR/lc)
Egyptian fruit bat, *Rousettus aegyptiacus* (LC)
Leschenault's rousette, *Rousettus leschenaultii* (R/lc)

Vespertilionidae

Lesser mouse-eared bat, *Myotis blythii* (LR/lc)
Whiskered myotis, *Myotis muricola* (LR/lc)
Geoffroy's bat, *Myotis emarginatus* (LC)
Kashmir cave bat, *Myotis longipes* (DD)
Eastern barbastelle, *Barbastella leucomelas* (LR/lc)
Botta's serotine, *Eptesicus bottae* (LC)
Gobi big brown bat, *Eptesicus gobiensis* (LR/lc)
Sind bat, *Eptesicus nasutus* (VU)
Serotine, *Eptesicus serotinus* (LC)
Lesser noctule, *Nyctalus leisleri* (LR/nt)
Mountain noctule, *Nyctalus montanus* (LR/nt)
Common noctule, *Nyctalus noctula* (LR/lc)
Desert long-eared bat, *Otonycteris hemprichii* (LR/lc)
Kelaart's pipistrelle, *Pipistrellus ceylonicus* (LR/lc)
Indian pipistrelle, *Pipistrellus coromandra* (LR/lc)
Java pipistrelle, *Pipistrellus javanicus* (LR/lc)
Kuhl's pipistrelle, *Pipistrellus kuhlii* (LC)
Common pipistrelle, *Pipistrellus pipistrellus* (LC)
Least pipistrelle, *Pipistrellus tenuis* (LR/lc)
Savi's pipistrelle, *Pipistrellus savii*
Desert yellow bat, *Scotoecus pallidus* (NT)
Greater Asiatic yellow bat, *Scotophilus heathi* (LR/lc)
Lesser Asiatic yellow bat, *Scotophilus kuhlii* (LR/lc)
Dormer's pipistrelle, *Scotozous dormeri* (LC)
Hutton's tube-nosed bat, *Murina huttoni* (LR/nt)
Scully's tube-nosed bat, *Murina tubinaris* (LR/lc)

Rhinopomatidae

Lesser mouse-tailed bat, *Rhinopoma hardwickei* (LC)
Greater mouse-tailed bat, *Rhinopoma microphyllum* (LC)
Small mouse-tailed bat, *Rhinopoma muscatellum* (LC)

Family: Emballonuridae

Naked-rumped tomb bat, *Taphozous nudiventris* (LC)
Egyptian tomb bat, *Taphozous perforatus* (LC)

Family: Megadermatidae

Indian false vampire bat, *Megaderma lyra* (LC)

Family: Rhinolophidae

Blasius's horseshoe bat, *Rhinolophus blasii* (NT)
Greater horseshoe bat, *Rhinolophus ferrumequinum* (LR/nt)
Lesser horseshoe bat, *Rhinolophus hipposideros* (LC)
Blyth's horseshoe bat, *Rhinolophus lepidus* (LR/lc)
Big-eared horseshoe bat, *Rhinolophus macrotis* (LR/lc)
Trident leaf-nosed bat, *Asellia tridens* (LC)
Ashy roundleaf bat, *Hipposideros cineraceus* (LR/lc)
Fulvus roundleaf bat, *Hipposideros fulvus* (LR/lc)
Persian trident bat, *Triaenops persicus* (LC)

Family: Molossidae

Egyptian free-tailed bat, *Tadarida aegyptiaca* (LC)

6.8.8 Order: Pholidota (Pangolins)

Eight species of pangolin represent this order. Pangolins have powerful claws to dig ant holes to extract ants. Their long snout and tongue help them to extract ants from narrow and deeper holes.

Family: Manidae

Indian pangolin, *Manis crassicaudata* LR/nt

6.8.9 Order: Cetacea (Whales)

This order consists of whales, dolphins, and porpoises. Their body is spindle shaped and hairless. They are fully adapted to aquatic life. A thick layer of blubber fully protects their body. They swim in water with the help of modified forelimbs. Modification of the tail provides them with propulsion under water.

Family: Balaenopteridae

Bryde's whale, *Balaenoptera edeni* (DD)
Pygmy blue whale, *Balaenoptera musculus brevicauda* (EN)
Fin whale, *Balaenoptera physalus* (EN)
Humpback whale, *Megaptera novaeangliae* (VU) (the world's most isolated, most endangered, and only non-migratory population)

Family: Platanistidae

Indus river dolphin, (national marine mammal of Pakistan) *Platanista minor minor* (EN)

Family: Phocoenidae

Finless porpoise, *Neophocaena phocaenoides* (DD)

Family: Kogiidae

Pygmy sperm whale, *Kogia breviceps* (LR/lc)
Dwarf sperm whale, *Kogia sima* (LR/lc)

Family: Physeteridae

Sperm whale, *Physeter macrocephalus* (VU)

Family: Ziphidae

Blainville's beaked whale, *Mesoplodon densirostris* (DD)
Ginkgo-toothed beaked whale, *Mesoplodon ginkgodens* (DD)

Family: Delphinidae (marine dolphins)

Rough-toothed dolphin, *Steno bredanensis* (DD)
Sousa chinensis (DD)

Bottlenose dolphin, *Tursiops aduncus* (DD)
Bottlenose dolphin, *Tursiops truncatus* (DD)
Spinner dolphin, *Stenella longirostris* (LR/cd)
Common dolphin, *Delphinus capensis* (LR/lc)
Fraser's dolphin, *Lagenodelphis hosei* (DD)
Risso's dolphin, *Grampus griseus* (DD)
Melon-headed whale, *Peponocephala electra* (LR/lc)
Pygmy killer whale, *Feresa attenuata* (DD)
Killer whale, *Orcinus orca (*DD)

6.8.10 Order: Carnivora (Carnivorans)

Two hundred and sixty species represent this order. Most of them eat meat as their primary food. They have typical dentition and skull shape.

Family: Felidae (cats)

Caracal, *Caracal caracal* (LC)
Jungle cat, *Felis chaus*
Sand cat, *Felis margarita* (LC)
Pakistan sand cat, *Felis margarita scheffeli*
Wildcat, *Felis lybica*
Asiatic wildcat, *Felis lybica ornata* (LC)
Pallas's cat, *Otocolobus manul* (NT)
Eurasian lynx, *Lynx lynx* (LC)
Leopard cat, *Prionailurus bengalensis* (LC)
Fishing cat, *Prionailurus viverrinus* (VU)
Leopard *Panthera pardus* (VU)
Indian leopard, *Panthera pardus fusca*, nationally (CE)
Snow leopard, *Panthera uncia* (EN)

Family: Viverridae (civets, mongooses, etc.)

Masked palm civet, *Paguma larvata* (LR/lc)
Small Indian civet, *Viverricula indica* (LR/lc)

Family: Herpestidae (mongooses)

Indian gray mongoose, *Herpestes edwardsii* (LR/lc)
Small Asian mongoose, *Herpestes javanicus* (LR/lc)

Family: Hyaenidae (hyaenas)

Striped hyena, *Hyaena hyaena* (LR/nt).

Family: Canidae (dogs, foxes)

Blanford's fox, *Vulpes cana* (VU)
Red fox, *Vulpes vulpes* (LC)
Bengal fox, *Vulpes bengalensis* (LC)
Rüppell's fox, *Vulpes rueppellii* (LC)

Golden jackal, *Canis aureus* (LC)
Common jackal, *Canis aureus aureus* (LC)
Gray wolf, *Canis lupus* (LC)
Indian wolf, *Canis lupus pallipes* (EN)
Dhole *Cuon alpinus* (EN)
Tien Shan dhole, *Cuon alpinus hesperius* (EN)

Family: Ursidae (bears)

Brown bear, *Ursus arctos* (LC)
Himalayan brown bear, *Ursus arctos isabellinus* (CR)
Asiatic black bear, *Ursus thibetanus* (VU)

Family: Mustelidae (mustelids)

Stoat, *Mustela erminea* (LR/lc)
Mountain weasel, *Mustela altaica* (NT)
Marbled polecat, *Vormela peregusna* (LR/lc)
Yellow-throated marten, *Martes flavigula* (LR/lc)
Beech marten, *Martes foina* (LR/lc)
Ratel, *Mellivora capensis* (LR/lc)
European otter, *Lutra lutra* (NT)
Smooth-coated otter, *Lutrogale perspicillata* (VU)

6.8.11 Order: Perissodactyla (Odd-Toed Ungulates)

Members of this family are either grazers or browsers with simple stomachs. They are odd-toed ungulates with a large middle toe. Their size is large to very large in some members. This order has two families which are given below.

Family: Equidae (horses etc.)

Asiatic wild ass, *Equus hemionus* (NT)
Indian wild ass, *Equus hemionus khur* (NT)
Persian onager, *Equus hemionus onager* (EN) (extirpated)

Family: Rhinocerotidae

Indian rhinoceros, *Rhinoceros unicornis* (VU) (extirpated)

6.8.12 Order: Artiodactyla (Even-Toed Ungulates)

Unlike perissodactyls which bear their weight on the third toe, the even-toed ungulates bear their weight mostly or entirely on their third and fourth toes. This order has about 220 species covered under various families. There are several species which are economically important for human beings.

Family: Suidae (pigs)

Wild boar, *Sus scrofa* (LR/lc)

Family: Moschidae

Himalayan musk deer, *Moschus chrysogaster* (LR/nt)

Family: Cervidae (deer)

Barasingha, *Rucervus duvaucelii* (VU) (extirpated)
Spotted deer, *Axis axis* (LC) (extirpated)
Indian hog deer, *Hyelaphus porcinus* (EN)
Barking deer, *Muntiacus muntjak*

Family: Bovidae (cattle, antelope, sheep, goats)

Blackbuck, *Antilope cervicapra* (NT)
Chinkara, *Gazella bennettii* (LC)
Goitered gazelle, *Gazella subgutturosa* (VU)
Nilgai, *Boselaphus tragocamelus* (LC)
Wild goat, *Capra aegagrus* (VU)
Sindh ibex, *Capra aegagrus blythi*
Chiltan ibex, *Capra aegagrus chialtanensis*
Markhor, *Capra falconeri* (EN)

There are five distinct types of Markhor. This is the national animal of Pakistan. Out of them all, the Chiltan Markhor looks very conspicuous due to its long horns. Since adaptation of its protection measures, its population has been multiplied.

Astor markhor, *Capra falconeri falconeri* (LR/nt)
Kabul Markhor, *Capra falconeri megaceros* (EN)
Siberian ibex, *Capra sibrica* (LR/lc)
Gray goral, *Nemorhaedus goral* (LC)
Argali, *Ovis ammon* (NT)
Marco Polo sheep, *Ovis ammon polii* (LR/nt)
Mouflon, *Ovis orientalis* (VU)
Blanford urial, *Ovis orientalis blanfordi*
Punjab urial, *Ovis orientalis punjabiensis*
Ladakh urial, *Ovis orientalis vignei*
Bharal, *Pseudois nayaur* (LC)

REFERENCES

Adams, J.M. & Woodward, F.I. (1989). Patterns in Tree Species Richness as a Test of the Glacial Extinction Hypothesis. Nature 339, 699–701.

Auffenberg, W. & Rehman, H. (1977). Geographic Variations in *Bufo stomaticus*, with Remarks on *Bufo olivaceus*: Biogeographical and Systematic Implications. In: Biodiversity of Pakistan, eds. Mufti, S.A., Woods, C.A., & Hasan, S.A., pp. 351–372. Museum of Natatural History Islamabad, Pakistan.

Begum, A., Ghalib, S.A., Khan, M.Z., et al. (2014). An Overview of the Current Status and Distribution of Amphibians in Sindh. African Journal of Science and Research 3, 21–23.

Blackburn, T.M. & Gaston, K.J. (1996). Spatial Patterns in the Species Richness of Birds in the New World. Ecography 19, 369–376.

Braulik, G.T., Smith, B.D. & Chaudhry, A.A. (2012). *Platanista gangetica* sp. minor. In The IUCN Red List of Threatened Species. Version 2014.3: http://www.iucnredlist.org. Downloaded on 23 April 2018.

Brown, J.H. & Lomolino, M.V. (1998). Biogeography. 2nd ed. Sinauer, Sunderland, MA.

Chown, S.L. & Gaston, K.J. (1999). Patterns in Procellariiform Diversity as a Test of Species-Energy Theory in Marine Systems. Evolutionary Ecology Research 1, 365–173.

Clarke, A. & Crame, J.A. (1997). Marine Biodiversity: Patterns and Processes. In: Biodiversity and Conservation, eds. Ormond, R.F.G., Gage, J.D. & Angel, M. V., pp. 122–147. Cambridge University Press, Cambridge.

Clements, J.F. (2000). Birds of the World: A Checklist. Cornell University Press.
Currie, D.J. (1991). Energy and Large-Scale Patterns of Animal- and Plant-Species Richness. American Naturalist 137, 27–49.
Eeley, H.A.C. & Lawes, M.J. (1999). Large-Scale Patterns of Species Richness and Species Range Size in Anthropoid Primates. In: Primate Communities, eds. Feagle, J.G., Janson, C. & Reed, K.E., pp. 191–219. Cambridge University Press, Cambridge.
Flather, C.H., Wilson, K.R., Dean, D.J. & McComb, W.C. (1997). Identifying Gaps in Conservation Networks: of Indicators and Uncertainty in Geographic-Based Analyses. Ecological Applications **7**, 531–542.
Frost, D.R. (2016). Amphibian Species of the World: An Online Reference. Version 5.2. Electronic Database. American Museum of Natural History, New York, USA.
Gabol, K. & Rahim, A. (2006). Obsevation on the Waterfowl of Jiwani Wetland Complex (Jiwani, Gawater Bay, Pishukan and Akara River Estuary) Balochistan. International Journal of Applied Biology and Biotechnology 3(4), 763–766.
Gaston, K.J. (1996). Biodiversity — Latitudinal Gradients. Progress in Physical Geography 20, 466–476.
Gaston, K.J. & Blackburn, T.M. (1999). A Critique for Macroecology. Oikos 84, 353–368 (1999).
Griffiths, D. (1997). Local and Regional Species Richness in North American Lacustrine Fish. Journal of Animal Ecology 66, 49–56.
Groomridge, B. & Wright, L. (1982). The IUCN Amphibia – Reptilia Red Data Book. Part 1, test dines, Crocodylia, Rhynchocephalia. IUCN: 426–428.
Hader, D-P., Kumar, H.D., Smithc, R.C. & Worrestd, R.C. (2007). Effects of Solar UV Radiation on Aquatic Ecosystems and Interactions with Climate Change. Photochemical and Photobiological Sciences 6, 267–285.
Hammer, K.C. & Hill, J.K. (2000). Scale-Dependent Effects of Habitat Disturbance on Species Richness in Tropical Forests. Conservation Biology 14(5), 1435–1440. Wiley.
Haq, S.M. & Rehman, Fu (1973). A New Calanoid Copepode, Centropages Karachiensis from the Inshore Waters of the Karachi Coast, West Pakistan. Zoologische Mededelingen 46,183–188.
Hashmi, M.U.A., Khan, M.Z., Safi, A. & Huda, N., 2013. Current Status, Distribution and Threats of Varanus Spp. (*Varanus bengalensis* & *Veranus griseus*) in Karachi & Thatta of Sindh. International Journal of Fauna and Biological Studies 1(1), 34–38.
Hawkins, B.A. & Compton, S.G. (1992). African Fig Wasp Communities: Under Saturation and Latitudinal Gradients in Species Richness. Journal of Animal Ecology 61, 361–372.
Helfman, G. (2007). Fish Conservation: A Guide to Understanding and Restoring Global Aquatic Biodiversity and Fishery Resources. Corpus ID: 127239904. 10.1899/0887-3593(2008)27[802:BR]2.0.CO;2.
Hiddink, J.G., MacKenzie, B.R., Rijnsdorp, A., Dulvy, N.K., Nielsen, E.E., Bekkevold, D., Heino, M., Lorance, P., & Ojaveer, H. (2008). Importance of Fish Biodiversity for the Management of Fisheries and Ecosystems. Fisheries Research 90, 6–8.
Ishaq, S. & Mustaquim, J. (1996). Ploychaetous Annelids(Order Sabellida) from the Karachi Coast, Pakistan. Pakistan Journal of Marine Sciences 5(2), 161–197.
IUCN (1990). 1IUCN Red List of Threatened Animals. Prepared by the IUCN Conservation Monitoring Centre, Cambridge, U.K., in association with the IUCN Species Survival Commision and the International Council for Bird Preservation.
Kazmi, Q.B. & Naushaba, R. (2004). Rediscovery of *Chiton peregrinus* Thiele, 1910 (Mollusca, Polyplacophora) from Pakistan coast. Pakistan Journal of Marine Science 13(1&2), 75–77.
Khan, M.S. (1997). A New Toad of Genus *Bufo* from the Foot of Siachin Glacier, Baltistan, Northeastern Pakistan. Pakistan Journal of Zoology 29, 43–48.
Khan, M.S. (2001). Notes on Cranial-Ridged Toads of Pakistan and Description of a New Subspecies (Amphibia: Bufonidae). Pakistan Journal of Zoology 33, 293–298.
Khan, M.S. (2006). Amphibians and Reptiles of Pakistan. Krieger Publishing Company, Malabar, Florida.
Khan, M.S. (2014). Amphibians of Pakistan and Their Conservation Status(Chapter 7). In: Conservation Biology of Amphibians of Asia, eds. Heatwole, H. & Das, I., 22 pages. Natural History Publications, (Borneo), Kota Kinabalu.
Khan, M.Z., Amtyaz, S., Fatima, F., Hashmi, M.U.A., Hussain, B., Siddiqui, S., Khan, S.I. & Ghalib, S.A. 2015. An Evaluation of Distribution, Status and Abundance of Freshwater Turtles in the Selected Areas of Sindh and Khyber Pakhtunkhwa Provinces of Pakistan. Canadian Journal of Pure & Applied Sciences 9(1), 3201–3219. https://web.archive.org/web/20140914014025/http://cjpas.net/instructions.htm
Koslow, A. (2007). The Silent Deep: The Discovery, Ecology and Conservation of the Deep Sea. University of New South Wales Press, Sydney.
Lawton, J.H. (1996). Patterns in Ecology. Oikos 75, 145–147.

Lepage, D. (2007). "Checklist of birds of Pakistan". Bird Checklists of the World. Avibase.
Lyons, S.K. & Willig, M.R.A. (1999). Hemispheric Assessment of Scale Dependence in Latitudinal Gradients of Species Richness. Ecology 80, 2483–2491.
Masroor, R. (2011). An Annotated Checklist of Amphibians and Reptiles of Margalla Hills National Park, Pakistan. Pakistan Journal of Zoology 43, 1041–1048.
Mazhar, Y. (2020). The Most Amazing National Reserves in Pakistan. Zameenblog.
Moazzam, M. & Osmany, H.B. (2024). A Review of Fishes Belonging to the Family Carangidae from Pakistan. International Journal of Biology and Biotechnology 21 (4), 527–596.
Murty, S.J., Bett, B. & Gooday, A.J. (2008). Megafaunal Responses to Strong Oxygen Gradients on the Pakistan Margin of the Arabian Sea. DeepSea Research II. https://doi.org/10.1016/j.dsr2.2008.05.029.
Platnick, N.I. (1991). Patterns of Biodiversity: Tropical vs Temperate. Journal of Natural History 25, 1083–1088.
Punjab Fisheries Manual (2014). Pages 196.
Rais, M., Abbassi, S. & Batool, T., et al. (2014). A Note on Recapture of *Nanorana vicina* (Anura: Amphibia) from Murree, Pakistan. Journal of Animal and Plant Sciences Lahore 24, 455–458.
Rais, M., Baloch, S. & Rehman, J., et al. (2012). Diversity and Conservation of Amphibians and Reptiles in North Punjab, Pakistan. Herpetological Bulletin 122, 16–25.
Ricklets, R.E. & Schluter, D. (1993). Species Diversity in Ecological Communities: Historical and Geographical Perspectives.pp 194–202. University of Chicago Press, USA.
Rizvi, N. & Moazzarn, M. (2006). Sessile Barnacles (Cirripedia) from the Pakistan Coast. Pakistan Journal of Marine Sciences 15(1), 91–118.
Rosenzweig, M.L. (1995a). Species Diversity Gradients: We Know More and Less than We Thought. Journal of Mammalogy 73, 715–730.
Rosenzweig, M.L. (1995b). Species Diversity in Space and Time. Cambridge University Press, Cambridge.
Rosenzweig, M.L. & Abramsky, Z. (1993). In: Species Diversity in Ecological Communities, eds. Ricklefs, R. E. & Schluter, D., pp. 52–65. University of Chicago Press, Chicago.
Roy, K., Jablonski, D., Valentine, J.W. & Rosenberg, G. (1998). Marine Latitudinal Diversity Gradients: Tests of Causal Hypotheses. Proceedings of the National Academy of Science USA 95, 3699–3702.
Safi, A. & Khan, M.Z. (2014). Distribution and Current Population Status of Freshwater Turtles of District Charsadda of Khyber Pakhtunkhwa, Pakistan. The Journal of Zoology Studies 1(4), 31–38. http://www.journalofzoology.com
Sarwar, M.K., Malik, M.F., Hussain, M., Azam, I., Iqbal, W. & Ashiq, U. (2016). Distribution and Current Status of Amphibian Fauna of Pakistan: A Review. Electronic Journal of Biology 12(3), 243–246.
Siddiqui, Q.A. (2006). The Ostracod Genus Paijenborchella and Some of Its Species in the Early Tertiary of Pakistan. Journal of Micropalaeontology 25, 65–72.
Stevens, G.C. (1989). The Latitudinal Gradient in Geographical Range: How so Many Species Co-exist in the Tropics. American Naturalist 133, 240–256.
Turley, C., Blackford, J., Widdicombe, S., Lowe, D. & Nightingale, P. (2006). Reviewing the Impact of Increased Atmospheric Carbon Dioxide on Oceanic pH and the Marine Ecosystem. Plymouth Marine Laboratory, Plymouth UK.
Turner, J.R.G., Lennon, J.J. & Lawrenson, J.A. (1988). British Bird Species Distributions and Energy Theory. Nature 335, 539–541.
Watson, R.T., Heywood, V.H., Baste, I., Dias, B., Gamez, R., Janetos, T., Reid, W. & Ruark, G. (1995). Global Biodiversity Assessment Summary for Policy-Makers. Published for the United Nations Environment Programme Cambridge University Press.
Wright, D.H. (1983). Species-Energy Theory: an Extension of Species Area-Theory. Oikos 41, 496–506.

7 Human Activities and Their Effects on Animal Populations

7.1 INTRODUCTION

The concept of the importance of biodiversity emerged in the minds of human beings when they thought about it and realized that biodiversity is being lost in one way or another. They further realized that either we would be facing the challenges of a biodiversity crisis or we have already faced several and we have reached the midpoint of this dangerous process (GBS, 1992). This concern grew in the human mind and is still growing, starting from the last about five decades ago when people started to observe losses in several species of animals and plants. This loss was not restricted to species only; rather, it extended to populations, even domesticated varieties of plants and animals, as well their natural habitats such as wetlands which host plants and animals and tropical rain forests where the majority of plants inhabit. In fact, people have been involved in the question of biodiversity and thinking about it since their emergence on this planet (GBS, 1992). The first definition of biodiversity might have evolved when the two terms "genetic diversity" (meaning the amount of genetic variability within species) and "ecological diversity" (meaning the number of species in a community of organisms) were coined together (Norse and McManus, 1980 cited in GBS, 1992). If elaborated further, ecological diversity tells us how many species are present in a defined area and what type of ecological roles and functions they perform. Ecological diversity further tells us the changes in species composition as we move across a region and different groupings of species or within an ecosystem which may be grassland or in the forest. Different processes going on within an ecosystem or across the ecosystems as well as interactions among animals and/or ecosystems across different parts of land or occurring in or among biomes are also linked to this definition of ecological biodiversity (GBS, 1992). It has recently been estimated that human beings have significantly altered more than half of the total land which could probably be inhabited by different plants and animal species. These concerns further ameliorate our realization that the required knowledge about biodiversity regarding plant varieties, animal varieties, microorganisms, and the ecosystems is not fully explored and requires further deliberation (Lovejoy, 1980). It is universally accepted that our definition of ecological diversity covers three main components (ecosystems, species, and genes) which are used commonly when biodiversity is studied or practiced for utilitarian purposes.

Over the past few decades, awareness about the environment has increased considerably. This increased awareness has suggested further understanding and comprehension of the interactions between biodiversity and society. Including human dimensions, this notion has further broadened the scope of biodiversity far greater than it was thought of a few decades earlier. Human activities are always on the rise to affect wildlife on all ecological scales. Human interferences start to display their effects in the form of short-term changes in the behavior of animal species, initially by removing the animal and/or plant species at local level and, if this process further continues, animal and/or plant species become totally extinct at the global level (Pimm et al., 1995; Chapin et al., 2000).

It is a real challenge for those who are involved to understand the effects of human beings on wildlife and its populations and to plan strategies to manipulate these changes in a positive and befitting manner. It is essentially important to assess the effects of human activities on wildlife and then monitor them continuously to ensure that all the resources are properly being taken care of and further required strategies have been designed to mitigate the effects of human activities for the simultaneous conservation and restoration of wildlife. Among all these interventions are usage of

DOI: 10.1201/9781003472001-7

land and its management. In general, these effects are land use or land management practices that alter the routes of succession in ecology, ultimately causing changes in abiotic factors of a particular site following changes in that plant community which are currently dominant in overall population (Robert & Brian, 2006). Human activities have greatly affected vertebrates ecologically and, accordingly, these effects have been well explored too (Spies et al., 1996; Lichstein et al., 2002). Those ecological effects of human beings that are less clear are from non-consumptive activities which do not alter the physical environment significantly but nevertheless have adverse effects on wildlife. Wildlife viewing, visiting parks, hiking, and boating are examples of all such activities which are non-consumptive in their essence. Sensitive species have a direct negative response with the rise of recreational activities in the wilderness and other protected areas (Steidl & Anthony, 2000). The type of recreational activity, frequency, magnitude, duration, location, and timing of the disturbance in a particular area are all factors which have a direct impact on wildlife, which may be of short or long duration (Fernandez & Azkona, 1993; Holmes et al., 1994; Swarthout & Steidl, 2001, 2003; Mann et al., 2002; Johnson et al., 2005). The observed effects include displacement from preferred environments (McGarigal et al., 1991), which increases animals' energetic stresses (Bélanger & Bédard, 1990), with consequent changes in the activity/energy budgets spent during daily functions (Swarthout & Steidl, 2001, 2003). All these changes ultimately result in the reduction of animal productivity, and in related outcomes like decreased survival of young ones and ultimate death in some severe cases of stress (White & Thurow, 1985). Overuse of particular areas including construction of buildings and roads and/or destruction of vegetation cause physical changes to parks, do have short as well as long term effects on wildlife.

7.2 LOSS OF BIODIVERSITY

Currently, degradation of the environment and consequent losses to biodiversity are at quite an accelerated rate than observed in the past. Further, there is no hope that these degradative processes will slow down at any time in the future. The loss of biodiversity and the related changes in the environment are now faster than ever before in human history and there is no sign of slowing down of this process. There are dramatic changes and distortions in earth's ecosystems due to human activities and there is continuous conversion of these ecosystems into agriculture lands and for other uses. Geographical spread of many animals and plants has been severely disturbed and their number has considerably declined. Although the extinction of animals is a natural process, human activities have increased its normal rate by up to 100 times (Gaynor et al., 2018). Loss of biodiversity is attributed to several drivers. Any factor that induces changes in an ecosystem which may be direct or indirect, natural, or human-induced is called a driver. A direct driver unambiguously affects the processes of ecosystems. Effects of indirect driver are more diffused and less obvious and manipulative and results in the alteration of direct drivers which affect biodiversity disturbing its natural balance as well existence. Such factors which may be natural or human induced interactively intensify the effects of each other.

7.2.1 Changes in Habitats of Plants and Animals and/or Total Destruction

In general at global level habitat alteration and destruction are the two major factors playing a key role in loss of biodiversity. In habitat destruction the appropriate available habitats where plants and animals are thriving successfully lose their function and become unable to support life. Resultantly habitat destruction reduces the number of organisms presently inhabiting there by dislocation or destruction (Ayoade et al., 2009; Agarwal et al., 2011). In the latter half of the 20th century there is considerable increase in the human destruction of the habitats. Human destroys the natural habitats either for harvesting the natural resources, for industrialization and/or for urbanization. Forests are removed for agriculture and riverine habitats are changed to lacustrine (reservoir) habitat when hydroelectric projects are constructed on the rivers (Agarwal et al., 2014). Other factors that result in fragmentation and destruction of habitats are logging, mining, construction of highways, and urban

Since the intrusion of both mosquito and parasite in conjunction with other threats caused decline and extinction of about 60 species endemic forest birds due to an ailment of avian malaria. Even some which appeared in the absence of this disease were so weak and intolerant to this disease that they could not cope with the debilitating effects of the novel parasite. Nevertheless, after the laps of more than 100 years now some birds like thrushes (*Myadestes* spp.) are now capable enough to resist this disease. It is sad that remaining species of birds which belong to the family Drepanididae, are still vulnerable. These birds are now restricted to altitudes where temperatures are so low that they cannot support survival of this mosquito vector. It is predicted that global warming will enhance the distribution ranges altitudinally which will be lethal for disease disease-susceptible birds with the disappearance of mosquito free habitats. Transmission of malaria can be conveniently reduced by elimination of vector mosquito populations by chemical applications or diminishing the places where mosquito larvae inhabit. The most least known example of lake Victoria (tropical east Africa) be quoted where an invasion catastrophe occurred due to infiltration of cichlid fishes. In addition to Tilapia this lake hosts several other fish species. During the last century exploitation of non-cichlid fisheries, increased human population pressure, removal of forests and conversion of biotas for agriculture purposes are some of the causes of mass extinction of animal fauna in and around this lake. Despite all these factors in fact the most devastating effect was the introduction of the predatory Nile perch (*Lates niloticus*) in the 1950s. After several decades of its introduction in Lake Victoria though the population of the Nile perch did not increase much but its population explosion in 1980s directly or indirectly eliminated 200–400 cichlid species endemic to the Lake which was in addition to the extinction of several non-cichlid fish species.

Among other factors human mediated climatic changes are the major contributors in biodiversity losses. There are five principal ways in which global warming can affect the future biodiversity losses. (1) alterations in the densities including alterations in the structure and composition of community (2) shifting of the ranges of the species, this shift may be upward in elevation or it can be toward poles (3) changes in behavior such as migration phenology, which is time of life cycle events of an animal, time and place of breeding, and flowering time; (4) variations and changes in the size of body means morphology, and (5) onset of inbreeding depression due to the reduction in genetic diversity.

Predicted loss of habitats by inundation due to rising sea levels is another threat for island and coastal biotas. Reliable information on the effects of different factors on animal fauna is scarce with several assumptions. Nevertheless, there are several local and regional examples and model based predictions that climate change in association with other drivers result in habitat degradation and species loss. In future this will be a major challenge for the conservation of biodiversity. One possible hint can be the disappearance crisis frog and toads(40%, 20 out of 50) the highland forests of Monterverde (Costa Rica), where 40% (20 of 50) of frog and toad species. This loss of amphibian fauna was the result of occurrence of population crashes (in 1987) synchronous with other factors. Most of these crashes were attributed to the changes in local climate when it progressively warmed up and became dry.

The mortality of endemic golden toad (*Bufo periglenes*) in this area was significant casualties. Some researchers suggest that when climate warmed, it withdrew the clouds, and dried up mountain habitats. These changes made amphibians more susceptible to fungal and parasite outbreaks which ultimately succumbed to death. The pathogenic chytrid fungus *Batrachochytrium dendrobatidis*, grows on the skin of amphibians and results in frog mortality. Previously it has been observed that this fungus have become the cause of loss of harlequin frogs (*Atelopus* spp.) in Central and South America. This fungus also reduced frog populations in other parts of the globe too. Presumably warm and dry conditions might have stressed the amphibians and made them more vulnerable to the fungal infection.

Although there is a decline in population size, uncommon events are more prominent when a population is not growing. Further, it has been observed that there is a considerable reduction in competition among individuals at low densities, which can enhance population growth but the ‘Allee effect'—also called the countervailing phenomenon—can push the population toward extinction with the disruption of behavioral patterns which are dependent on the number of individuals present in that particular area or by inbreeding depression, which is one of the genetic threats. Chance events and Allee effects dominate small populations, which reduce the population to their minimum viable size. On degradation of habitat, many species take decades until they perish completely. It has also been observed that some species may recover from the preliminary shock of land clearance and development. However, the lack of food for the animals, breeding sites, and the fact that the population is scattered will ultimately result in the extinction of that species. From this phenomenon, the concept of the living-dead species emerges, which means that these species have to undergo extinction in one way or another. With deforestation and habitat loss, species become extinct, known as extinction debt, which can never be recuperated even if there is a reduction in the rate of deforestation or it is completely halted.

7.3.2 Vulnerability to Extinction

The behavior of an animal, morphology, and its physiology are some of the characters which make a species more susceptible to extinction drivers than others. For example, larger species whose distribution is in a limited area with a specialized habitat are more vulnerable to extinction due to loss of habitat, e.g., Javan rhinoceros, *Rhinoceros sondaicus*. Animals with low densities with specific habitats, which means they are rare in nature, may be more prone to extinction than common species. Species range is another determinant to species loss; larger habitat areas are less prone to extinction than those found in smaller areas. Such trends in species propagation and extinction are troublesome, because those species whose area of domain and distribution is shrinking due to human activities become more prone to climatic changes. Habitat loss reduces the overall space available for animals to live. It diminishes the connectivity between different sub-populations and dispersals, which ultimately lowers the viability of the population. If extinction threats are more rapid and intense, then large-bodied organisms are more prone to extinction than small-bodied organisms.

It is a fact that heavier animals are more threatened than lighter animals. This trend can be further explained in that body size is inversely correlated with population size, which means that larger animals are less abundant in population and are more prone to environmental perturbations because bigger animals need larger areas, higher habitat specificity, greater food intake, and lower reproductive rates. Smaller animals are more vulnerable to habitat loss, while bigger animals are more vulnerable to human pursuits such as hunting. For example, larger parrots, because of their slower reproductive rate, are more vulnerable to overexploitation than finches, which are smaller in size, although fewer numbers of parrots are captured for the pet trade. Other than size of an animal, some morphological features of an animal are also prone to extinction. For instance, in dimorphic species due to the better development of secondary sexual characteristics their adaptation to a changing environment decreases and hence they become more attractive to specimen or pet trade collectors. When abrupt or gradual changes occur in climatic conditions at a rate higher than adaptation capacity of an animal via natural selection then an animal may succumb to death. Animals with specific habitats to live in and special food to eat means they have narrow ecological niches and cannot tolerate the effect of these changes, leading eventually to extinction. Breeding and feeding sites may decrease. These animals are heavily predated, soil is eroded disturbing the habitat, species cannot disperse, nutrients are lost, and all these factors induce stress to the species and result in their total elimination. Those species which are living in open plains contrary to forests are better capable to tolerate disturbed environment with fewer competitors. Evolution favors producing both specialist and generalist species so that both can coexist. A rapid change in habitat and climate renders specialization a modern "curse" in evolutionary terms. The process of foraging can enhance the ability of animals to adapt themselves to changing habitats. There are several studies which say that frugivorous and insectivorous birds are more vulnerable to extinction due to lack of year round availability of their required food. Several hypotheses explain the possible reasons for the disappearance of insectivorous birds with the onset of deforestation and fragmentation of areas. First, deforestation can reduce the required microhabitats (e.g., dead leaves) for these birds, ultimately impoverishing the insect fauna. Second, poor dispersal of insectivorous birds is another reason for their decline. They build their nests near the ground, which are easily approachable by nest predators. Poor dispersal rather than a lack of food is the reason for elimination of insectivorous birds from small fragments. The dispersal in birds and insects is very much dependent on the morphology of these organisms. These are wing loading and intolerance to sunlight when moving out of the forest's umbrella. These species then cannot replace their population within a given area, which means they cannot support sink habitats and produce sub-populations which are not viable. Poor dispersal ability, generally low population densities, and patchy distributions lead to poor maintenance of genetic diversity. The absence of genetic diversity results in inbreeding depression, which further paves the way toward extinction. Nevertheless, the role of life history traits in the decline and extinction is still not yet clear.

7.3.3 Consequences of Extinctions

The consequences of extinction depend on the type of animal species which have become extinct. For example, if pollinators and large predators have become extinct, they have more devastating ecological consequences than those animals which do not fall under this category. It is unfortunate that when large predatory species become rare in tropical communities, it exacerbates the vulnerability of avian fauna to predation. The following example further explains this phenomenon. Predation of large cats—also called jaguars (*Panthera onca*)—is not direct; rather, smaller predators such as medium-sized and small mammals (meso-predators) come under the limiting force exerted by jaguars. These animals then become more abundant when there is a decline in bird populations.

Therefore, if meso-predators become abundant in population they predate eggs and nestlings of the birds. Due to their abundance, this effect is above the average and hence the bird population declines. This 'meso-predator-release' hypothesis is predominantly applicable to mammals, e.g., the release of Australian dingoes, *Canus lupus*, for suppression of the population of cats and foxes. Similarly, coyotes were released in California to control cats. There is a possibility that when the harpy eagle (*Harpia harpyja*) (a large predatory bird) is lost, this might inflict similar effects on the ecosystem. Like terrestrial fauna, similar effects have been observed in the marine environment too. For example, when pelagic sharks were overexploited they in turn increased the population of rays and skates, which ultimately repressed the population of scallops (which have commercial importance).

When extinction disrupts different ecological processes, in turn it induces catastrophic alterations in co-extinctions. Fruiting plants and frugivorous birds are directly linked to each other. Plants support animal nutrition, while animals help dispersal of gametes and pollination in plants. Therefore, habitat degradation jeopardizes two interdependent taxa. The same is true with forest invertebrates. When there is a loss of habitat and degradation, their population declines or they are totally eliminated, hence their essential ecosystem functions are totally halted. Figs are a keystone species in Southeast Asian rainforests. These plants rely heavily on tiny (1–2 mm) species-specific wasps for their pollination. Seed dispersal ability is limited in some wasps. If forests are disturbed, then this will further reduce the density of wasps, with ultimately limited or no pollination in figs. Similarly, dung beetles contribute heavily to the nutrient-recycling processes, dispersal of plant seeds, and risks of diseases from the accumulation of dung. Therefore, extinction of the dung beetle will result in the total cessation of the ecological function that it plays. Animals pollinate almost all flowering plants in tropical rainforests. It has been estimated that one-third of all human food comes from insect pollinated plants in tropical countries. If the population of plant pollinators declines in the forest, this will not only impede plant reproduction in the forests but will also affect reproduction in the adjacent areas. A study in Costa Rica showed that the presence of forests has increased the production of lowland coffee (*Coffea canephora*) by 20% in fields within 1 km of the forest edge due to pollination by honeybees. It is estimated that pollination provided services worth US$60,000 to a 1,100 ha farm in 2000-2003, which emphasizes the importance of conservation of forests directly and conservation of pollinators, e.g., honey bees indirectly.

7.4 POPULATIONS, INDIVIDUALS, AND GENETIC DIVERSITY

A balanced diversity of species hinders the ability of invasive species to intrude. Biodiversity further increases productivity and stabilizes the natural communities or ecosystems. In general, this effect of biodiversity on the ecosystem is called the 'biodiversity effect' (Cadotte, 2013). It has been observed that cumulative productivity of the community that has diverged functions is higher than that in communities where components of that community function are independent of each other. Why does this happen? There are two possible reasons behind this. The First is complementarity, which means that different species living in the same community interact with each other positively. The second is a selection effect, which states that inclusion of the most productive species in the community enhances the production of that community (Cadotte et al., 2009). In this context, it has also been further observed that even diversity within the same species is very useful and improves the fitness of the population (Forsman et al., 2008). The presence of diversity within a population or species is the result of the presence of polymorphisms and personalities (Karpestam et al., 2016). This diversity enhances productivity and the growth rate, and thus stabilizes the community, ultimately enhancing the capacity of the community to continue. This diversity within species further reduces competition for resources within the population. Further, dispersion of species lowers the predation risks, reduces harmful effects, and also reduces the threat of extinction from fluctuating environmental conditions (Wolf & Weissing, 2012).

Nevertheless, it is not yet clear how polymorphism affects the fitness of the population in a certain community. In a population, genetic polymorphism is a common phenomenon, which is maintained in nature by balancing a selection in a population or by a selection – migration balance (Slatkin, 1973). It is predicted that polymorphisms maintained by balancing selection will enhance population fitness (Karpestan et al., 2016). The reason behind this phenomenon is that individuals present in a polymorphic population enjoy the advantage of those which have rare morph advantages, while in case of a monomorphic population none of the individuals can enjoy rare morph advantages. Depending on the ecological and evolutionary context, there is a variation in the structural consequences (Farkas et al., 2016). Further, practical support for the effects of genetic polymorphism on population fitness is lacking and hence the presence of a diversified effect of genetic polymorphism is disputed (Forsman, 2016). Interestingly, when the food is nutrient limited, negative frequency-dependent selection is observed, which intensifies the resource competition among the larvae. Nevertheless, when the food is nutritious, this competition diminishes immediately. This suggests that foraging behavioral polymorphism is maintained by a negative frequency-dependent selection (Fitzpatrick et al., 2007). If the polymorphic conditions prevail in populations of organisms, they will increase the growth rate of the population beyond a single generation because of the number of females present and their size in that particular population group. Both the number and size of females can increase the number of progeny and biomass of the preceding and/or following generations (Tantawy & Vetukhiv, 1960). In the females of damsel fly species, polymorphic populations increased population density and productivity, making females more fecund and increasing the rate of oviposition in females (Takahashi et al., 2014). It has further been observed that there is an increase in the survival of polymorphic population of *D. melanogaster,* which is due to efficient utilization of the available resources. It has been yet further suggested that polymorphic conditions increase the carrying capacity of a population and persistence of the population in the environment with temporary fluctuations.

7.4.1 Species of Special Concern

The biota of Pakistan are also on the decline and there is no reason for their exemption from this phenomenon. At least four mammals, including tigers (*Panthera tigris*), swamp deer (*Cervus duvauceli*), lions (*Panthera leo*), and the Indian one-horned Rhinoceros (*Rhinoceros unicornis*) have disappeared within the last 400 years. In recent decades, there is the possibility that four other species of mammals, namely the Asiatic cheetah (*Acinonyx jubatus venaticus*), the Hangul (*Cervus elaphus hanglu*), and the Indian wild ass (*Equus hemionus khur*) have become extinct in recent decades (Roberts 1977). The fourth of these species, the blackbuck (*Antelope cervicapra*), has been bred in captivity, although it is extinct in the wild and has been listed as extinct.

7.4.2 Animal Species Which Are Internationally Threatened

Thirty-seven species and 14 sub-species which are internationally threatened or near-threatened mammals found in Pakistan have been included in the latest Red List of THE IUCN (1996).

The status of these species can be categorized as follows: critically endangered, 2; endangered, 9; vulnerable, 11; near threatened, 23; data deficient, 5; conservation dependent, 1. The critically endangered mammals are the Baluchistan black bear (*Ursus tibetanus gedrosianus*) and the Chiltan goat (*Capra aegagrus chiltanensis*). Species-wise, the list of threatened and near-threatened mammals includes bats (Chiroptera), 4 species; primates, 2 species; dogs (Canidae), 3 species; cats (Felidae), 3 species and 2 sub-species; otters (Mustelidae), 1 species; and bears (Ursidae), 1 specie s and 1 sub-species. This list has been further expanded to other mammals too which are: cetacean, 3 species; Artidactyla, 1 species and 1 sub-species; Cervid, 1 sub-species; Bovidae (antelopes, goats, sheep, etc.), 11 species and 9 sub-species; pangolin (Manidae), 1 species; and Rodentia, 7 species. Some researchers (Collar et al., 1994) and the IUCN (1996) state that there are 25 species of birds which are declared internationally threatened. The status of these birds can be categorically described as: critically endangered, 1; endangered, 2; vulnerable, 22; and near-threatened, 17 (Steidl & Anthony, 1996).. The lesser florican (*Eupodotis indica*) is a critically threatened bird.

There are 10 species of reptiles found in Pakistan which have been declared internationally threatened. Their status can be described as follows: endangered, 3; vulnerable, 3; near-threatened, 3; and data deficient, 1.

According to the IUCN (1996) data, none of the amphibian species found in Pakistan is internationally threatened. The latest Red Data Book IUCN (1996) lists of some of those species which are data deficient. They are one species of fish, the spiny eel (*Macrognathus aral*) and one species of invertebrate, a butterfly (*Hyles hippophaes*).

7.4.3 Animal/plant Species of National Concern

The availability of lists of those animals which have been declared internationally threatened is very scanty. For example, of birds present in Europe, 5% of them have the status of global conservation concern. Another 33% are unfavorable, although they are of not much importance and hence do not matter from a conservation point of view (Tucker & Heath, 1994). Insufficient information on wildlife of Pakistan is available, and so it is not clear which species/populations are declining or have already declined considerably. One thing is very clear: Degradation and fragmentation of habitats are the causes of this decline. When a habitat is fragmented, the population of species faces isolation and scattering. This activity decreases genetic diversity and exposes them to increased risk of extinction (UNEP, 1995). Preparation of a list of nationally threatened animal species has been attempted. One list tells us that 500 plant species are nationally rare or threatened. For Pakistan, there is no comprehensive compilation of different species of national concern. If prepared, this list would cover animals/plants which are: (1) nationally rare and declining; and (2) nationally rare but not declining but are also at risk due to population fluctuations, natural catastrophes, indiscriminate killing, etc. In addition to the species mentioned above, this list will also include those animals which are highly locally distributed and those animals which are widely distributed and common in presence although on a significant decline.

7.5 THREATS TO BIODIVERSITY

Deforestation poses specific threats to biodiversity which have been estimated at 1% annually. Agriculture practices, fishing, overgrazing, soil erosion, and rampant hunting are some other threats which result in biodiversity decline and loss. It has been estimated overall that 12% of plant fauna is threatened. There is insufficient data to make realistic estimates of the actual status of flora and fauna. Major threats are always from human beings, which involves a variety of activities. These threats are described below.

7.5.1 Population Growth

When the population of human beings increases, it adds pressure on natural resources in order to maintain living standards, which jeopardizes the stability of biodiversity. When the economy of any country develops, it widens the inequality gap and with the consequence that the poor then have to depend heavily on natural resources for a living. Different development models have limited applications and can become incompatible with the factual circumstances. Sustainability of biodiversity can become unstable and tends toward decline.

7.5.2 Irrigated Agriculture

After a population explosion, agricultural initiation and development is another threat to the viability of biodiversity. Riverine habitats are destroyed, mangrove forests are removed, and animal fauna quickly disappear. Riverine forests that hosted plants like obhan and animals like jungle cats, fishing cats, gray and black partridges, and hog deer have now lost their ability to support such flora and fauna due to destructive human activities. Similarly, many fish live in mangroves and their elimination means the death of several fish species. Riverine forests and mangroves have been considered as dangerous ecosystems because their absence translates into the disappearance of unique plant and animal species. River habitats demand the release of water at regular intervals and if that water is diverted, then the habitat will be lost, which will take several plant and animal species with it.

7.5.3 Hunting and Poaching

Culturally, Pakistan has a rich population of hunters, which started with the Moghuls. Hunting is considered to be an important recreational activity, especially for the rich. The consequences are that wildlife is diminishing at an accelerated rate. Some animals are hunted for their skins, e.g., reptiles including various lizards, crocodiles, and snakes. In addition to reptiles, larger mammals are also hunted for the same purpose. With the decline in the population of large cats, the numbers of porcupines, jackals, and wild boars have been boosted. The decline in numbers of large cats has disturbed the ecosystem balance and hence these vertebrate pests have flourished uncontrollably, inflicting heavy losses to agriculture. The excessive population of wild boars has trampled snake nests and fauna, which in turn has promoted the population of mice, which became a nuisance for agriculture and its products.

7.5.4 Deforestation and Loss of Habitat

The disappearance of habitat and competition for grazing animals is a far greater threat to wildlife than hunting and poaching. There is an annual 1% decrease in the closed canopy forest in KPK

(Pakistan). The clearing of different areas and commercial logging have increased pressure on forests, which in turn exerts pressure on the survival of the wildlife present in these forests. Excessive grazing by animals, indiscriminate cutting of trees, and clearing of forests for development are some of the other causes of deforestation.

7.5.5 The Scale of the Human Enterprise

All the above-mentioned factors which affect wildlife are just an estimate or an approximation. In fact, the size of the existing human population and the growth rate of this population, known as the scale of human enterprise, are the major drivers toward the decline of biodiversity (Ehrlich, 1995).

7.5.5.1 Population Size and Growth

It was estimated that human population touched the figure of 6.1 billion in 2000. Since then, the world's population continues to increase at an accelerated rate. However, the population of great apes, our closest relatives, is between 10,000 and 25,000, the bonobo Pan paniscus, 100,000–150,000, the chimpanzee *Pan troglodytes*, 40,000–65,000, the gorilla *Gorilla gorilla* and the orangutan *Pongo pygmaeus* being approximately 38,500 (Cincotta & Engelman, 2000). Due to the pivotal role of human population on the sustainability of biodiversity, it has been observed that there is a marked positive correlation between the human population present at a particular place and the animal species found there. It has been observed in sub-Saharan Africa (Balmford et al., 2001) that both animal species and human beings closely relate in primary productivity and seek similar places and habitats for living and multiplication. If we look at the human distribution pattern, 1.1 billion human individuals have been distributed in 25 places on the globe which are indeed biodiversity hotspots. These areas are very risky and threatening for other forms of life to thrive (Cincotta & Engelman, 2000; Cincotta et al., 2000).

Average population density of human beings is 42 per km^2, while in the areas which are also called population hotspots the density of people is about 73 per km^2. There is a direct relationship between the number of people and the intensity of the degradation of habitats. With the rise in population, the conflict between people and biodiversity continues to intensify. If the population continues to increase, then this relationship will break down completely. In 1970, it was established that the non-availability of data regarding native plant species was due to the increase in human population in these areas (Thompson & Jones 1999). It was further observed that there was a decrease in large vertebrates where human beings were in abundance, despite their presence in protected areas (Walsh et al., 2003). It is expected that nature and the intensity of this conflict will continue to grow as the human population is increasing at an annual rate of 1.2%, directly impacting animal populations.

There is an annual increase of 13,380 million individuals or, to put it another way, the number of people in the world increases by 250,000 every day. It is expected that by the year 2050 the human population will be between 7.9 billion and 10.9 billion (with a medium variant of predictions of 9.3 billion) (United Nations, 2001). This dramatic increase in human population has been observed over the last 200 years, although it was very slow for most of human existence.

Looking at population increases globally, population growth was higher in 19 of the total biodiversity hotspots. It has further been observed that in the hotspots of developing countries human population will grow further for several decades (Cincotta et al., 2000). Silphion, *Ferula historica*, a herb in the carrot family (Cincotta & Engelman, 2000), is a good example of interaction between human population growth and this herb. It grew in abundance on the coast of Libya near Cyrene and was highly valued because of its oral contraceptive quality. In Cyrene's trade, it was once considered as a principal commodity. People tried to cultivate this herb but it failed to thrive. Consequently, it was overharvested and it vanished in the second or third century AD.

7.5.5.2 Primary Production

Human beings are responsible for consumption or destruction of potential terrestrial net productivity up to approximately 35–40% of use (Pimm, 2001). Total productivity in aquatic systems is only 8% of the primary production. Nevertheless, if we also consider both freshwaters and nearshore areas, the productivity is much higher and approaches the value of terrestrial productivity (Pauly & Christensen, 1995).

7.5.5.3 Water

Humanity uses water from two sources: (1) terrestrial evapotranspiration; and (2) runoff. The quantity from terrestrial evapotranspiration is more than a quarter of the 69,600 km^3 yr^{-1}, while from runoff it is half of the total runoff (12,500 km^3 yr^{-1}). This share is from those water sources which are geographically and temporally accessible (Postel et al., 1996). At a global level, 42% of water consumption goes to agriculture and 14% goes to industry. Freshwater is limited but its requirements are on the rise for agricultural and economic developments. Thu, it can be seen that this decrease in the availability of water also inflicts hugely detrimental effects on biodiversity.

REFERENCES

Agarwal, N.K., Singh, G. & Rawat, U.S. (2014). Present Status and Threats to the Ichthyofaunal Diversity of a Snow Fed River Nandakini in Central Himalaya (Garhwal), India. In: Uttarakhand Disaster: Contemporary Issue of Climate Change and Development with Holistic Approach, eds. Rawat, U.S. & Semwal, V.P., pp. 173–182. Winsar Publication, Dehradun, India.

Agarwal, N.K., Singh, G. & Singh, H. (2011). Present Status of Icthyofaunal Diversity of Garhwal Himalayan River Bhilangna and Its Tributaries with Reference to Changing Environment. Environment Conservation Journal 12(3), 101–108.

Ayoade, A.A., Agarwal, N.K. & Chandola-Saklani, A. (2009). Changes in Physico-Chemical Features and Plankton of Two Regulated High Altitude Rivers, Garhwal Himalaya, India. European Journal of Scientific Research 27(1), 77–92.

Balmford, A., Moore, J.L., Brooks, T., Burgess, N., Hansen, L.A., Williams, P. & Rahbek, C. (2001). Conservation Conflicts across Africa. Science 291, 2616–2619.

Balmori, A. & Hallberg, Ö, (2007). The Urban Decline of the House Sparrow (Passer domesticus): A Possible Link with Electromagnetic Radiation. Electromagnetic Biology and Medicine 26(2), 141–151. https://doi.org/10.1080/15368370701410558.

Bélanger, L. & Bédard, J. (1990). Energetic Cost of Man-Induced Disturbance to Staging Snow Geese. Journal of Wildlife Management 54, 36–41.

Cadotte, M.W. (2013). Experimental Evidence That Evolutionarily Diverse Assemblages Result in Higher Productivity. Proc. Natl Acad. Sci. USA 110, 8996–9000. https://doi.org/10.1073/pnas.1301685110.

Cadotte, M.W., Cavender-Bares, J., Tilman, D. & Oakley, T.H. (2009). Using Phylogenetic, Functional and Trait Diversity to Understand Patterns of Plant Community Productivity. PLoS ONE 4, e5695. https://doi.org/10.1371/journal.pone.0005695.

Chapin, F.S. III, Zaveleta, E.S., Eviner, V.T., Naylor, R.L., Vitousek, P.T., Reynolds, H.L., Hooper, D.U., Lavorel, S., Sala, O.E., Hobbie, S.E., Mack, M.C. & Diaz, S. (2000). Consequences of Changing Biodiversity. Nature 405, 234–242.

Cincotta, R.P. & Engelman, R. (2000) Nature's Place: Human Population and the Future of Biological Diversity. Population Action International, Washington, DC. (An interesting analysis of the relationship between biodiversity hotspots and human population.)

Cincotta, R.P., Wisnewski, J. & Engelman, R. (2000). Human Population in the Biodiversity Hotspots. Nature 404, 990–992.

Collar, N.J., Crosby, M.J., & Shattersfield, A.J. (1994). Birds to Watch 2: The World List of Threatened Birds: The Official Source for Birds on the IUCN Red List. Birdlife Conservation Series no. 4. Cambridge: Birdlife International, Washington, D.C.

Dumont, E., (2012). Estimated Impact of Global Population Growth on Future Wilderness Extent. Earth System Dynamics Discussions 3, 433–452. https://doi.org/10.5194/esdd-3-433-2012

Ehrlich, P.R. (1995). The Scale of the Human Enterprise and Biodiversity Loss. In: Extinction Rates, eds. Lawton, J.H. & May, R.M., pp. 214–226. Oxford University Press, Oxford.

Farkas, T.E., Mononen, T., Comeault, A.A. & Nosil, P. (2016). Observational Evidence That Maladaptive Gene Flow Reduces Patch Occupancy in a Wild Insect Metapopulation. Evolution 70, 2879–2888. https://doi.org/10.1111/evo.13076

Fernandez, C. & Azkona, P. (1993). Human Disturbance Affects Parental Car of Marsh Harriers and Nutritional Status of Nestlings. Journal of Wildlife Management 57, 602–608.

Fitzpatrick, M.J., Feder, E., Rowe, L. & Sokolowski, M.B. (2007) Maintaining a Behaviour Polymorphism by Frequency-Dependent Selection on a Single Gene. Nature 447, 210–212. https://doi.org/10.1038/nature05764.

Forsman, A. (2016). Is Colour Polymorphism Advantageous to Populations and Species? Mol. Ecol. 25, 2693–2698. https://doi.org/10.1111/mec.13629.

Forsman, A., Ahnesjo, J., Caesar, S. & Karlsson, M. (2008). A Model of Ecological and Evolutionary Consequences of Color Polymorphism. Ecology 89, 34–40. https://doi.org/10.1890/07-0572.1.

Gaynor, K.M., Hojnowski, C.E., Carter, N.H. & Brashares, J.S. (2018). The Influence of Human Disturbance on Wildlife Nocturnality. Science 360(6394), 1232–1235. https://doi.org/10.1126/science.aar7121.

Global Biodiversity Strategy (GBS). (1992). Guidelines for Action to Save, Study and Use Earth's Biotic Wealth Sustainable and Equitably. World Resources Institute (WRI), World Conservation Union (IUCN) United Nations Environmental program (UNEP), IUCN, Gland, Switzerland, pp. 30–100.

Green, R.E., Newton, I., Shultz, S., Cunningham, A.A., Gilbert, M., Pain, D.J. & Prakash, V. (2004). Diclofenac Poisoning as a Cause of Vulture Population Declines across the Indian Subcontinent. Journal of Applied Ecology, 41(5), 793–800. https://doi.org/10.1111/j.0021-8901.2004.00954.x.

Holmes, T.L., Knight, R.L., Stegall, L. & Craig, G.R. (1994). Responses of Wintering Grass-Land Raptors to Human Disturbance. Wildlife Society Bulletin 21, 461–468.

Johnson, C.J., Boyce, M.S., Case, R.L., Cliff, H.D., Gau, R.J., Gunn, A. & Mulders, R. (2005). Cumulative Effects of Human Developments on Arctic Wildlife. Wildlife Monographs 160, 1–36.

Karpestam, E., Merilaita, S. & Forsman, A. (2016). Colour Polymorphism Protects Prey Individuals and Populations against Predation. Scientific Reports 6, 22122. https://doi.org/10.1038/srep22122.

Lichstein, J.W., Simons, T.R. & Franzreb, K.E. (2002). Landscape Effects on Breeding Song-Visitor Bird Abundance in Managed Forests. Ecological Applications 12, 836–857.

Lovej, T.E. (1980) A projection of species extinctions. In: Council on Environmental Quality (CEQ), The Global 2000 Report to the President, vol CEQ. Washington, DC, pp. 328–331.

Prance, G.J. (1991), What Is Ethnobotany Today? Journal of Ethnopharmacology 32, 209–216.

Mann, S.L., Steidl, R.J. & Dalton, V.M. (2002). Effects of Cave Tours on Breeding Cave Myotis. Journal of Wildlife Management 66, 618–624.

McGarigal, K., Anthony, R.G. & Isaacs, F.B. (1991). Interactions of Humans and Bald Eagles on the Columbia River Estuary. Wildlife Monographs 115. Allen Press. URL. www.jstor.org/stable/3830569.

Millennium Ecosystem Assessment. (2005). Ecosystems and Human Well-Being: Biodiversity Synthesis. World Resources Institute, Washington, DC.

Muralidharan, S., Dhananjayan, V., Risebrough, R., Prakash, V., Jayakumar, R. & Bloom, P.H., (2008). Persistent Organochlorine Pesticide Residues in Tissues and Eggs of White-Backed Vulture, Gyps bengalensis from Different Locations in India. Bulletin of Environmental Contamination and Toxicology 81(6), 561–565. DOI 10.1007/s00128-008-9529-z.

Norse, E.A. & McManus, R.E. (1980). Ecology and living resources biological diversity. In: Environmental Quality 1980: The Eleventh Annual Report of the Council on Environmental Quality, pp. 31–80. Washington, D.C.: Council on Environmental Quality.

Pauly, D. & Christensen, V. (1995). Primary Production Required to Sustain Global Fisheries. Nature 374, 255–257.

Pimm, S.L. (2001). The World According to Pimm: A Scientist Audits the Earth. McGraw-Hill, New York.

Pimm, S.L., Russell, G.J., Gittleman, J.L. & Brooks, T.M. (1995). The Future of Biodiversity. Science 269, 347–350.

Prakash, V. (2007). Recent Changes in Populations of Resident Gyps Vultures in India. Journal of Bombay Natural History Society 104(2), 129–135.

Rawat, G.S. (1998). Temperate and Alpine Grassland of the Himalaya: Ecology and Conservation. Parks 8(3), 27–36.

Rawat, U.S. & Semwal, V.P. (2014). Uttarakhand Disaster: Contemporary Issue of Climate Change and Development with Holistic Approach, Winsar Publication, Dehradun, India.

Robert, J.S. & Brian, F.P. (2006). Assessing the Effects of Human Activities on Wildlife. The George Wright Forum 23(2), 50–58.

Roberts, T.J. (1977). The Mammals of Pakistan. Ernest Benn Limited, London.

The high latitudes of northern hemisphere show the highest warming. Compared to the ocean, this level of warming is significantly higher on the land. With the passing number of days, the heat index above 100°F might increase if the temperature is increasing persistently.

By the end of the century, central America may experience 60 to 90 additional days a year with a heat index of more than 100°F. Prevalent heat waves will last longer if there is an increase in the average global temperatures. There is also a possibility that, with the rise in global temperatures, illnesses and deaths related to heat will also rise. Accordingly, it can confidently be said that there will be a decrease in cold extremes and so with the cold-related deaths. A comparison of recorded high temperatures with recorded low temperatures shows a ratio of 2:1. The ratio of mid-range emission scenarios can possibly rise in the middle of the century by 20:1 and up to 50:1 at the end of the century.

8.1.3 Expected Changes in Precipitation

It is projected and expected that global warming will intensify the regional differences in precipitation which already exist. It is expected that dryness will increase in dry areas and they will become even dryer. However, precipitation will increase in wet areas and they will become even wetter than their present wet level. The reason for these changes is that warmer temperatures will mean that more water is evaporated from the soil, plants, lakes, and oceans. Both theory and observations agree at a common point that there will be an increase in water vapors of up to 7% for each degree rise in temperature 1°C (1.8°F).

Evaporation of water raises the moisture level in the air. This increased moisture provides the basis for snow and rain. Nonetheless, this activity results in the drying of the land's surface. This drying of the land will negatively impact these regions and will result in drought conditions. It has further been observed and is expected that possibly extreme rainstorms will intensify by 5–10% for each 1°C (1.8°F) rise in the temperature of the globe. These rainstorms will intensify more in the Tropics where rains are always the heaviest in the world. Annual streamflow will be affected by changes in precipitation. This streamflow will be equal to the amount of water coming from runoff (which usually comes from rain or melting of snow which leads to rivers and creeks) (Fig. 8.1).

This decline in water resources is mainly attributed to the increased rates of evaporative losses which are always due to higher temperatures. These changes cannot be due to increased evaporation caused by higher temperatures. Nevertheless, this effect cannot be counterbalanced by low and high concentrations of precipitation. Nevertheless, it has been observed that if the rate of evaporation is increased, it will dry out the land surface, which in turn will also result in a reduction in precipitation in that area.

8.1.4 Climate Change Impacts Sea Ice and Snow

As most of the population is living in low elevation regions, if there is a rise in ocean levels that becomes a matter of concern. Sea-level rise is directly related to human-induced increases in the emission of greenhouse gases. Hence, it is projected that, due to this effect, by 2100 there will be an estimated increase of 0.5–1.0 meter (20–39 inches) of mean sea-level rise. Scientists are also of the view that melting of the sea ice could increase sea levels further. The reason for this unexpected rise in sea levels is that melting of glaciers and ice sheets is happening at a more accelerated rate than expected.

Presently, the sea level rises are at the top ranges of those observed and projected in 1990). When we talk about the magnitude of threats which rising seas and floods can pose to particular coastlines, it has been observed that if compared with 1990 ocean rises on average 0.5 meters (20 inches), then flooding due to this rise globally means that between 5 million and 200 million people will be displaced.

There could be permanent displacement of up to 4 million people, which will erode more than 250,000 square kilometers of wetland and 98,000 square miles of dry land. Relocation of people in

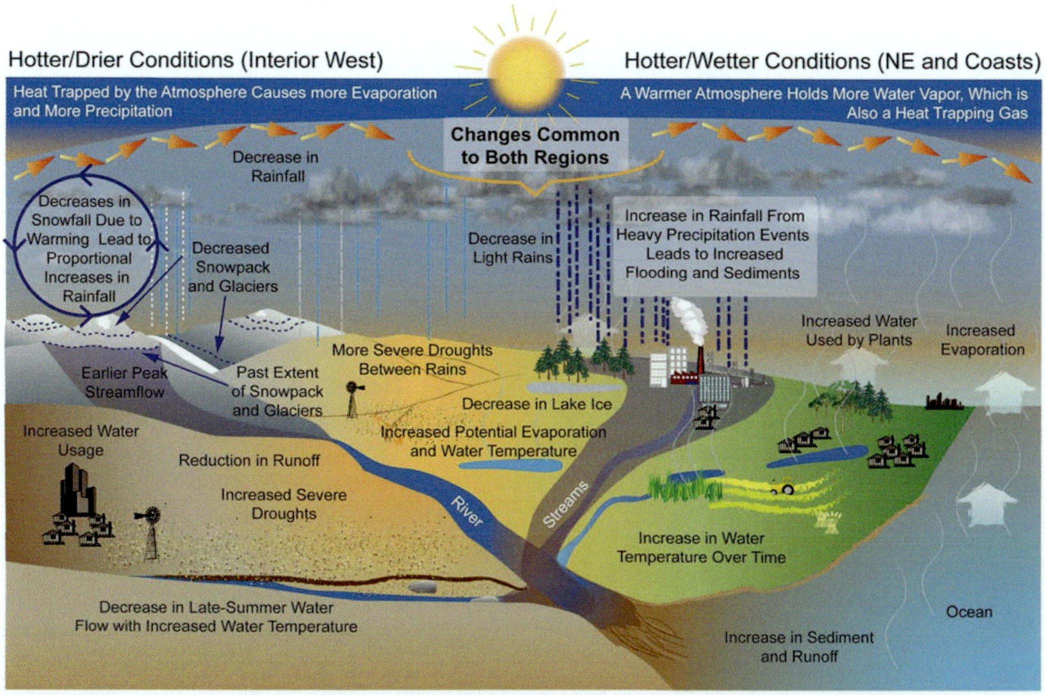

FIGURE 8.1 Climate effects on water resources. (*Source:* USGCRP 2009: 19january2017snapshot.epa.gov (open source).)

towns along the coast of Alaska is already underway and the reduction in sea ice and the melting of permafrost is already battering and eroding the shoreline. It is expected that if there is rise in seal level of up to 1 meter, then it will threaten coastal erosion of many parts of the U.S. coastline.

8.1.5 Projected Effects of Sea Level

If there is a rise in sea level of up to one meter (3.3 feet), coastal flooding can be expected in these areas. If sea level rises up to six meters (19.8 feet), those areas shown in red will also be susceptible.

Climate change also has a bearing on other aspects in addition to oceans, etc. Under the changing climatic conditions, all the organisms (marine or terrestrial; cold or warm-blooded) will try to adapt themselves to the changing conditions and if they fail to do so then they will move to those locations which are more favorable for them. Nevertheless, harsh conditions will force the animals to move out of their jurisdiction. Those animal species which are living on the mountain tops and are adapted to cold conditions are more prone to changing climatic conditions. Seasonal changes disrupt the life cycle events of plants and animals. These events, including breeding, hatching, and blooming, will cause mismatches in the timing of the seasons. This activity will in turn disrupt the feeding levels and behavior of animals, as well as to the intensity and nature of pollination of plants and important aspects of food which play a pivotal role in the timing of the seasons (Fig. 8.2).

Roads, land use, and sometimes competition and interaction with other species (human infrastructural barriers) restrict the movement of species and their ability to adapt to prevailing changes. Circulation changes in the ocean are the key factors which impact ecosystems. Less mixing of warm waters has been observed in cooler waters. In this surface, marine life is separated from the nutrient rich media found underneath. Resultantly, this reduces phytoplankton productivity, which is the basis of food present in oceans.

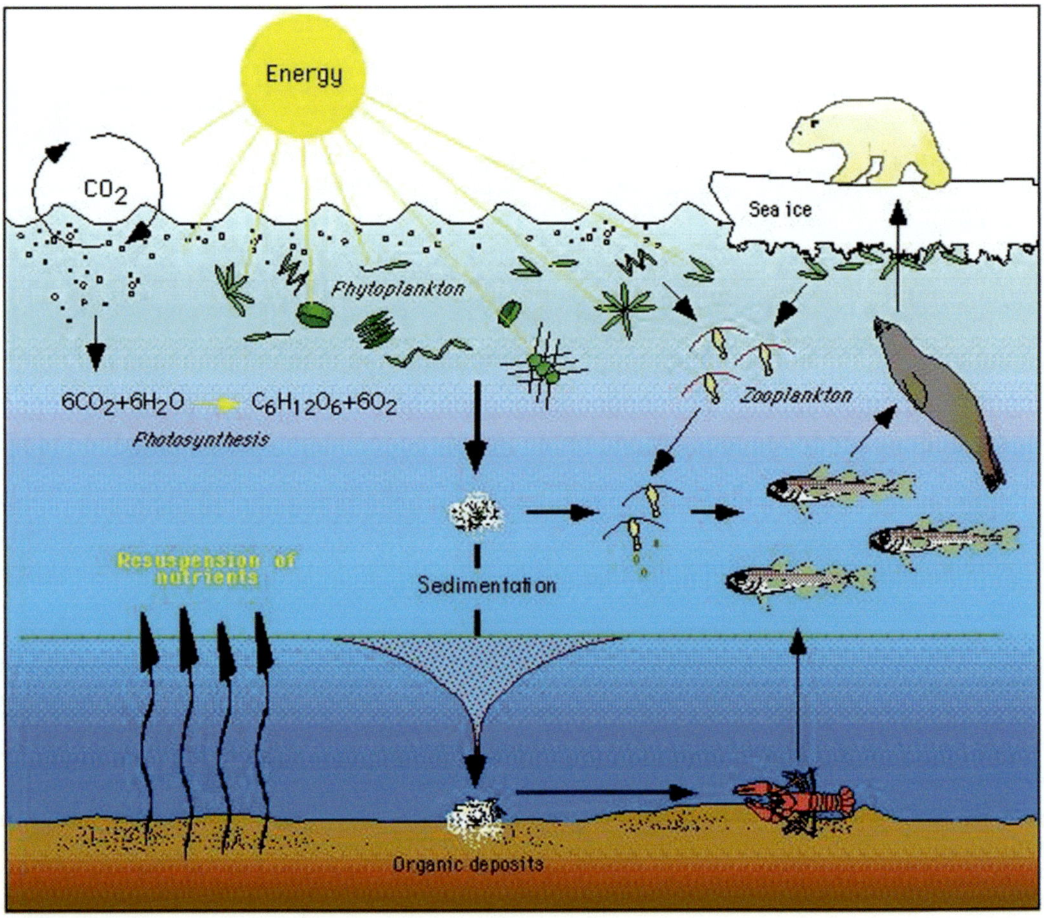

FIGURE 8.2 Climate impacts on ecosystems. (Courtesy of National Oceanic and Atmospheric Administration (NOAA, 2011). (open source).)

This problem caused by changes in climate will be more severe in the Tropics and subtropics. It is expected that there will be losses in sea ice and vertical mixing of water could increase in temperate and polar waters. This persistent warming of oceans will simultaneously push the habitat ranges of different species toward the poles. This warming of oceans will bring about chemical changes in the water with a decline in oxygen at surface waters, creating dead zones. Those species which are high on food chains will perish in this dead zone due to a lack of oxygen.

At a global level, climate change is stressing farming and ultimately threatening food security. In the atmosphere, although higher levels of carbon dioxide facilitate plant growth, they do not ensure additional food production. There is quick growth of crops under higher temperatures. Under such conditions, however, plants grow quickly but they have a limited time to bear enough grain. This reduces the growth period of plants, giving less time to plants to produce adequate grain. In addition, elevated temperature peaks will damage crops and will bring other hazards too. Different crops are affected by changes in climatic effects differently, so the variations will be present in different regions too. If there are increases in carbon dioxide but temperatures are moderate it will increase precipitation. This precipitation could be beneficial to pasture lands and crops located in the middle to higher latitudes. Nevertheless, there will be a reduction in production in those areas which are located in low-latitude areas. Further, those regions which are heavily dependent on the waters from glacier melt the heavy snow packed underneath.

In the current scenario, policy-makers need to address, first, what is the acceptable level of risks due to the warming effects of climate and, second, what is the safe level of emissions of gases to keep the environmental temperature within an acceptable limit.

8.1.6 Science Inform Emission Choices

To stabilize the environment, there should be a considerable reduction in the emissions of carbon dioxide gas.

Although there are several gases and/or other factors which affect climate, among all of them carbon dioxide bears the greatest importance and leaves long-lasting effects on climate changes-. It is not a question of what we should do without the emission of gases but what matters is how smartly we produce and use energy. Take the USA, for example. Its population is only 5% of the total world population, but it is already emitting half of the total CO_2 in the atmosphere caused by human activities. Currently, the USA's share in the emission of greenhouse gases is roughly 20% of global CO_2 emissions.

8.1.7 Choices for the Emission of Greenhouse Gases

Our major choice to reduce greenhouse gases is an efficient use of energy. There are several ways in which we can improve energy efficiency, e.g., devising and using efficient methods of heating, cooling, lighting, and insulating our buildings. Upgrades and improvement in industrial equipment will increase its efficiency and will produce fewer greenhouse gases. We should purchase vehicles and home appliances which are energy efficient. Use of low carbon dioxide producing sources is another way to reduce greenhouse gas emissions. In this case, we can use natural gas instead of coal and oil. Similarly, use of nuclear power and other renewable energy sources like solar, wind, geo-thermal, and hydropower should be encouraged. Carbon dioxide can be directly captured and sequestered from the atmosphere. Capturing and confiscation of carbon dioxide directly from factories and power plants can also control excessive carbon dioxide from the air. Management of forests and land should be exercised in such a way that they use more carbon for their internal processes. Control of emissions of carbon dioxide or to control its concentration in the air is very much dependent on, first, the behavior and consumer choices of individual households, and, second, the nature and levels of investments employed in the private sector.

All governments at different levels can play their role to influence key players involved in emissions and the control of gases by their use of incentives and effective policies. Speaking generally, there are four major areas where we can work to control the emission of greenhouse gases, ultimately saving our environment. First, we must devise mandates and regulations to control gas emitters directly and by adopting automobile fuel economy standards, standards involved in the efficiency of appliances, fulfilling the requirements of labels, allotting definite codes to buildings, development of low carbon or renewable carbon portfolio standards for generation of electricity etc. Second, we must devise the policy of imposition of carbon taxes or the adoption of a cap and trade system. Third, we must initiate a loaning process and provide subsidies permitting users to reduce emissions, and educate people about the harms of these emissions and promote incentives and initiatives that people should reduce emissions voluntarily, Fourth, and finally, we should fix prices and costs for each and every emission into the atmosphere.

8.1.8 Reduction of Other Warming Agents (instigated by Human Beings)

There is a possibility of reduction of other greenhouse gases (e.g., nitrous oxide, methane, and industrial gases such as hydrofluorocarbons), except for carbon dioxide. These gases comprise about 15% of the total gases emitted from different sources and they are worse for the climate than carbon dioxide. Carbon dioxide contributes much than other gases because it is abundant in nature and has a long life time.

8.1.9 Choices for the Mitigation of Climate Impacts

In several countries, adaptation measures are underway to protect their resources, people, and infrastructure. All these measures are heavily dependent on the history of variations in climate when it was stable. Adaptation efforts are hampered by a lack of solid information about limitations of different responses, their potential, costs, and benefits. Choices for mitigation are always adapted because the impacts of climate change are very diversified and most of the animals and plants are vulnerable to its effects. Second, there is little research which can help and guide us which focuses on the ways and measures to adapt to the changing climatic conditions. Presently, low cost strategies are employed which can offer short-term benefits that can amicably reverse poor adaptation practices and policies.

8.1.10 Uncertainties About the Risks of Climate Change and Need to Take Actions

Comprehensive and productive dealing with uncertainties and its risks of the climatic change will not be possible for future research. Inherent complexities that a climate system carries in it and many other prevalent factors like behavioral, economic, and technological whose prediction is difficult in future are associated with uncertainties and risks. Nevertheless, uncertainty does not mean we should stop all further actions; much has been done on the effects of climatic changes on agriculture, livestock, forests, and other animals, and much is in progress. A few important areas are described below.

8.1.10.1 Food Security

In the coming three or four decades, the Himalayan and Karakorum glaciers will deplete if more water flows into the Indus River system. There will be A reduction of 30–40% in its current potential. This depletion will threaten food and the livelihoods of those who depend on the Indus Basin irrigation system, which will be in addition to the 95% of the total irrigated land under the direct impact of this disaster (Porter et al., 2014). Since 1952, 4,544 ha of agricultural land along Humbas Wali creek has already been destroyed. Up to 1960s when Indus River was flowing regularly this area which includes Haji Mauruf Jat and Goth Qabul Dablo was irrigated through Ochito River.

8.1.10.2 Public Health

Climate change may increase the incidence of several diseases and many other vector-borne diseases. The reason is that climate change increases the temperature of the atmosphere following heavy rainfall, which becomes the cause of these diseases. Heavy rains and the accumulation of unnecessary water has facilitated uncontrolled breeding of mosquitos, which affected 20,000 people in 2011. Facing that challenge in 2011, the province of Punjab is well aware of that menace and has managed to control the spread of mosquitos. However, the situation in other provinces is not so good. Though Malaria declined significantly but a related mosquito caused dengue fever surged up swiftly (UN OCHA Pakistan, 2022). During 2022, a total of 25 932 confirmed dengue cases and 62 deaths (CFR 0.25%) were reported in Pakistan, with 74% of these cases reported in the month of September alone.

8.1.10.3 Water Security

According to a World Bank report, Pakistan is a water-stressed country and it is predicted that water availability will be less than 1,000 m^3/y per person by 2035. This indicates that there is an extreme shortage of freshwater resources in Pakistan. Rainfall (50 million acre feet) by monsoon, westerly winds, and river inflows (141 million acre feet) is the primary source of water in Pakistan. However, the Indus River system also receives water from glaciers and snowmelt from the Hindukush-Karakoram-Himalayas. By 2030, the shortage of water due to climate change will affect agriculture production and aggravating the insecurity of food. This state of affairs will intensify transboundary conflicts.

8.1.10.4 Coastline's Inundation

When there is rise in sea level this will inundate the coastline. Salt water will encroach into the Indus delta. It will threaten mangroves and coral reefs, with the prediction that fish breeding grounds will also be disrupted. An increased intensity of cyclones combined with a rise in sea levels will threaten life and property in coastal areas. This state of affairs will be a high risk for Karachi, the southern coastal mega city, and other coastal areas of the Sindh-Makran coast.

8.1.10.5 Rise in Sea Level

Rising sea levels have been observed along the coast of Pakistan. This rise in sea levels will erode regions located in and around the coast and will create flooding in the coastal plains and wetlands. In addition, it will inundate deltaic plains and will salinize aquifers and soils, ultimately destroying habitat for plants, fish, and wildlife.

8.1.10.6 National Security

Intense weather, drought, storms, and floods can destabilize a country in several ways. In recent decades, recurring droughts and floods have been experienced in Pakistan. Moreover, harsh weather can incapacitate the managerial functions of a government or state.

8.2 CLIMATE CHANGE AND ECOSYSTEM STRUCTURE AND FUNCTIONING

Climate primarily controls the rate of different processes on earth, ranges of different species, and distribution of ecosystems. Ecosystem patterns and processes (such as rates of primary productivity or the input–output balance of chemical elements) are controlled by multiple factors. Due to the complexities of these factors, the response of ecosystems toward a changing environment will also be complex. Respiration in an ecosystem and balance of primary production determine whether a particular forest is serving as a source of carbon sink or not. This pattern of forests determines the response to climate change. The occurrence of physical changes in an ecosystem (e.g., flood and drying regimes in streams and rivers, thermal stratification patterns in lakes and oceans, or intensification of hydrological cycles across large basins) ultimately affects the structure and function of an ecosystem. This change in ecosystem in turn affects the economy of the country, as well as its inhabitants. It has been observed that changes in overall average conditions of the ecosystem are less severe than changes in timing, which can impose greater impacts on society and costs. Therefore, to address such issues action plans and management strategies are designed in such a way that they can handle predicted extreme environmental changes, as well as seasonal changes.

8.2.1 Climate Change Impacts on Plant and Animal Biomes

The universal impacts of climate change affect the productivity of ecosystems and their ability to process chemical elements. Higher organic inputs, rapid warming, and loss of sea ice reduce the productivity of lakes. Similarly, insect outbreaks and wildfires reduce forest output. These processes are most common in the arid and semi-arid regions of the country. Under wet conditions, warming promotes forest productivity. Due to climate change, animal species are moving away from their previous habitats and there is a possibility that by 2100 biome composition will change (which is currently 5–20% of the land area). Both winter warming and intensification of the hydrological cycle will collect nutrients from the terrestrial ecosystems and will shift them to running water, resulting in huge nutrient wastage. Emissions of carbon dioxide and methane (from wetlands) and subsequent melting and thawing of permafrost will exacerbate the current rate of changes in the climate.

8.2.2 Broadleaf Evergreen Forest (Tropical and Sub-Tropical)

Lakes warm up twice as fast as the air that surrounds them (Schneider & Hook, 2010). This pattern of warming decreases the duration of ice cover for up to 12 days in each 100 years (Livingstone et al., 2010). This change will instigate the emergence of harmful algal blooms in lakes (Paerl & Huisman, 2008) and higher rates of precipitation will reduce the transparency of water to ultraviolet radiations. Sunlight, therefore, cannot disinfect surface waters, which proliferate the insect population (e.g., parasites which cause epidemics). Parasitic epidemics include zooplankton (Overholt et al., 2012), amphibians (Ortiz-Santaliestra et al., 2011) and affects human beings very quickly (Connelly et al., 2007). This category of species plays a key role in different ecosystems. In terrestrial vegetation, the abundance of insects, the occurrence of wildfires, and other climate dependent disturbances alter dissolved organic matter, which results in the production of by-products which are carcinogenic in nature (Beggs & Summers, 2011). There is also an increase in the thermal stratification of oceans and large lakes, which does not permit mixing surface waters with nutrient-rich deeper waters. Due to these changes, a reduction in primary productivity of up to 1% has been observed in eight out of ten major oceans basins (Boyce et al. 2010).

8.2.3 Change of State of Ecosystem

If there are changes in the state of an ecosystem, this can dramatically affect the functions of an ecosystem. It will change net plant productivity, will induce changes in nutrient and water cycling, and changes will be made in trophic interactions, as well as in regulation of regional climate (Zavaleta, 2006). These changes in an ecosystem will impact the consequences of the services which an ecosystem provides to society. If the changes observed in the state of an ecosystem are noticeable, then it will substantially impact Earth's system. Presently, climatic models and earth's system are capturing the impacts of climatic changes on the ecosystems (Euskirchen et al., 2009). Paleo-ecological studies tell that over the time period broad-scale patterns of state of changes has been observed in the ecosystems during last couple of decades. These changes or patterns seldom reveal the fundamental mechanisms going on in an ecosystem.

Most global-change experiments are short term and hence model predictions about climate change are normally standardized against current vegetation distributions. Nevertheless, there is no empirical testing of these models. Smith et al. (2009) proposed the hierarchical response framework and postulate that sequential hierarchy of mechanisms can be triggered by a long-term global-change experiment. These hierarchal mechanisms are actually the physiological changes. These changes reorganize species within communities, and also become the cause of their immigration and finally their loss. There is a considerable increase in insect outbreaks due to climate changes, mainly because of elevation in resistance of plants against drought stress, increased survival of insects during the winter season, and, finally, shortening of the cycle for reproduction and development of insects (Raffa et al., 2008). If the changes are occurring during the winter season, this will enhance the total amount of nutrients going to water and affecting water quality negatively. The sensitivity of the release of the total concentration of greenhouse gases is very much linked to hydrologic changes; nevertheless, this sensitivity is not certain in the permafrost zone. Forests have played a critical role in the absorption of CO_2 for most of the past century, in the present, as well as in the future (Zhang et al., 2012). This absorption of CO_2 results in regrowth after the forests are disrupted, including in fertilization by elevated CO_2 and N deposition as well. Nevertheless, when there are insect outbreaks, this can result in wildfires when forest dieback weakens the carbon sink in the terrestrial environment. Dry areas found at regional and continental scales witness the source of carbon (Zhao & Running 2010). Terrestrial ecosystems can play a dual role as producers of CH_4 and N_2O and as a sink of these two gases. Temperature and precipitation increase the emission of CH_4 from wetlands. However, deposition of more nitrogen will decrease CH_4 uptake by forests.

Both these processes contribute more CH_4 to the atmosphere (Xu et al., 2010). At the continental scale, higher temperatures and precipitation events, when combined with the conversion of forest to agricultural land, fertilization and deposition of nitrogen increase the emission of gases (Guenther et al., 2012). History tells the impacts of climate change and future risks and vulnerabilities at the ecosystem level are predicted by several researchers. Resource agencies should then address future challenges instead thinking about a past that no longer exists.

8.3 OCEAN PROCESSES, AND PLANKTON REGIME SHIFTS: ROLE OF CLIMATE CHANGE

Plankton is the first step in the overall food chain and hence plankton serve as an important source of food for the whole marine fauna. The carrying capacity of the ecosystem is therefore heavily dependent on the abundance, timing of production, and composition of plankton. The carrying capacity of ecosystems here is constituted of various fish species, their abundance, and finally their recruitment to the individual stocks. These organisms export CO_2 (an important greenhouse gas) to the deep ocean by biological pump (or carbon sequestration) ultimately playing an important role in bringing about changes in the environment of that area.

If the export process of carbon dioxide does not work, concentrations of carbon dioxide will be much higher in the atmosphere, which in turn will warm up the climate of the world. Since the mid-1980s, marine ecosystems have experienced substantial stepwise changes and shifting of regimes which are very much obvious in all components of the ecosystem. Nutrients and current fluxes, birds, fish, benthos, and plankton are important examples (Reid et al., 2001; Beaugrand, 2004a). Simultaneously, warmer water plankton has moved to the north by 10° latitude in 40 years, with similar backward movement of coldwater to the north (Beaugrand et al., 2002). The mechanisms which are directly involved in the causation of these changes are not fully known at present but it is presumed that variations in hydro-climate have served as an important modulator in these changes and movements. In any future assessment of the ecological state of coastal waters, this dominant natural variability (which may be forced by climate change at a global level) demands consideration.

8.4 CLIMATE CHANGE AND FISHERIES

Rises in the levels of greenhouse gases like CO_2, CH_4, and nitrogen dioxide (NO_2) are believably the largest contributor to current global warming. Our planet is surrounded by a thermal blanket of greenhouse gases which indirectly allow life on earth to continue to live (IPCC, 2014). Since the industrial revolution, there has been an exponential increase in the production of greenhouse gases, which remained really an extraordinary during the last 800 000 years. In the middle of the 18th century, atmospheric CO_2 concentration was 278 ppm, which has now increased up to 400 ppm (Ciais et al., 2013).

The hydrological cycles are greatly affected by the warming of the climate. When melting of ice and snow changes precipitation, temperature, and climatic patterns, it alters the quantity and quality of water, which ultimately affects its flow and recycling in the natural ecosystems. These factors lead to unavoidable changes in aquatic ecosystems. Permafrost is already warming due to climate change and the thawing of glaciers in high-latitude regions is already in progress. Glaciers are shrinking in high-elevation regions, which will consequently result in downstream water resources (IPCC, 2014). In the marine systems, when the Arctic sea ice melts, it will potentially slow down or disrupt the global ocean conveyor belt (Liu et al., 2017). Since 1901, the changes observed in precipitation are variable across the regions such as the mid-latitude land areas of the northern hemisphere which have shown increments, while decreases have been observed in others (Hartmann et al., 2013).

8.4.1 Aquatic Systems and Climatic Change

Global warming has not affected fisheries to the extent that has been observed in other agriculture sectors. Nevertheless, when we consider human influences like construction of dams on the rivers, removal of water, and its regulation, then which factor is inducing changes in the status of fisheries is hard to conclude. Although future possibilities are uncertain, it is expected in the near future that snowmelts will increase the flow of water in rivers (Pervez & Henebry, 2015). There will be substantial alterations in ecologically important qualities of water flow regimes in many rivers and wetlands due to changes in precipitation. These changes in precipitation will increase the level and intensity of human impacts in water usage in developed river basins. It is expected that in the 21st century warming in the upper portion of the sea will continue, especially in the Tropical and northern hemisphere sub-Tropical regions. However, substantial warming is also expected in the Southern Ocean. The increase in the temperatures of the surface of the sea has already been observed and crossed the limit of natural seasonal variations in the sub-Tropical areas, as well as those observed in an arctic (Ding et al., 2017). It has also been observed that rising water temperatures will increase existing issues in water quality, especially those systems which are facing excessive loading of nutrients due to anthropogenic activities. These changes in water quality will shift freshwater species and their distributions (IPCC, 2014) (Fig. 8.3).

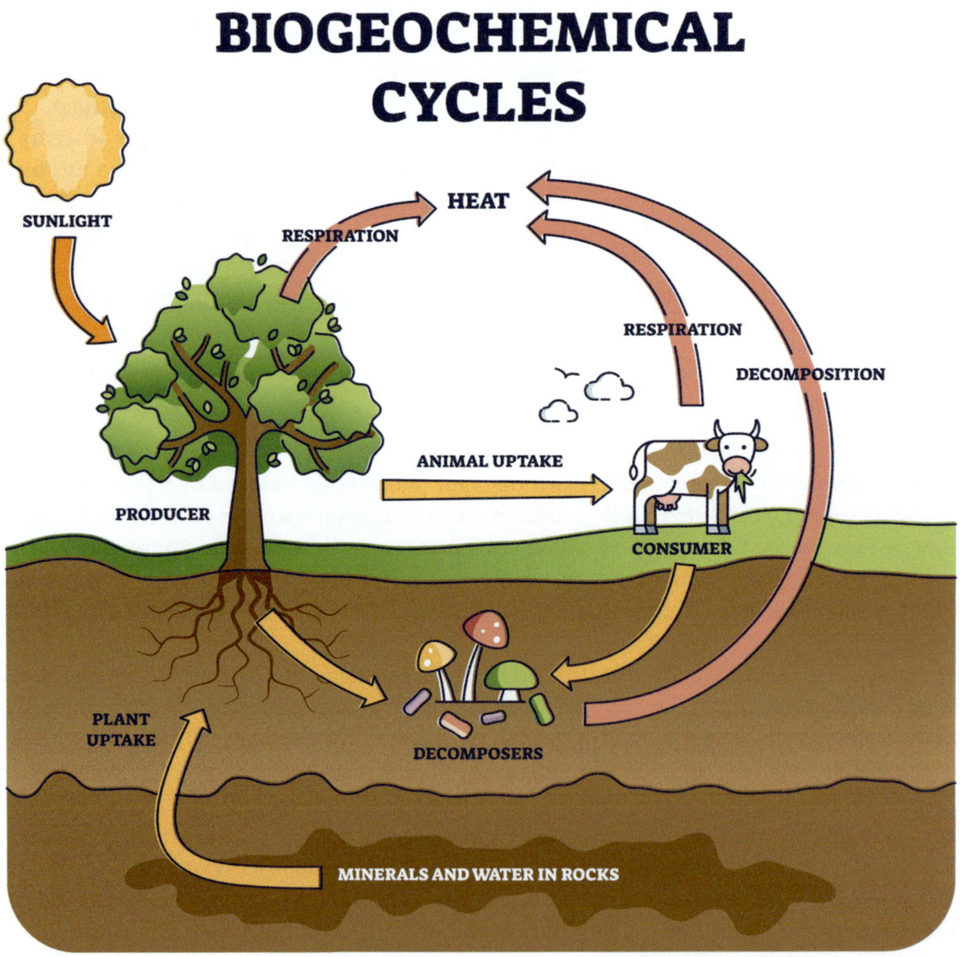

FIGURE 8.3 Organisms movement, trophic interactions, and biogeochemical cycles. (*Source*: www.shutterstock.com.)

In aquatic ecosystems, dissolved oxygen has a pivotal role. The activity of oxygen is very much linked and heavily impacts the global carbon and nitrogen cycles (IPCC, 2014). There is a great deal of variation in the concentration of oxygen in oceans and freshwaters, which ranges from supersaturated levels to zero levels in places of the water sources where aquatic life cannot exist. Oxygen levels are zero in costal sediments of Antarctic waters due to higher consumption than over longer periods of time. There are several such systems where oxygen is at zero level with no or restricted circulation. These oxygen minimum zones are coastal upwelling zones, deep basins of semi-enclosed seas, deep fjords, and some other areas. Since the 1960s, widespread decreases in oxygen concentrations were discovered in coastal areas and oxygen minimum zones have been in expanded recent decades (IPCC, 2014). At their most fundamental levels, both biological and biogeochemical processes are affected by oxygen but these effects vary with the variations in tolerance limits of oxygen in various species and taxonomic groups. Specifically, it has been observed in some fish species including tuna and bill fishes that their vertical migration depths decrease when oxygen is present in low concentrations (Eby & Crowder, 2002).

8.4.2 Ice Coverage

A decrease in Arctic ice has been observed at a rate of 3.5 to 4.1% per decade over the period from 1979 to 2012. Similarly, a reduction in perennial ice has been observed up to 11.5% per decade. Over the same period, average annual increase in Antarctic sea ice is 1.2–1.8% per decade (Vaughan et al., 2013). Considering this at a global level from 2003 to 2009, almost all glaciers have shrunk. Major losses have been observed in the glaciers of Alaska, the Canadian Arctic, the periphery of the Greenland ice sheet, the Southern Andes, and the Asian Mountains (Vaughan et al., 2013). It is possible that with the continuous rise in global temperature there will be further shrinking and thinning of Arctic sea ice cover with a decrease in the northern high-latitude springtime snow cover and near surface permafrosts in the near future. Further, some unconfirmed short-term reductions are observed in Antarctic sea due to climatic changes (Bindoff et al., 2013). When ice and snow coverage melts following a reduction in mountain glaciers, it contributes to water levels and flows into aquatic systems. Higher sea levels are the result of ice melting, while reduction in glaciers will impact the flow of water in the rivers over the short term.

8.4.3 Primary Production

The marine food web is based on the phytoplankton productivity. It controls energy and makes the food available to higher tropic levels and ultimately to fish. Due to climate changes, Earth system models project uncertainty in the production of primary food for aquatic fauna. Some models nevertheless project up to 20% increases (Taucher & Oschlies, 2011) and declines by 2100. Primary production integrates changes in light, temperature, and nutrients, but due to uncertainties induced by climatic changes, primary productivity is on the decline, creating uncertainties for other aquatic fauna. Considering and comprehending all of these factors, it is estimated that global marine primary production will decline by 6% by 2100 (Kwiatkowski et al., 2017). Increase in primary production has been observed in freshwater lakes like the Arctic (Michelutti et al., 2005) and boreal. However, Lake Tanganyika in the Tropics has shown decreases in primary production (O'Reilly et al., 2003).

8.5 CLIMATE CHANGE AND WILDLIFE

With the rise in the surface temperature of the world by 0.6 ± 0.2°C since 1900, the global climate has changed over the last century. These increases in temperature have led to retreats in glaciers and sea ice, has warmed the oceans, has increased the number of heat waves, has caused sea levels to rise (10–20 cm), increased the rainfall, induced droughts in some areas, and, finally, has reduced the chances of frosts (IPCC, 2001). All of these environmental changes are due to the increased concentrations of greenhouse gases in the atmosphere, which is all produced by human activities. Future forecasts say that by 2100 temperatures will rise up to 4°C. These increased levels of temperature will change patterns of precipitation. These conditions will compel them to shift their boundaries and habitat. By a 1°C change in temperature, the ecological zone will shift in latitude or altitude by 160 km and there will also be a change in precipitation for each degree of change in temperature. Other than changes in temperature, habitat destruction is another important factor that threatens biodiversity.

Temperature and rainfall determine where particular species of plants and animals can live, grow, and reproduce. For example, environmental temperature determines the sex of the developing embryos of alligators and turtles. Similarly, an abundance of different species in a particular area and their geographical distribution directly depends on the level of precipitation. For African elephants, living under longer dry periods and in restricted space makes them more vulnerable to climate change. Systematic data was collected on bird nesting in the United Kingdom for a period of 60 years. These studies revealed that 78% of 65 species of birds studied during this research effort started breeding earlier than their normal breeding period.

It has been observed that with a rise of only 0.5°C in Western Australia, a significant number of frogs, mammals, and plant species are restricting themselves to smaller areas (Pouliquen-Young & Newman, 2000). Habitat fragmentation isolates the species population geographically with the reduction of genetic variations. Migration and dispersal of species to suitable habitats is also attributed to climatic change. The marine turtles of Pakistan are an important group of reptiles. They visit almost all of the sandy beaches starting from the coast of Sindh to the coast of Baluchistan. They migrate from deeper oceans to the coastal areas during their breeding season to lay eggs. The Olive Ridley turtle (*Lepidochelys olivacea*) and the Green turtle (*Chelonia mydas*) (two species of marine turtles) are common along the coastal areas of Pakistan. The other two species of turtles (the Loggerhead and Hawksbill turtle) are rarely observed (Firdous, 2001).

Wildlife department in its long-term study has studied the distribution and biological aspects from the Hawkes Bay and Sandspit areas of Karachi (Firdous, 2001). The IUCN has listed the green turtle in its red list of threatened species. Under the wildlife regulatory provisions, Sindh Wildlife Protection Ordinance 1972 provides protection to turtles. Pakistan has signed the CITES (Conservation on International Trade in Endangered Species of Wild Fauna and Flora) agreement and has approved clause 5 in the Pakistan Fish Inspection and Quality Act 1997 of the Federal Ministry of Food, Agriculture and Livestock, Government of Pakistan (MINFAL). Both these documents forbid the domestic consumption and export of all aquatic turtles.

From 1981 to 1997, a study was conducted on the distribution of nesting incidences of these turtles. Females of both *C. mydas* and *L. olivacea* (web address) show a decreasing trend recorded at the Sandspit area of Karachi. Possible reasons for this could be habitat destruction of these animals and anthropogenic impacts. Similarly, while talking about birds some workers claim that Pakistan serves as Palearctic to Asia bird migration routes because of strategic and geographic location. All the rivers in Pakistan serve as major migratory routes for migratory birds.

When we consider the importance of bird migration routes at a global level, it is the fourth major bird migration flyway in the world. Seasonal migratory birds come and go in winter and summer. The Uchalli, Khabbaki, and Jahlar lakes, when combined, make a complex of lakes called the Uchalli Wetland Complex. In 1996, this site was declared a Ramsar site. Different migratory birds and others regularly visit this lake complex (Ali, 2005).

In response to annual rainfall in combination with variations in climatic conditions year after year, there are many fluctuations in the water levels of these lakes. Drought conditions started after 1997, which resulted in a 50% decline in rainfall. The reduction in rainfall reduced the area of the Uchalli lakes, which remained only 27% (336 ha) of the total (1,243 ha). This reduction in overall water surface area impacted many migratory bird populations (Ali, 2005). In the Uchalli wetland complex, annual precipitation and bird counts are significantly positively correlated. Yearly (short term) and decade wise (long term trends) creates wet and dry states in wetlands. When surface interacts with groundwater; underneath rocks and soil affects salinity and chemistry of wetlands. Climate change due to high temperatures affects wetlands physically, with higher rates of evapotranspiration, extended ice-free periods, and a long growing season. Increases in precipitation levels increase water levels of lakes, while an increase in evapotranspiration decreases water levels of lakes. If higher levels of precipitation and evapotranspiration increase simultaneously, there is no significant impact on the water levels of lakes, although it shortens the water resident time in the lakes (Ali, 2005, 2011). Those aquatic systems which are associated with the Indus River are highly vulnerable to the different changes in the climate and variations in the weather. The release and deposition of greenhouse gases in the atmosphere affects evaporation, snowpack, patterns of precipitation, flood, and drought. All these factors severely affect the supply of freshwater and its quality (IPCC, 2001). The Indus River depends heavily on glacier water and hence it is very much prone to climate changes. Starting from the Mansarovar Lake located in Tibet at an elevation of about 5,182 m, the Indus River flows for about 2,880 km down and finally passing through the Indus Delta before it falls into the Arabian Sea. 450,000 square miles of its total area expands into India, contributing an average annual inflow of water in the range of 175 million acre feet. There is seasonal fluctuation in the flow of Indus River with the melting of Himalayan glaciers (70–80%). This glacier water flows in Indus river which is the highest percentage compared to any river in Asia (Kiani, 2005).

8.6 CLIMATE CHANGE AND AGRICULTURE

With every passing day, climate changes are the biggest challenges that the world is facing and has yet to face. At a global level, agriculture is directly susceptible to climate change. Being an agricultural country, the economy of the country is very much dependent on various forms of agriculture and its products. 70% of Pakistani agriculture directly depends upon the annual rainfall. Therefore, annual rainfall and temperature always have a pivotal role in the establishment and promotion of agriculture. In different districts of Pakistan, climate changes have displayed significant effects (Shakoor et al., 2011). For example, one-third of the total agriculture and cultivated land depends on annual rainfall, hence its stoppage drastically affects the productivity of this area. Therefore, natural factors like rainfall, temperature, and humidity play a pivotal role (up to 65%) in Pakistani agriculture. A strong interaction has been observed between existing agriculture and population density with emissions of carbon dioxide, agriculture nitrous oxide, and greenhouse gases.

Among greenhouse gases, methane emissions have limited effects on agriculture if exposed for a short duration of time, while carbon oxide emissions and greenhouse gas emissions have direct bearing on agriculture. However, nitrous oxide emission has a short-term relationship between annual growth and production of agriculture. Nevertheless, both independent and dependent variables of climatic changes directly affect the production of agriculture.

All over the world temperature and environment are continuously changing, which has been experienced with every passing day. During the last five years the onset of floods, drought, and other changes are good indictors of climatic effect in Pakistan. As stated above, the severe impacts of changing climate on agriculture and its productivity cannot be ignored and maximum efforts should be exhausted to take remedial measures to mitigate these effects. Under this situation, the following measures can be adopted to overcome and/or to neutralize these effects.

1. Careful monitoring of climatic changes should be done by government agencies. Whatever the changes and effects that they find they should immediately inform farmers so they can take precautionary measures to mitigate these effects.
2. Agriculture experts should exclusively focus their research activities in line with the climatic changes. They should develop such varieties of crops which can tolerate changing climatic conditions according to the regions where these effects are observed and/or expected to be observed in the future.

3. Farmers may be provided with soft loans by the government so that they can purchase agriculture implements which match their requirements and in accordance with the changing environment and prevailing conditions so that they can increase their productivity.
4. Water shortage is a persistent problem for Pakistan in general and for agriculture specifically. The government should take such decisions and initiatives to make sure that sufficient water is available to the farmers during the sowing and growing periods of the crops.

Mangrove forests develop extensively on and along shorelines. They grow in tropical intertidal habitats. Mangrove trees constitute 34 species and three hybrids. Their center of diversity is in Southern Papua New Guinea. The 20°C isotherm controls the global distribution of mangroves. Their diversity and forest height decreases with increasing latitude away from the equator (Ellison, 2015). When the sea level rises, it erodes sediments, inundates the required area for their growth, and increases salinity at landward zones, which in turn threatens the survival of the mangroves. If landward migration is restricted by topography or by human developments, then the mangroves face the problem of coastal squeeze or shrinkage. Rising temperatures and concentrations of carbon dioxide in the atmosphere expand the ranges of mangroves into higher latitudes with the increase in their productivity, and changes in their phenological patterns. Mangroves provide many useful products to human societies and important services to the adjacent ecosystems. First, they protect inland sites from storm surges and flooding. Second, they serve as nursery habitats for fish species and invertebrates, although they spend their adult life in coral reefs. Third, they provide building materials. Fourth, they sustain offshore water quality for coral reefs by trapping sediments. Fifth, they provide firewood. Sixth, they provide food. Seventh, they are an important source of traditional medicine. With the expansion of the human population, there is a shrinkage of productive land in many developing countries of the world. Therefore, people have started to clear mangroves for agriculture and aquaculture production. Increases in the demand for timber and coastal developments are also injurious for mangroves. Overexploitation by humans has reduced mangroves to such an extent that only mere remains are found in some places. In addition to these pressures, annual sea level rises of up to 0.9 and 8.8 mm/year are also threatening mangroves severely. Studies have shown that mangroves are closely controlled by sea level elevation at their seaward margin, although there are several important factors which determine the patterns of their advance or retreat. There is a distinct zoning of mangrove species from low to high water. This zoning depends on the frequency of inundation and their exposure to salinity. The elevation of the substrate surface relative to mean sea level also controls this pattern of mangrove spread. Part of this accumulation is from the local production of organic matter but external inputs of sediment from rivers augment this accumulation. From the rates of deposition it can be assessed that projected rates of sea level rises have important effects on mangrove systems. Those mangroves which are present in low relief islands in carbonate settings where there are no rivers are probably more sensitive to sea level rises because they do not have too much sediment in their environments and hence there is poor deposition too. Those mangroves which are present on low islands seem more sensitive to the expected rises in the sea levels. Those mangroves which are present on continental margins also face retreat and disruption in their patterns. At stable sea level, mangroves have extensive capacity to establish themselves but if there are changes in sea level they are highly prone to retreat and disruption.

REFERENCES

Ali, Z., (2005). Ecology, distribution and conservation of migratory birds at Uchalli Wetlands Complex, Pakistan. PhD Dissertation, University of the Punjab, Lahore, Pakistan.

Beaugrand, G. (2004). The North Sea Regime Shift: Evidence, Causes, Mechanisms and Consequences. Progress in Oceanography 60, 245–262.

Beaugrand, G., Reid, P.C., Ibanez, F., Lindley, J.A. & Edwards, M. (2002). Reorganisation of North Atlantic Marine Copepod Biodiversity and Climate. Science 296, 1692–1694.

Beggs, K.M.H. & Summers, R.S. (2011). Character and Chlorine Reactivity of Dissolved Organic Matter from a Mountain Pine Beetle Impacted Watershed. Environmental Science & Technology 45, 5717–5724.

Bindoff, N.L., Stott, P.A., AchutaRao, K.M., Allen, M.R., Gillett, N., Gutzler, D. & Hansingo, K. et al. 2013. Detection and Attribution of Climate Change: From Global to Regional. In Climate Change 2013: The Physical Science Basis. Contribution of Working Group I to the Fifth Assessment Report of the Intergovernmental Panel on Climate Change, eds. Stocker, T.F., Qin, D., Plattner, G.-K., Tignor, M., Allen, S.K., Boschung, J., Nauels, A., Xia, Y., Bex V., & Midgley, P.M. pp. 867–952. Cambridge, UK and New York, USA, Cambridge University Press.

Boyce, D. G., Lewis, M. R. & Worm, B. (2010). Global Phytoplankton Decline over the Past Century. *Nature* 466, 591–596.

Bush, M. (1997). Ecology of a Changing Planet. Prentice Hall, Upper Saddle River, NJ.

Ciais, P., Sabine, C., Bala, G., Bopp, L., Brovkin, V., Canadell, J. & Chhabra, A. et al. (2013). Carbon and Other Biogeochemical Cycles. In Climate Change 2013: The Physical Science Basis. Contribution of Working Group I to the Fifth Assessment Report of the Intergovernmental Panel on Climate Change, eds. Stocker, T.F., Qin, D., Plattner, G.-K., Tignor, M., Allen, S.K., Boschung, J., Nauels, A., Xia, Y., Bex V., & Midgley, P.M. pp. 465–570. Cambridge, UK and New York, USA, Cambridge University Press.

Connelly, S.J., Wolyniak, E.A., Williamson, C.E. & Jellison, K.L. (2007). Artificial UV-B and Solar Radiation Reduce in vitro Infectivitiy of the Human Pathogen *Cryptosporidium parvum*. Environmental Science & Technology 41, 7101–7106.

Ding, Q., Schweiger, A., L'Heureux, M., Battisti, D.S., Po-Chedley, S., Johnson, N.C., Blanchard-Wrigglesworth, E., Harnos, K., Zhang, Q., Eastman, R. & Steig, E.J. (2017). Influence of High-Latitude Atmospheric Circulation Changes on Summertime Arctic Sea Ice. Nature Climate Change 7, 289–295.

Eby, L.A. & Crowder, L.B. (2002). Hypoxia-Based Habitat Compression in the Neuse River Estuary: Context-Dependent Shifts in Behavioral Avoidance Thresholds. Canadian Journal of Fisheries and Aquatic Sciences, 59(6): 952–965. https://doi.org/10.1139/f02-067.

Ellison, J.C. (2015). Vulnerability Assessment of Mangroves to Climate Change and Sea-Level Rise Impacts. Wetlands Ecology and Management 23, 115–137.

Euskirchen, E.S., McGuire, A.D., Chapin FS III , *et al.* (2009). Changes in Vegetation in Northern Alaska under Scenarios of Climate Change, 2003–2100: Implications for Climate Feedbacks. Ecological Applications 19, 1022–1043.

Firdous, F., (2001). Some bio-ecological studies on green turtles (*Chelonia mydas*) and olive ridley (*Lepdochyles olivacea*) from Karachi coast. PhD Thesis. University of Karachi. pp. 1–648.

Guenther, A.B., Jiang, X., Heald, C.L., Sakulyanontvittaya, T., Duhl, T., Emmons, L.K. & Wang, X. (2012). The Model of Emissions of Gases and Aerosols from Nature Version 2.1 (MEGAN2.1): An Extended and Updated Framework for Modeling Biogenic Emissions. Articles 5(6) 1471–1492. https://doi.org/10.5194/gmd-5-1471-2012.

Hansen, J., Ruedy, R., Sato, M. & Lo, K. (2010). Global Surface Temperature Change. Reviews of Geophysics, 48(4), RG4004. https://doi.org/10.1029/2010RG000345.

Hartmann, D.L., Klein Tank, A.M.G., Rusticucci, M., Alexander, L.V., Brönnimann, S., Charabi, Y. & Dentener, F.J. et al. (2013). Observations: Atmosphere and Surface. In Midgley, eds. Climate Change 2013: The Physical Science Basis. Contribution of Working Group I to the Fifth Assessment Report of the Intergovernmental Panel on Climate Change, eds. Stocker, T.F., Qin, D., Plattner, G.-K., Tignor, M., Allen, S.K., Boschung, J., Nauels, A., Xia, Y., Bex V., & Midgley, P.M. pp. 159–254. Cambridge, UK and New York, USA, Cambridge University Press.

IPCC. (2001). Climate Change: Impacts, Adaptations and Vulnerability. Cambridge University press, Cambridge.

Kiani, K. (2005). Water-Related Crisis Feared in 20 Years. *Dawn*: the Internet Edition. 04 January 2005. The Dawn Group of Newspapers, Islamabad, Pakistan. http://www.dawn.com/2005/01/04/nat12.htm

Kwiatkowski, L., Bopp, L., Aumont, O., Ciais, P., Cox, P.M., Laufkötter, C., Li, Y. & Séférian, R. (2017). Emergent Constraints on Projections of Declining Primary Production in the Tropical Oceans. Nature Climate Change, 17, 355–359. https://doi.org/10.1038/NCLIMATE3265.

Liu, W., Xie, S.P., Liu, Z. & Zhu, J. (2017). Overlooked Possibility of a Collapsed Atlantic Meridional Overturning Circulation in Warming Climate. Science Advances, 3(1), e1601666. https://doi.org/10.1126/sciadv.1601666

Livingstone, D.M., Adrian, R. & Blenckner, T., et al. (2010). Lake Ice Phenology. In: The Impact of Climate Change on European Lakes, ed. George, G. (Ed.), pp. 51–61. Springer, New York, NY..

Michelutti, N., Wolfe, A.P., Vinebrooke, R.D., Rivard, B. & Briner, J.P. (2005). Recent Primary Production Increases in Arctic Lakes. Geophysical Research Letters, 32(19), L19715. https://doi.org/10.1029/2005GL023693.

National Academy of Sciences (2012). Climate Change, Evidence, Impacts and Choices; Answers to Common Questions about the Science of Climate Change, National Research Council of the National Academies. Pages 40.

National Oceanic and Atmospheric Administration (NOAA) (2011). The Arctic web is complex. The loss of sea ice can ultimately affect the entire food web, from algae and plankton to fish to mammals. In: climate impacts ecosystems. Under subheading: Food web disruptions.

O'Reilly, C.M., Alin, S.R., Plisnier, P.-D., Cohen, A.S. & McKee, B.A. (2003). Climate Change Decreases Aquatic Ecosystem Productivity of Lake Tanganyika, Africa. Nature, 424(6950), 766–768. https://doi.org/10.1038/nature01833.

Ortiz-Santaliestra, M.E., Fisher, M.C. & Fernandez-Beaskoetxea, S., et al. (2011). Ambient Ultraviolet B Radiation and Prevalence of Infection by *Batrachochytrium dendrobatidis* in Two Amphibian Species. Conservation Biology 25, 975–982.

Overholt, E.P., Hall, S.R. & Williamson, C.E., et al. (2012). Solar Radiation Decreases Parasitism in Daphnia. Ecology Letters 15, 47–54.

IPCC. (2014). Climate Change 2014: Synthesis Report. ContriBution of Working Groups I, II and III to the Fifth Assessment Report on the Intergovernmental Panel on Climate Change. Core Writing Team, eds. Pachauri, R.K. & Meyer, L.A. Geneva, Intergovernmental Panel on Climate Change. 151 pp. http://www.ipcc.ch/report/ar5/syr/.

Paerl, H.W. & Huisman, J. (2008). Blooms Like It Hot. Science 320, 57–58.

Pervez, M.S. & Henebry, G.M. (2015). Assessing the Impacts of Climate and Land Use and Land Cover Change on the Freshwater Availability in the Brahmaputra River Basin. Journal of Hydrology: Regional Studies, 3, 285–311. https://doi.org/10.1016/j.ejrh.2014.09.003.

Porter, J.R., Xie, L., Challinor, A.J., Cochrane, K., Howden, S.M., Iqbal, M.M., Lobell, D.B. & Travasso, M.I. (2014). Food Security and Food Production Systems. In: Climate Change 2014: Impacts, Adaptation, and Vulnerability. Part A: Global and Sectoral Aspects. Contribution of Working Group II to the Fifth Assessment Report of the Intergovernmental Panel on Climate Change, eds. Field, C.B., Barros, V.R., Dokken, D.J., Mach, K.J., Mastrandrea, M.D., Bilir, T.E., Chatterjee, M., Ebi, K.L., Estrada, Y.O., Genova, R.C., Girma, B., Kissel, E.S., Levy, A.N., MacCracken, S., Mastrandrea, P.R. & White, L.L., pp. 485–533. Cambridge University Press, Cambridge, United Kingdom and New York, NY, USA.

Pouliquen-Young, O. & Newman, P. (2000). The Implications of Climate Change for Land-Based Nature Conservation Strategies. Final Report 96/1306. Australian Greenhouse Office, Environment Australia, Canberra; Institute for Sustainability and Technology Policy, Murdoch University, Perth.

Raffa, K.F., Aukema, B.H. & Bentz, B.J., et al. (2008). Cross-Scale Drivers of Natural Disturbances Prone to Anthropogenic Amplification: the Dynamics of Bark Beetle Eruptions. BioScience 58, 501–17.

Reid, P.C., Borges, M.F. & Svendsen, E. (2001). A Regime Shift in the North Sea Circa 1990. Fisheries Research 50, 163–171.

Schneider, P. & Hook, S.J. (2010). Space Observations of Inland Water Bodies Show Rapid Surface Warming Since 1985. Geophysical Research Letters 37, L22405.

Shakoor, U., Saboor, A., Ali, I. & Mohsin, A.Q. (2011). Impact of Climate Change on Agriculture: Empirical Evidence from Arid Region. Pakistan Journal of Agricultural Sciences 48(4), 327–333.

Smith, M.D., Knapp, A.K. & Collins, S.L. (2009). A Framework for Assessing Ecosystem Dynamics in Response to Chronic Resource Alterations Induced by Global Change. Ecology 90, 3279–89.

Taucher, J. & Oschlies, A. 2011. Can We Predict the Direction of Marine Primary Production Change Under Global Warming? Geophysical Research Letters 38(2), L02603. https://doi.org/10.1029/2010GL045934.

UN OCHA PAKISTAN (2022). Monsoon Floods Situation Report No. 6 As of 16 September 2022. https://reliefweb.int/report/pakistan/pakistan-2022-monsoon-floods-situation-report-no-6-16-september-2022

Vaughan, D.G., Comiso, J.C., Allison, I., Carrasco, J., Kaser, G., Kwok, R. & Mote, P. et al. (2013). Observations: Cryosphere. In Climate Change 2013: The Physical Science Basis. Contribution of Working Group I to the Fifth Assessment Report of the Intergovernmental Panel on Climate Change eds. Stocker, T.F., Qin, D., Plattner, G.-K., Tignor, M., Allen, S.K., Boschung, J., Nauels, A., Xia, Y., Bex V., & Midgley, P.M., pp. 317–382. Cambridge, UK and New York, USA, Cambridge University Press.

Watson, R.T., Zinyowera, M.C., Moss, R.H. & Dokken, D.J. (eds) (1996). Climate Change 1995, Impacts, Adaptations and Mitigation of Climate Change: Scientific–Technical Analyses. IPCC. Cambridge University Press, Cambridge, UK.

Xu, X.F., Tian, H.Q. & Zhang, C., et al. (2010). Attribution of Spatial and Temporal Variations in Terrestrial Methane Flux Over North America. Biogeosciences 7, 3637–55.

Zavaleta, S. (2006). Shrub Establishment under Experimental Global Changes in a California Grassland. Plant Ecology 184, 53–63.

Zhang, F., Chen, J.M. & Pan, Y., et al. (2012). Attributing Carbon Changes in Conterminous US Forests to Disturbance and non-Disturbance Factors from 1901 to 2010. Journal of Geophysical Research 117, G02021.

Zhao, M. & Running, S.W.(2010). Drought-Induced Reduction in Global Terrestrial Net Primary Production from 2000 Through 2009. Science 329(5994), 940–943. https://doi.org/10.1126/science.1192666.

Zwerver, S., Van Rompaey, R.S.A.R., Kok, M.T.J. & Berk, M.M. (1995). Climate Change Research: Evaluation and Policy Implications. Studies in Environmental Sciences 65 (B). Elsevier, Amsterdam.

9 Conservation of Biodiversity

9.1 INTRODUCTION

Biodiversity conservation means the protection of life present on this planet in whatever form it is present. It also entails maintenance of natural ecosystems. Elaborating this concept further, it stresses to preserve it, maintain it, and then use it in a sustainable manner to enhance its viability. The recovery of various components of biological diversity and its further proliferation also come under this concept. Here, the two terms conservation and preservation are used interchangeably but the way of working with both terms and interpretation varies. "Conservation" emphasizes the use of the resources sustainably, which means exploiting them but also ensuring their protection. Preservation nonetheless says that we must conserve the available natural resources without any alteration or change. Another aspect of biodiversity conservation is sustainable development of various natural resources in such a way that "biodiversity is fulfilling the requirements of the current generation at the same time it is not harming the resources for future generations to fulfill their needs and necessities." Sustainability therefore refers to impartial treatment within and among the generations. An appropriate balance between society, environment, and development can only ensure biodiversity. This whole process requires maximum involvement of environmental institutions, which should ensure that laws and policies are properly enforced and implemented when and where needed.

Throughout the evolutionary history of life, biological diversity present on earth has faced expansion and contraction similar to other natural processes. Different levels of genetic, population, species, and ecosystems are the major areas which face this expansion and contraction. Currently, due to anthropogenic forces of environmental change we are facing a rapid loss of biodiversity at a global level (Barnosky et al., 2011). Presently, global climate changes and their associated effects are influencing biodiversity negatively. These trends are destruction and conversion of marine, freshwater, and terrestrial ecosystems; degradation of environment by pollution or other factors; extensive alterations and amendments in hydrological systems are too few to count and are believed to be responsible for biodiversity losses. In addition to these factors, the intrusion of invasive species can create havoc for local fauna, disturbing biodiversity and ecosystem balances. Further, if the natural resources are not harvested sustainably a severe imbalance occurs, which puts both flora and fauna present within that ecosystem at the verge of extinction. These are actually social and economic forces which drive these trends to build relationships between landscapes, ecosystems, and people. Both of these forces are old ones. History witnesses that human communities and economies which associated them always affected the character, composition, and function of ecosystems. Behavioral norms which evolved from some of the societies contributed to relatively sustainable and robust environmental conditions, while some others failed to achieve this. As the human population grew and distributed itself geographically—which resulted in the expansion and intensification of their economies—this whole scenario generated concerns for sustainability of biodiversity at a global level. Recent decades have witnessed that healthy and resilient ecosystems are the key to the long-term vitality and security of human socioeconomic systems. Indirectly, this vitality is attributed to the Earth system where all the ecosystems are embedded. Co-evolution of diversity is a complex of several factors which strengthen the resilience of ecosystems. Human activities inflict the major losses to biodiversity. Essentially, conservation of biodiversity can promote and benefit ecosystems which are resistant and robust in nature, which can ensure sustainable future of living organisms present on this globe. The majority of the population have strong ethical arguments which support the preservation of biodiversity. Because the

 DOI: 10.1201/9781003472001-9

general public has a strong foundation in cultures, religions, and philosophies, hence they accept it quickly, setting aside the economic value of these efforts. These arguments and phenomena are: (1) each living organism has a right to live and exist comfortably; (2) all species depend on each other; (3) other animals should have the same facilities and ecological limitations which human beings enjoy; (4) people should behave responsibly and accept their mistake if they commit any; (5) people should think about future generations and should consider what they are leaving for generations to come after them; (6) wastage of resources should be completely prohibited; (7) each individual should be respected; (8) spiritual and aesthetical values present in nature should exceed the economic values of biodiversity; and, finally, (9) biological diversity is the source for determination of the origin of life.

Millstein (2015) describes this concept and says that any activity is appropriate when it tries to conserve and maintain the integrity, stability, and beauty of biotic community, which means where the life is present; otherwise, it can be disregarded.

Species help to maintain relatively stable ecosystems providing a variety of functions (Shashidharan & Quadros, 2013). In turn, when demand for fuel, fiber, and food increases, it inflicts losses on biodiversity and ecosystem services. Biodiversity constitutes a variety of plants, micro, and macroorganisms. At a global level, forests, wetlands, and coral reefs, where the majority of animals live, are on the decline. Therefore, it can be confidently said that, at a global level, loss of habitats—which also involves habitat degradation and fragmentation—are the most important causes of biodiversity loss. In addition to these causes, forestry, fisheries, and agricultural practices also impose losses to biodiversity. Degraded and fragmented habitats cannot support that level of life and cannot provide ecosystem services of that level, which they could if they were intact. Over the history of life, as the scientific knowledge advanced and there was a shift in social values, conservation evolved continuously and proportionally. Although advocates for the preservation of biodiversity and its use for utilitarian purposes discouraged early efforts toward the conservation of biodiversity, researchers working in this filed managed to nullify these efforts and controlled their negative approaches. Nevertheless, they were successful in the presentation of a wider and integrated set of values for biodiversity and its importance for living organisms. That critical process took in the growing and appreciable role of biodiversity in the maintenance of the resilience of the ecosystem. Now, conservation of biodiversity faces several challenges in its effective evolution to cope with the social and environmental changes at a global level. Parallel to the evolution in conservation and to enhance its efficiency, ethical foundations must also undergo the process of evolution and upgradation.

9.2 IMPORTANCE OF CONSERVATION

Biodiversity supports the life present on our planet. Biodiversity supplies us oxygen to breathe, provides food to eat, and pure and clean water for drinking. Medicines like penicillin, aspirin, quinine, and taxol are alleviating human suffering and are saving millions of lives. All these medicines we get from plants. Wetlands filter out pollutants from water and minimize pollution. Trees and plants absorb carbon dioxide and reduce global warming. Bacteria and fungi fertilize the soil by breaking down organic matter present in it. Only a healthy ecosystem supports the native species and enhances their density, which improves the quality of life for human beings. It can confidently be said there is a close relationship between biodiversity and our sustainable future, which means that our lives depend on biodiversity which in turn demands that it should be conserved in any way possible.

The World Bank accepted the National Conservation Strategy (NCS), also called the National Environmental Action Plan, for implementation for the conservation of biodiversity. The NCS has 14 core programs which address issues of biodiversity. Provincial conservation strategies have been formulated to conserve biodiversity. Conservation strategies have also been initiated at the district level, which deals with biodiversity at the local level more explicitly (GOP, 1999).

The Biodiversity Action Plan (BAP) was also taken to conserve biodiversity. It was adopted in 1999 and in this plan all the ecological regions of the country have been recognized with a specific focus on semi-arid and arid regions of the country where native communities exist. In the ecological regions of the country native communities are under pressure regarding natural resources due to their limited availability. All the provinces of the country are working on mitigation of the threat to biodiversity and to improve its conservation by adopting national conservation strategies. Specifically, the KP Provincial Conservation Strategy (SPCS) presented a chapter on biological diversity, with recognition of protected areas and parks (GONWFP, 1997). Nevertheless, the requirements of the Convention on Biodiversity (CBD, 1992) are not comprehensively addressed in this document. The local authorities have designed and finalized conservation strategies at the district level for Chitral and Abbottabad, with local availability of approachable hierarchically nested conservation plans. Presently existing conservation plans and policies from forestry, fisheries, wildlife, and agriculture are under the process of implementation (GOP, 1999; GOP, 2007).

9.2.1 Participation of Community in Biodiversity Conservation

It is well established that if the local community does not participate in planning, management, and monitoring conservation programs, none of the legal provisions can be effective. This can be archived by several initiatives taken both by government organizations and non-governmental organizations (NGOs). For example, continuous stress persists from the Joint Forest Management philosophy that degraded forest land in the vicinity of villages, which should be regenerated and protected. This effort cannot be fruitful without the participation and involvement of village communities. Therefore, it is mandatory that successful conservation should earn the confidence and invite the participation of the local communities (Dobhal et al., 2011).

9.2.2 Conservation of Biodiversity: International Efforts

The issue of conservation of biodiversity does not pertain to any single country or community living anywhere in world. Conservation of biodiversity is a critical issue of the globe and hence demands that it be addressed globally. The United Nations Conference on Human Environment (UNEP) (1972) realized the importance of the sustainable use of various components of biodiversity and its conservation and in 1973 stressed it as a priority area. Therefore, to achieve the conservation of biodiversity and world resources, the acquisition of a legal mandate is required for any further initiative. Negotiations could then be started as a legally binding instrument to address biological diversity, its losses, and its fair and equitable distribution among the masses. These negotiations became the cause of the opening of the Convention on Biological Diversity in 1992 and later the CBD itself in 2011. This convention motivated and enhanced the concern for sustainable development all over the world. The convention's objectives, although briefly mentioned above, were: (1) conservation of biological diversity; (2) utilization of components for biodiversity sustainably; and, finally (3) sharing of its benefits following equivalent and fair practices.

At a global level, the CBD was the first comprehensive convention that could be considered to address all the aspects of biological diversity, including ecosystem diversity, species diversity, and genetic resources. To strengthen the international participation and commitment towards the conservation of biodiversity several international treaties and agreements are in place. These include (1) the African Convention on Conservation of Nature and Natural Resources; (2) the UNESCO Program on Man and Biosphere; (3) the Convention on International Trade for Endangered Species (CITES); (4) the Convention on Conservation of Migratory Species of Wild Animals; (5) the International Convention for the Protection of Birds, (5) the United Nations Conference on Environment and Development (UNCED)/Earth Summit; (6) the Ramsar Convention on Wetlands of International Importance; (7) the International Union for the Conservation of Nature (World

Conservation Union); (8) the International Board for Plant Genetic Resources; (9) the World Resources Institute; (10) the Worldwide Fund for Nature; and (11) the International Convention for the Regulation of Whaling.

9.2.3 Initiatives Taken by Pakistan for Conservation and Promotion of Biological Diversity

Pakistan recognizes the importance of biodiversity as a beneficial entity (Baig & Al-Subaiee, 2011). Therefore, Pakistan is focusing seriously on the following key areas for meaningful conservation of biodiversity.

9.2.3.1 Establishment of Protected Areas

For conservation of the maximum number of species and ecosystems, to date the government has established 190 protected areas. The detail of these protected areas covers national parks (15), wildlife sanctuaries (83), game reserves (85), and 14 unclassified areas. If we take the total percentage of these areas then 9% of the total area of the country is included, which is about 71,649 square kilometers. Under the Ramsar Convention, nine wetlands have been designated.

9.2.3.2 Institutional Measures

In 1968, the Wildlife Enquiry Committee (WEC) was established and for the first time legislation was introduced to protect biodiversity. The WEC drafted and formulated this legislation which was enacted with various provincial acts and ordinances for its adoption in letter and spirit. With the facilitation and support of these laws (statutes), national parks were created. These statutes also resulted in the establishment of wildlife sanctuaries and game reserves. In addition to these facilities, private game reserves were also established for the conservation of wild animals. In 1974, the National Council for Conservation of Wildlife (NCCW) was established within the Ministry of Food, Agriculture, and Livestock and within the NCCW an advisory board was constituted. This advisory board coordinates, formulates, and implements wildlife policies at the provincial and federal levels and promotes the conservation of wildlife while coordinating activities with international agencies. The first piece of legislation was the Environmental Protection Ordinance (EPW), which was enacted in 1983. Later on, the NCS shifted its original motto of simple regulation and protection measures and focused more on an holistic view of environmental issues and problems.

9.2.3.3 International Conventions and Agreements

Pakistan has signed all the major international agreements executed for the conservation of wildlife. Convention on International Trade in Endangered Species of Wild Flora and Fauna (CITES), the Convention on the Conservation of Migratory Species of Wild Animals (Bonn), the Convention on Wetlands of International Importance (Ramsar), and the World Heritage Convention are the conventions where Pakistan is a signatory. The World Conservation Union (IUCN) and the International Waterfowl and Wetland Research Bureau (IWRB) are other organizations where Pakistan is a member. Rio de Janeiro is also a convention on biological diversity with 125 member countries, of which Pakistan is one. The NCS, and the Ministry of Environment (Urban Affairs) are two other bodies designated for conservation of wildlife. Forestry, and Wildlife, Islamabad is considered an apex body in Pakistan and is responsible for the resolution of environment-related issues of Pakistan. The NCS was formulated by Forestry and Wildlife, Islamabad. National conservation strategy is responsible for all issues of biodiversity in Islamabad. In addition, 13 other core areas fall under its jurisdiction. This strategy provides linkages among various core areas in both forward and backward directions. The NCS realizes that there are several flaws in the existing protected areas system and that it is not comprehensive at all. These flaws are: (1) the devised protected areas system does not represent all the ecosystems, plants, and animal communities; and (2) in most cases the boundaries drawn for the existing designated protected areas do not fulfill the ecological criteria. Many designated protected areas are too small and poorly located.

The NCS also says that: (1) plans for specific management requirements for rare and endangered species are lacking; and (2) plans are also missing for the identification of species hotspots. The area where conservation is process, local people living and relate to that particular area are quite important entities and can be of great help in the implementation of conservation strategies. The NCS does offer benefits and incentives which can alleviate any constraints which are presently hampering the pace of biodiversity in the country. There are several factors which are the main causes of the severe threat to biodiversity of Pakistan and demands immediate and critical attention for their sustainable removal and remediation and for further advancements in the conservation of biodiversity. Due to these important factors, much of Pakistan's biodiversity is severely threatened, and in need of critical attention. For the conservation of biodiversity, the NCS recommends the implementation of the following seven key programs with an investment of Rs. 1.08 billion. These seven programs are: (1) management of protected areas and national parks; (2) new national parks be developed; (3) new wetland reserves be developed; (4) execution of medical applications from plants and a program for the preservation of germplasm; (5) implementation of management programs for game reserves by local communities which are actually private entities; (6) programs for rehabilitation and conservation of endangered species to obviate their extinction; and finally (7) programs which encourage and facilitate the private sector for the captive breeding of rare and endangered wildlife animals.

9.3 ROLE OF BIOTECHNOLOGY IN THE CONSERVATION OF BIODIVERSITY

Biotechnology makes a product from living organisms or modifies the product which has been produced. This technology is also used to improve plants or animals and also develops microorganisms for specific uses. This is quite a wide-ranging technology. It starts from traditional biotechnology to much more advanced and novel techniques like cell and tissue-culture methods and rises to the development of transgenic organisms. Plant biotechnology helps to improve economically important plants, ultimately resulting in higher production. With the rise in production, it also has great impacts on the conservation of biodiversity. Specifically, transgenic technology produces a category of germplasm with new and advanced features. Biotechnology produces cell lines of species characteristics and can also transform the material genetically (Engelmann, 1991). If the germplasm is collected from rapid, medium, and slow-growing multiplication processes, following its storage at slow-growth processes, then cryopreservation works wonders with widespread applications and helps to lower the risk of loss of genetic resources of the plants. Generally, these germplasms are well regarded both for growth and storage conditions when required (Pence et al., 2002). At the laboratory scale, the use of slow growth techniques is a routine matter. Slow growth germplasm technique is useful for conservation of germplasm collected from numerous endangered animal species, both from temperate and tropical areas of origins. If desirable and important somaclonal variations are selected and preserved, then among the growing populations it can serve as an important source of genetic variability if it is persistent in following generations. A number of biotechnological approaches are in place to determine the genetic variability among germplasms. For determination of this variability, either biomolecular markers are used or its determination is DNA-based. Biomolecular markers which are used to determine this variability restriction fragment length polymorphism (RFLP), random amplified polymorphic DNA (RAPD), amplified fragment length polymorphism (AFLP), sequence characterized amplified regions (SCAR), simple sequence repeat (SSR), and inter simple sequence repeat (ISSR). All these techniques are either based on PCR or are not based on PCR (Glaszmann et al., 1987; Khan et al., 2012). Gene banks are developed after full analysis of DNA procured from different methods. The structure of a gene bank requires storage conditions, the provision of an inventory, and evaluation of the viability and verification of genetic stability. Ex-situ approaches along with gene banking are used for identification and conservation of biodiversity.

9.4 CONSERVATION AND SUSTAINABLE USE OF BIODIVERSITY: GENERAL MEASURES

9.4.1 Conservation Measures

9.4.1.1 Early Western Approaches to Conservation

Environmental deterioration due to human action has been observed since classical times. Plato observed the deforested mountains of Attica and stated that they were like bones of a body who's softer and richer parts have been wasted away and only the skeleton is left (Jewett, 1892). Historically, management and supervision of natural resources has been practiced by both Western and non-Western cultures. These countries particularly focused on the protection of wild animal and plant species. They extended their efforts to the protection of land to sustain the production of different agroecosystems present locally.

9.4.1.2 Early Conservation Efforts in North America

European colonists faced the challenges of poor landscape with plenty of diseases, which decimated their population when they entered in the new world. Early settlers exploited the natural resources recklessly. Although there was much destruction and depletion of natural resources, they still tried very hard to protect different natural resources, including soil, water, wild animals, and forests. After the establishment of the new American states through the 1800s, they intermittently passed laws and prohibited the hunting of non-game birds. At the same time, they strictly directed hunters to observe the closed hunting season. In addition, they tried to control predators by various methods to save life and tried to keep pests in balance with their resources.

9.4.1.3 Emergence and Evolution of Biodiversity Conservation

Although at that time there was little scientific knowledge or techniques for sustainable use of natural resources, what there was remained in practice. Sometimes they had to use cultural and artistic expressions. There were revolutionary changes in these practices with the emergence of knowledge and techniques. During the 18th and 19th centuries, Alexander von Humboldt, John and William Bartram, Meriwether Lewis, and George Rogers Clark (and others) embarked on many voyages to explore new and hidden ecosystems. Linnaean introduced binomial nomenclature in 1700 for the naming of plants and animals. Taxonomic research emerged and flourished when this system was adopted. Alfred Russel Wallace and Charles Darwin formulated an evolutionary theory and developments in the field of geology laid down the foundations of evolution. At that time, insufficient biological methods were developed or available, yet field techniques were fully comprehended and people explored the ways in which life diversifies and then maintains itself in that particular area of its inhabitation. At the same time, geologists and other earth scientists started to work on modern climate science and what changes it might exert on the fauna and flora of the earth.

9.4.1.4 Foundations of Conservation Philosophy and Policy

With advancements in knowledge of science, Western concepts of order, beauty, and value of the natural world were revolutionized. Natural philosophers emphasized the smooth functioning of natural systems and advised people to comprehend the natural objects and processes with their own presentation of different concepts of nature. They further highlighted the possible effects of human activities on the natural world. Books written by early economic philosophers in general and Thomas Malthus in particular provided the basics for understanding and considering the intricate complexities of human populations, economies, and the ultimate effects of natural resources on them or vice versa.

9.4.1.5 The Progressive Conservation Movement

Under the banner of the progressive movement in the United States, policy reforms advanced. In that broad platform conservation was one component. Later on, conservation of natural resources appeared as a new platform before the development of the progressive conservation movement. Following this, people started to value the natural world and changed their attitudes toward it. Muir (1901) introduced the importance of natural resources and their conservation. His book appeared about 15 years before an establishment of U.S. National Parks Service. He lamented that beauty in Asia was swiftly deteriorating and, in particular, he noted that plant fauna, although very charming, was also very vulnerable to destruction.

9.4.1.6 Ecology and Conservation

When the institutions started to host the concept of natural resources then the creed of utilitarian conservation was followed. When 1930 approached, application of those utilitarian principles were not only applied to forests but rather it was expanded to other areas of natural resources. For example, river systems, recreational areas, sports, fisheries, game and wildlife populations, agricultural soils, and scenic lands were some other important natural resources. New academic disciplines and professional societies then focused on these various natural resources and started to promote sustainability of production in and from these sources.

9.4.1.7 Biodiversity Conservation and the Modern Environmental Movement

Evolutionary developments occurred in genetics to evolutionary biology. These developments expanded later on to atmospheric chemistry and geology. All these developments made us think and understand the global biosphere and to attempt to determine the place of human beings in this biosphere. New tools were developed for further explorations. These tools facilitated full comprehension of the interrelations within and among the various systems of earth. Terrestrial, atmospheric, and aquatic systems may be quoted as examples. A landmark book entitled *Silent Spring* came from Rachel Carson in 1962. In her book, she examined the biological effects of various synthetic chemical pesticides, e.g., DDT. These studies instigated and motivated us to realize the modern concerns about the prevalent environment and how these concerns might be resolved and mitigated. Although Rachel Carson's book has mainly focused on the indiscriminate use of synthetic pesticides, diverse natural systems have also been discussed in detail. She expressed in her book that nature has created great variety of life on earth but human beings is always in the desire to simplify it. She further elaborated that nature be controlled appropriately and conserved earnestly.

9.4.1.8 Conservation Biology and Environmental Ethics

When changes took place in the environment and foundational sciences, these changes served as a driving force for us to rethink the need for conservation across various disciplines of study. At a global scale, scientists and biodiversity conservationists were alarmed by the rising continuous losses of ecosystems and species by the late 1970s. Destruction of the species-rich forests at an accelerated rate, located in the humid Tropics, was of special concern. The principles of island biogeography stated that human activity had degraded and fragmented the habitats occupied by biota. Further, it has been observed that these areas are small in size and are isolated from each other and are hence at risk. Conservation biologists understood that disruption of ecological functions following the loss of biodiversity present in these ecosystems were closely associated to each other. Therefore, to think about preservation of those ecosystems as well as the life within them would not be an adequate strategy. Hence, in those landscapes which are heavily populated with human beings if the ecosystem processes are disrupted and biodiversity is lost, then it becomes a great and increasing concern for natural resource managers.

9.4.1.9 Conserving Dynamic Systems

The proper balance of nature and recovery of lost species has raised the need to reframe the strategies for biodiversity protection and sustainable use. Biodiversity conservationists have always focused on sustaining biodiversity within a stable landscape but they further emphasize the functions of larger ecosystems, as well as their resistance capability to the changing conditions. Although it can comfortably be said that protected areas are an important tool for conservation, it is hard to ensure protection per se if the medium which is holding and hosting these protected areas is non-sustainable. Therefore, if we want to sustain ecosystems and their services for the benefit of human beings, then diversity of these ecosystems should be fully considered.

9.4.1.10 Ecological Restoration

In the last several decades, it was thought that if we intentionally and willfully renew and rehabilitate the ecosystems which have been previously degraded and destroyed, then scientists were of the view that it will be considered a new intentional renewal and rehabilitation of degraded, damaged, or destroyed ecosystems, which has been considered a new aspect of conservation of biodiversity. This practice is a good alternative which can help protection and sustainability of biodiversity with the recovery of lost diversity of an ecosystem and its function. With the increase in human activities following habitat losses and loss of species, restorationists must face complex challenges under unexpected and rapidly changing climatic conditions (Woodworth, 2013).

9.4.1.11 Range of Variability and Historic Baselines

For the planning and conservation of biodiversity, historical conditions have been a helpful guide for land supervisors, restoration ecologists, and natural resource managers. These conditions provide a gauge to measure progress of conservation and any setbacks faced during the course of conservation. The variations observed in the past are a good record of the nature of the ecosystems which persisted at that time. Hydrologists, foresters, and others used a range of variability observed throughout the history of life for understanding the reasons behind the successful functioning of these ecosystems. With the increase in intensity of anthropogenic activities, these measures need reconsideration because past experiences are no longer a simple and straightforward guide to future aims and actions of conservation. Now, the scientists are of the view that to fully comprehend the past trajectory and present conditions of ecosystems they need to shift previously fixed basslines as a helping guide.

9.4.2 Sustainable Use of Biodiversity

If sustainable resources are used in such a way that an earth can renew them, this is called sustainable use of biodiversity. Sustainability means that we are fulfilling the needs of the present generation without inflicting any harm to resources meant for the future generations. The increase in population means more pressure on the available resources. We have limited resources on our planet and if in general we will not change our consumption patterns and behavior, the resources will deplete swiftly. In the past, humans always tried to recharge the resources and tried to meet the deficiency whenever felt necessary, but now our planet has reached its maximum capacity. It is very hard to renew it or recharge and hence its sustainable use and proper management has become mandatory. When we talk about management of resources, this covers many facets of diversified natural resources, including forestry, agriculture, tourism, fisheries, and water management.

For the sustainable use of biodiversity, in 2004 the parties to the CBD adopted the Addis Ababa Principles and Guidelines. These principles and guidelines are actually a set of 14 principles which are equally applicable to all activities and areas. These principles were reviewed and reaffirmed in 2010 and, accordingly, their implementation status was studied in depth. All governments are required to adopt the guidelines of the CBD in letter and spirit and integrate matters regarding conservation of biodiversity and its sustainability into their national policies. There are several sectors where we can apply the sustainability which are: (1) environmental impact assessment; (2) certification and eco-labeling; (3) quotas for fisheries and prohibited use of large nets; (4) organic farming; (5) management of protected areas; (6) reduction of bush fires; and (7) sustainable and limited tourism which current resources can sustain. It is expected that fish demand will grow proportionate to the increase in the human population. This issue will be more stringent in developing countries.

The 2010 Living Planet Report explains that since 1966 humanity's ecological footprint has approximately doubled. Destruction of mangroves has been observed, which has caused erosion of coastlines, a lack of nurseries for fish and shellfish, and siltation of coral reefs. The continuous decline in mangroves is their conversion into agricultural land.

Accordingly, a huge population of people (300 million) directly or indirectly depends on forests. Another 60 million people of indigenous and tribal groups depend completely on forests. Illegal harvesting and logging of forests and their products is a serious problem which costs an estimated amount of US$15 billion per year. Those tree species which are rare in nature and those which have higher value for timber and non-timber products are seriously in danger of extinction.

9.5 IN-SITU CONSERVATION

The process of recovery and conservation of natural habitats and ecosystems and maintenance of the viable population in their natural habitats is called in-situ conservation of biodiversity. In-situ ('onsite', 'in place') conservation uses several conservation techniques to conserve population without its displacement. These techniques first designate the population to be conserved, manage, and finally monitor its survival and propagation pace at the place where that population inhabits. In contrast to ex-situ conservation—which is always away from the site—in-situ conservation is at the place where that particular species is living with no displacement. In ex-situ ('offsite') conservation, implementation of conservation techniques is always away from the targeted natural habitat. Further, there is no ambiguity in the initiatives taken for ex-situ conservation. The initiatives taken in this method of conservation are: (1) the establishment of zoos; (2) captive breeding programs; and (3) development of seed banks. The in-situ conservation concept has a wider scope. It starts from the designing of sustainable management strategy for a particular habitat up to the establishment of a protected area.

9.5.1 Purpose and Use of In-Situ Conservation Strategies

In-situ management approaches can be either an ecosystem-based approach when conservation strategy is applied to the whole ecosystem or it can be a species-centered approach, when it is applied to a population which is constituted of selected species. The objectives of both these approaches are the same. The population of organisms present in any ecosystem is allowed to self-replicate for maintenance of its potential and to provide the material for perpetual ongoing evolutionary processes. Both these approaches require that all components of the natural system, namely species, populations, communities, and biophysical systems—as well as the ecological and evolutionary processes which are happening within that system—be conserved too. The focus of conservation measures is always surrounding of the target species where it is living, which can be an environment of the species with extensive modifications by human activities or be it a natural habitat for that particular species. In human-dominated environments, agricultural and domestic species have evolved with distinguishing traits. Therefore, when we conserve agriculture biodiversity, farmers should conserve agro-ecosystems through the use of conventional farming practices.

9.5.2 Examples of In Situ Conservation Initiatives

In-situ conservation is based on protected areas. The protected area network maintains target species and their habitats and conserves natural or semi-natural ecosystems which ultimately contribute towards the achievement of conservation targets. Awareness is growing for the extension of in-situ conservation beyond the protected areas by virtue of its importance. If a protected area is surrounded by socio-economic and political forces, then establishment of that protected area and conservation of its threatened flora and fauna will be difficult to achieve. In these situations, alternative "in-situ conservation management approaches" can be more useful. There is a possibility of enhancement of the Indonesian coral reefs by the implementation of management plan which is an opening and closing of reef harvesting resources within a prescribed location. There are several other in-situ initiatives which may be implied for conservation of biodiversity. These are: (1) planning strategically how to use biological resources sustainably following their efficient management; (2) habitat restoration, which may recover animal/plant species and rehabilitate the disturbed habitat; and (3) designing of various recovery programs for recovery of threatened and endangered animal/plant species at the national or sub-national level. Agriculture diversity conservation is applied with similar measures. Genetic resources are conserved for traditional crops and their wild counterparts, including both animals and plants. After application of genetic resources their performance is regularly monitored. Similarly, it is tried to alleviate other threats like control of alien species. At the same time efforts are made to preserve the species and to maintain traditional knowledge. Implementation of regulatory practices and legislation and proper management and exercising other frameworks to safeguard species and their habitats are some other measures used for the conservation of species and populations.

9.6 EX-SITU CONSERVATION

When zoos, botanical gardens, museums, gene banks, etc. are used for the conservation of components of biodiversity, it is called ex-situ conservation because conservation measures are being used outside of their natural habitat. In principle, these facilities are used for the protection of those species which are endangered and are on the verge of extinction. This type of conservation is also called captive conservation. Exhaustive efforts have been undertaken for the collection and preservation of genetic material of several animals in general and birds, fish, and plants in specific (Rawat & Agarwal, 2015). This work is performed in universities or very advanced research institutions. In addition to these methods, several other measures are also executed to conserve biodiversity. These measures include captive breeding programs, reintroduction of species which are on decline, and artificial feeding to food deficient animals to make up their health and facilitate their propagation. If we reintroduce any plant or animal to a place where it is declining or has become extinct, to meet its deficiency is also an example of ex-situ conservation. Introduction of the *Hobara bustard* in Rahimyar Khan can be quoted as an example of ex-situ conservation. Ex-situ conservation is also practiced through recreational, horticultural, and botanical gardens and seed banks. Ex-situ conservation measures are complementary to in-situ conservation.

9.6.1 The Relationship between In-Situ and Ex-Situ Conservation

Over the history of life it has been observed that the onsite (in-situ) method of conservation of biodiversity is more appropriate, holistic, and real. This method allows applying and involving those measures which the ex-situ approach even can hardly sense (Wikipedia, 2018). There are some exceptions to ex-situ conservation. These are monitoring and conservation of evolutionary processes, soil microbial processes, and specific ecosystems such as coral reefs. Ex-situ conservation cannot be applied to those species which have specialized needs to survive and propagate. The CBD recognizes that both ex-situ and in-situ conservation approaches complement each other in their principles and functions. In particular, the CBD specifies that the ex-situ conservation technique is applied for the complementation and support of in-situ conservation measures.

9.7 CRYOBIOLOGY, EMBRYO TRANSFER, AND ARTIFICIAL INSEMINATION IN EX-SITU ANIMAL CONSERVATION

When habitat of wild animals is degraded and destroyed, this threatens the preservation of genetic diversity of these animals (Croner, 1984). It is expected that if the present pace of habitat destruction persists, then 1 million species of plants and animals will become extinct (AAZPA, 1983). Continuous extinction of wild animals has generated interest in the field of wildlife production. Accordingly, zoo researchers and wildlife experts have started to study different methods of reproduction in wild animals. In this context, for the improvement of reproductive potential of wild animals, experts have also started to use technologies like embryo transfer and artificial insemination. Zoos now also have started their role as conservators rather than merely displayers, as has been seen in the past. Due to inbreeding of animals there is a loss of genetic diversity in zoos. This problem is also prevalent in those areas where animal numbers have been severely depleted and they cause a genetic bottleneck. An example can be quoted of a primate from Southeast Asia called the douc langur (*Pygathrix nemaeus*), which has almost become extinct in the wild. In captivity, their number is not enough to ensure the genetic diversity which is actually required to keep the species alive for the future (Gorman, 1980). For successful survival of the species, genetic diversity is a must. Enough variation is required in the genome of a species enabling it to adapt to the changing environment. Inbreeding has decreased fertility, and has caused high juvenile and birth defects in zoos. Scientists have realized now that, for animal species which are closer to extinction, advanced technologies like embryo transfer, cryopreservation of gametes, and artificial insemination may maintain the genetic diversity which is applicable only in captivity because the applications of this technology are still in their infancy in the wild.

9.7.1 Embryo Transfer

This technology follows the removal of fertilized ova and early embryos from the reproductive tract of a donor female, also called the genetic mother, and its transfer into the reproductive tract of a recipient female, also called the foster mother. This embryo develops into a fetus in the foster mother and the foster mother gives birth to this developed young one. Heape (1891) carried out the first successful attempt of the transfer of an embryo in a rabbit. Specific methods are used for each species. Nevertheless, the general principles are the same, whether it is applied for laboratory animals like rabbits and mice, or applied for large domestic animals like horses and cattle, or in wild animals like baboons and antelopes. In cattle and mice, embryo transfer has become routine.

In embryo transfer, injections of prostaglandins like Lutalyse and Estrumate are precisely timed. This synchronizes the timing of both the donor and the recipient female. On the injection of these hormones, the ovaries begin a new cycle. Injections of follicle stimulating hormones (FSH)—also called fertility hormones—accomplish superovulation in the donor. The process of superovulation has been optimized for ungulates but dosages for other animals still need to be refined. It has been observed in felines, e.g., cats which ovulate following a reflex action. This means that they do not ovulate until the process of copulation stimulates the process of ovulation. Therefore, unlike other animals, in cats synchronization of superovulation in donors and recipient females is a complex process. Research on embryo transfer in cats is in progress but is still limited (Kraemer et al., 1979; Goodrowe et al., 1986; Dresser et al., 1987), and in the case of wild cats research work on this aspect is very scanty (Reed et al., 1981; Bowen et al., 1982). If this technology becomes successful in domestic cats, then it can be applied to their wild counterparts called the black-footed cat, *Felis nigripes*. Nevertheless, much research work is still required for the determination of the best time when cat ovaries are stimulated to produce more follicles than are routinely reproduced. Also, information is lacking on the exact timing of ovulation with ultimate embryo production. If we want to maximize genetic diversity in domestic animals, then the development of embryo transfer techniques is essential. Zoo researchers are working to develop interspecies embryo transfers in such a way that collection of an embryo is taken from the female of an endangered species and then transferred to a more common species, which can act as its surrogate for development of that particular embryo. This can greatly enhance the reproductive potential of the donor species (endangered species). It has also been observed that there is a drastic reduction in disease transfer in embryo transfer technology (Hare, 1985). The possible reason for this beneficial effect can be that the intact embryo present in the donor female might be free of microbial or viral disease agents and the recipient female may confer passive immunity to her offspring. Although there are several advancements in the field of biological sciences, nevertheless whenever an embryo is transplanted from one location to another, be it in a domestic female or a wild female, the consequences are admirable. However, the transfer of embryos between species has had limited success and is far more limited in wild species. Intergeneric transfer of embryos like eland to cow (Dresser et al., 1982a) or water buffalo to cow (Drost, 1983) has not been successful. However, interspecies transfer, like tiger to lion (Reed et al., 1981) for example, has had encouraging results. Embryos can develop up to the blastocyst stage in the oviducts of those species which are not related to each other (Daniel, 1981). Further, for the proper development of embryo developments, donor and recipient need to be closely related. Similarity is required in the rate of ovum transport, implantation, and placenta formation. In addition to that, the length of gestation, weight at birth of the young one, and neonatal and maternal postpartum behavior of both donor and recipient are also important for the successful development of the embryo. Recently, a domestic horse mare was injected with donkey white blood cells and she gave birth to a donkey foal. Antczak (1985) reported that a domestic horse mare also gave birth to donkey foal even when she received an embryo from a donor female. Prior to this, none of the attempts was successful. For the collection of embryos from non-human primates, various attempts have been made but very few reports are available in the scientific literature regarding these cases. In a little over six years, dramatic progress has been observed in the micromanipulation of embryos specifically in domesticated animals. Willadsen (1979, 1980) and Willadsen et al. (1981) talked about the most important application of micromanipulation to exotic animals. In this technique, an embryo is surgically divided into two or more pieces, which are genetically identical twins or triplets. Although this technique could be relatively successful with exotic animals, nevertheless it could quickly increase the number of endangered and rare animals. In this context, it is further added that there should be no confusion in the transfer of embryos between species and/or inner cell mass transfer and hybridization. There is a possibility that the methods used for the production of chimera can be utilized for the rescue of endangered species. Possibly, chimeras could be produced in which an embryo is collected from an endangered species and is transferred to a common species, but this common species is related to an endangered species. Embryo transfer technique in particular has become an important technique for the production of young in domesticated animal species. In the wild, it is still a novel technique. Extensive research might help in the successful transfer of embryos in wild species, as has been practiced in domestic animal embryos for many years.

9.7.2 Cryopreservation of Embryos

Whittingham et al. (1972) reported the first successful freezing of mammalian embryos. Embryos of mice, rabbits, and cows have been routinely frozen and to date embryos from 11 species of mammals have been frozen with success. Examples of mammal species whose embryos have been successfully frozen include (1) horses (Yamamoto et al., 1982); (2) goats (Bilton & Moore, 1976); (3) sheep (Whittingham et al., 1972); (4) antelopes (Dresser et al., 1984); (5) cows (Wilmut & Rowson, 1973); (6) rabbits (Bank & Maurer, 1974); (7) mice (Whittingham et al., 1972); (8) rats (Whittingham, 1975); (9) cats (Dresser et al., 1987); (10) human beings (Trounson & Mohr, 1983); and (11) baboons (Pope et al., 1984). Among these freezing of mouse embryos have been the most successful. To date, thousands of mouse embryos have been frozen for a sufficient period of time to preserve valuable genetic stocks. In the case of mice, rabbits, and cows, freezing of embryos is accomplished with very advanced and sophisticated methods that when frozen embryos were thawed even up to 80–90% embryos develop. In this way, thousands of live young have been produced from frozen and subsequently thawed embryos in domestic species (Gupta et al., 2016). Studies are still limited in wild animals for embryo transfer to the recipient female. It is hoped that cryopreservation will support the reproductive research in non-domestic animals. Myers (1984) has reported that a group of 250 animals theoretically persevered 95% of their genetic material after laps of 50 generations. It means that it took them 400 years for conservation of their genetic material.

At a global level, all zoos have limited facilities for cryopreservation There is also a possibility of collecting new bloodlines from wild animal populations. This can be frozen in the field and then transported to zoos, before being inserted into the captive population to improve their bloodlines. In this way, the captive population will be improved without depletion of wild populations of different species.

9.7.3 Artificial Insemination

When semen is introduced artificially into the vagina or cervix of a female animal, this technique is called artificial insemination. Betteridge (1981) has reported that it is assumed that ancient scientists used this method of insemination but in the modern world it first occurred in 1784, where dogs were artificially inseminated. In the 1930s, Russia used this technique quite extensively in the insemination of livestock. Betteridge (1981) has further reported that the potential of artificial insemination was further revealed when in the 1920s and 1930s fresh semen of rabbits, sheep, and bulls was shipped from England to various other European countries. The discovery of the cryoprotective action of glycerol facilitated the successful freezing of semen. Today, this technique is widely and successfully used in the agriculture industry. One estimate tells that, in 1977, using the technique of artificial insemination about 90 million head of cattle were produced (Foote, 1981), where previously frozen semen was used. This technique is still very limited to birds and mammals in the wild, specifically. Examples of animals where this technique has been used successfully are birds, reptiles, cranes (several species), waterfowl (ducks), tortoises, albino cockatiels, pheasants, raptors, brown brocket deer, guanaco reindeer, red deer, wolves, squirrel monkeys, blackbucks, Speke's gazelles, Persian leopards, chimpanzees, bighorn sheep, giant pandas, Puma gorillas, and Ferret Rhesus monkeys. Although sufficient semen was collected from wild animals for artificial insemination, its evaluation and use has remained very limited. The wolf was the first animal which was inseminated in 1973 (Seager, 1981). A gorilla then successfully followed this (Douglass & Gould, 1981). Several attempts

have been made to inseminate cats artificially (family Felidae), but most of these attempts were fruitless (Dresser et al., 1982b). In 1980, London Zoo was successful in the production of a puma using a fresh semen sample (Moore et al., 1981). Production of a Persian leopard using fresh semen in Cincinnati zoo also followed this development (Dresser et al., 1982b). Semen is collected by three methods: (1) the male reproductive tract is stimulated manually; (2) an artificial vagina is used; and (3) the ejaculation of semen is facilitated by electric pulses (Cherfas, 1984). Among all of the three methods, electroejaculation is the most common mode of collection of semen for wild animals. The male animal is anesthetized and a lubricated probe is inserted into its rectum. This probe creates electrically driven stimuli resulting from pulsation which stimulates the reproductive tract of the animal, which results in ejaculation. The fertility of sperm collected in this manner is debatable. Several problems have been faced when this technique is extended to wild animals. These issues and problems are: (1) anesthesia is required to accomplish this, which is risky for both male and female for semen collection and for insemination respectively; and (2) the fertility of semen collected by electro-ejaculation is always questionable as compared to that which is collected naturally. Soon after collection, the sperm start to die. Even if a fresh sample is used, there is still doubt about the loss of a certain amount of sperm.

9.7.4 Semen Cryopreservation

The freezing process is not as safe for eggs as it is for sperm. Sometimes freezing can be detrimental to sperm and there is a possibility that after thawing they will not be as motile as they were before thawing (Cochran et al., 1985). For exotic animals, a great deal of work still remains to be done. Seager (1981) reported that the sperm of about 200 animal species were frozen but not many have been thawed and tested for fertility. Production of offspring is the ultimate test of the viability of the sperm. From the few semen samples which cryobiologists tested, they came up with the suggestion that each species needs slightly different conditions for extension and freezing than is currently practiced. A buffered solution is actually used as an extender that contains a cryoprotectant, e.g., glycerol, antibiotics, and either egg yolk or milk. In the basic recipe of semen extender, several variations have been made. The agriculture experts say that the extender is species specific in domestic animals and the same is also applicable in wild animals. However, it is probable that experts have to work with reptiles which have the capability to keep sperm viable in the abdomen of a female even up to six years after mating (Cherfas, 1984). Determination of the exact timing of artificial insemination is another problem and it is a fact that artificial insemination cannot be practiced frequently as the animal is inseminated naturally during the estrous cycle. Taking it positively, artificial insemination can greatly improve the genetic diversity of captive animals. Compared to embryo transfer, the risk and expense are far less than transporting animals from place to place for breeding purposes. Artificial insemination can alleviate the limitations of quarantine of animals, thus minimizing the risks of disease outspread.

9.8 SUSTAINABLE USE OF COMPONENTS OF BIOLOGICAL DIVERSITY

The various components of biodiversity should be used in a sustainable manner. Categorically, sustainable use can be expressed as follows.

Sustainable use means that all species, ecological processes, and genetic variability present in that particular ecosystem do not touch the threshold levels and remain adequately above them. Thus, biodiversity can be viable for a longer period of time and can also support the systems dependent on it. Total depletion of resources will completely halt the services it is providing. With this setback, it will also lose the option of recovery. Ecosystems and ecological processes operating in these ecosystems may change over time in nature and mechanism. Existing species can vary and/or new one can develop. Genetic variability present in these ecosystems may change over time. All these changes are spontaneous, irrespective of whether these resources are used or not. Therefore, it is mandatory for resource managers and governments responsible for maintenance of biodiversity to take in account the possibility of accommodating these changes for the viability of life in an ecosystem. In a given region, population is normally randomly distributed that might have adverse effects on biodiversity and its sustainability. These distributions should be considered during the use of biodiversity for utilitarian purposes.

There are circumstances where the chances of conversion of natural landscapes for other purposes is quite high. Under these circumstances, proper sustainable use of biodiversity can provide sufficient opportunities and incentives for the maintenance of ecosystems and habitats for various organisms. Sustainable use of biodiversity also facilitates the maintenance of species present in that particular system and genetic variability within them. To take an example, the crocodile is a dangerous animal and is a threat to human beings. However, when its utilization was sustainable it provided sufficient incentives for its conservation both in captivity and in the wild.

Biological diversity is producing a great deal—directly or indirectly—for the wellbeing of human beings and other organisms living in that vicinity. These products mainly constitute food, shelter, clean air, and freshwater, which are basic necessities of life. In many countries, it has been observed that the millions of the people (specifically poorer people) depend heavily on harvested animals and plants for their livelihoods. Biological diversity also provides pharmaceuticals for the prevention of diseases and to cure human beings and sustain their health. It has further been observed that indigenous and local communities and cultures where they are living are directly dependent on biodiversity for their sustenance. Under such heavy utilization of natural resources, concerned governments should devise appropriate policies and strategies following their strict implementation. They should further ensure that that all these policies are in place for the sustainable use of biodiversity and its future viability. There are possibilities when intrinsic biological characteristics of both species and ecosystem limit biological products and ecosystem services. These characteristics also restrict productivity of the system, its resistance to the ecological conditions, and stability of the system. Those biological systems which totally depend on the recycling of infinite resources are limited in the goods they can provide, as well as the services they are providing. There are some technological breakthroughs which can extend these limits to some degree, nevertheless constraints exist which limit the availability and accessibility that endogenous and exogenous resources can impose. Therefore, it is compulsory for all resource users to apply caution in their management decisions to obviate the potential negative effects on biodiversity encouraging sustainability of ecosystems and usage of their services.

REFERENCES

AAZPA (American Association of Zoological Parks and Aquariums). (1983). Species Survival Plan Handbook Publication. Oglebay Park, Wheeling, W. Va. 24 pp.

Antczak, D.F. (1985). Cute Cross-Species Birth. Science News 128, 8.

Baig, M.B. & Al-Subaiee, F.S. (2011). Biodiversity in Pakistan: Key Issues. Biodiversity 10(4), 20–29.

Bank, H. & Maurer, R.R. (1974). Survival of Frozen Rabbit Embryos. Experimental Cell Research 89(1), 188–196.

Barnosky, A.D., Matzke, N., Tomiya, S., Wogan, G.O., Swartz, B., Quental, T.B., Marshall, C., McGuire, J.L., Lindsey, E.L., Maguire, K.C., Mersey, B. & Ferrer, E.A. (2011). Has the Earth's Sixth Mass Extinction Already Arrived? Nature 471(7336), 51–57.

Betteridge, K.J. (1981). An Historical look at Embryo Transfer. Journal of Reproduction and Fertility 62, 1–13.

Bilton, R.J. & Moore, R.W. (1976). In *Vitro* Culture, Storage, and Transfer of Goat Embryos. Australian Journal of Biological Science 29, 125–129.

Bowen, M.J., Platz, C.C. Jr, Brown, C.D. & Kraemer, D.C. (1982). Successful Artificial Insemination and Embryo Collection in the African Lion *(Panthera leo)*. American Association Of Zoo Veterinarians. 57–59.

CBD (1992). Convention of Biological Diversity, United Nations Environment Program.

Cherfas, J. (1984). Test-Tube Babies in the Zoo. New Scientist, 104(1433).

Cochran, R.C., Judy, J.K., Parker, C.F. & Hallford, D.M. (1985). Prefreezing and Post-Thaw Semen Characteristics of Five Ram Breeds Collected by Electroejaculation. Theriogenology 23, 431–440.

Croner, S. (1984). An Introduction to the World Conservation Strategy. Prepared for the International Union for Conservation of Nature and Natural Resources (IUCN) by its Commission on Education. IUCN, Gland, Switzerland; United Nations Environmental Program, Nairobi, Kenya. 28 pp.

Daniel, J.C. (1981). Preserving the Genome of Endangered Species. SWARA—East African Wildlife Society 4, 16–18.

Dobhal, P.K., Kohli, R.K. & Batish, D.R. (2011). Impact of *Lantana camara* L. Invasion on Riparian Vegetation of Nayar Region in Garhwal Himalayas (Uttarakhand, India). Journal of Ecology and the Natural Environment 3, 11–22.

Douglass, E.M. & Gould, K.G. (1981). Artificial insemination in lowland gorilla *(Gorilla g. gorilla)*. pp. 128–130 in Annual Proceedings—American Association of Zoo Veterinarians. Hills Division, Riviana Foods, Topeka, Kansas.

Dresser, B.L., Kramer, K., Dahlhausen, R.D., Pope, C.E. & Baker, R.D. (1984). Cryopreservation followed by successful transfer of African eland antelope *(Tragelaphus oryx)* embryos. pp. 191–193 in Proceedings of the 10th International Congress on Animal Reproduction and Artificial Insemination. Standing Committee of the International Congress of Animal Reproduction and Artificial Insemination, University of Illinois at Urbana-Champaign, Illinois.

Dresser, B.L., Kramer, L., Pope, C.E., Dahlhausen, R.D. & Blauser, C. (1982a). Superovulation of African Eland *(Taurotragus oryx)* and Interspecies Embryo Transfer to Holstein Cattle. Theriogenology 17, 86.

Dresser, B.L., Kramer, L., Reece, B. & Russell, P.T. (1982b). Induction of Ovulation and Successful Artificial Insemination in a Persian Leopard *(Panthera pardus saxicolor)*. Journal of Zoo Biology 1(1), 55–57.

Dresser, B.L., Sehlhorst, C.S., Keller, G., Kramer, L.W. & Reece, R.W. (1987). Artificial Insemination and Embryo Transfer in the Felidae. In: Tigers of the World: The Biology, Biopolitics, Management and Conservation of an Endangered Species Au, eds. Tilson, R. & Seal, U. pp. 287–293. Noyes Publications, Park Ridge, NJ.

Drost, M. (1983). Reciprocal embryo transfer between water buffaloes and cattle. pp. 63–71 in Proceedings of the Annual Meetings of the Society for Theriogenology. Society for Theriogenology, Hastings, Nebr.

Engelmann, F. (1991). *In Vitro* Conservation of Tropical Plant Germplasm – A Review. Euphytica 57, 227–243.

Foote, R.H. (1981). The Artificial Insemination Industry. In: New Technologies in Animal Breeding, eds. Brackett, B.G., Seidel, G.E. & Seidel, S.M., pp. 13–39. Academic Press, New York.

Glaszmann, J.C. (1987). Isozymes and Classification of Asian Rice Varieties. Theoretical and Applied Genetics 74, 21–30.

GONWFP. (1997). Sarhad Provincial Conservation Strategy. Peshawar: Sarhad Program Office, IUCN-Pakistan, 248 PP.

Goodrowe, K.L., Howard, J.G. & Wildt, D.E. (1986). Embryo Recovery and Quality in the Domestic Cat: Natural versus Induced Estrus. Theriogenology 25, 156.

GOP (1999). Biodiversity Action Plan of Pakistan. Ministry of Environment/IUCN/WWF. Islamabad.

GOP (2007). Third National Report on Implementation of Convention on Biological Diversity (CBD) in Pakistan Directorate of Biodiversity, Ministry of Environment, Government of Pakistan, Islamabad, Pakistan.

Gorman, J. (1980). Brave New Zoo. Discover 1, 46–49.

Gupta, V., Sengupta, M., Prakash, J. & Tripathy, B.C. (2016). Animal Cell Culture and Cryopreservation. In: Basic and Applied Aspects of Biotechnology. Springer Nature, p. 527.

Hare, W.C.D. (1985). Diseases Transmissible by Semen and Embryo Transfer. Office Internationale des Epizost, Paris, 28–33.

Heape, W. (1891). Preliminary Note on the Transplantation and Growth of Mammalian Ova within a Uterine Foster-Mother. Proceeding of Royal Society of London 48, 457–458.

Khan, S., Al-Qurainy, F. & Mohammad, N. (2012). Biotechnological Approaches for Conservation and Improvement of Rare and Endangered Plants of Saudi Arabia. Saudi Journal of Biological Sciences 19, 1–11.

Kraemer, D.C., Flow, B.L., Schriver, M.D., Kinney, G.M. & Pennycook, J.W. (1979). Embryo Transfer in the Nonhuman Primate, Feline and Canine. Theriogenology 11, 51–62.

Millstein, R.L. (2015). Is Aldo Leopold's "Biotic Community" "Land Community" an Individual? Presented at ISHPSSB 2015, July 2015, Montreal, QC Department of Philosophy; Science and Technology Studies University of California, Davis. http://www.RLM.net/.

Moore, H.D.M., Bonney, R.C. & Jones, D.M. (1981). Induction of oestrus and successful artificial insemination in the cougar, *Felis concolor*. pp. 141–142 in Annual Proceedings—American Association of Zoo Veterinarians. Hills Division, Riviana Foods, Topeka, Kans.

Muir, J. (1901). "Hunting Big Redwoods," Atlantic Monthly 88(528).

Myers, N. (1984). Cats in Crisis! International Wildlife 14(Nov-Dec), 42–52.

Pence, V.C., Clark, J.R. & Plair, B.L. (2002). Wild and Endangered Species. In: *In Vitro* Collecting Techniques for Germplasm Conservation. IPGRI Technical Bulletin, Vol. 7, eds. Pence, V.C., Sandoval, J.; Villalobos, V.; Engelmann, F., pp 76–82, IPGRI, Rome.

Pope, C.E., Pope, V.Z. & Beck, L.R. (1984). Live Birth Following Cryopreservation and Transfer of a Baboon Embryo. Fertility and Sterility 42, 143–145.

Rawat, U.S. & Agarwal, N.K. (2015). Biodiversity: Concept, Threats and Conservation. Environment Conservation Journal 16(3), 19–28.

Reed, G., Dresser, B., Reece, B., Kramer, L., Russell, P., Pindell, K. & Berringer, P. (1981). Superovulation and artificial insemination of Bengal tigers *(Panthera tigris)* and an interspecies embryo transfer to the African lion (*Panthera leo*). pp. 136–138 in Annual Proceedings—American Association of Zoo Veterinarians. Hills Division, Riviana Foods, Topeka, Kans.

Seager, S.W.J. (1981). A Review of Artificial Methods of Breeding in Captive Wild Species. The Dodo: Journal of the Jersey Wildlife Preservation Trust 18, 79–83.

Shashidharan, N. & Quadros, G. (2013). Bridging gaps: Open Source Geospatial Technology as a Public Participatory Tool for Landscape Assessment. National Conference on Biodiversity: Status and Challenges in Conservation FAVEO 224–230.

Trounson, A.O. & Mohr, L. (1983). Human Pregnancy Following Cryopreservation, Thawing, and Transfer of an Eight-Cell Embryo. Nature 305, 707–709.

Whittingham, D.G. (1975). Survival of Rat Embryos after Freezing and Thawing. Journal of Reproduction and Fertility 43, 515–518.

Whittingham, D.G., Leibo, S.P. & Mazur, P. (1972). Survival of Mouse Embryos Frozen to −196° and −269°C. Science 178, 411–414.

Wikipedia. (2018). "Ex Situ Conservation." Wikipedia, Wikimedia Foundation, 9 July 2018.

Willadsen, S.M. (1979). A Method for Culture of Micromanipulated Sheep Embryos and Its Use to Produce Monozygotic Twins. Nature 277, 298–300.

Willadsen, S.M. (1980). The Viability of Early Cleavage Stages Containing Half of the Normal Number of Blastomeres in the Sheep. Journal of Reproduction and Fertility 59, 357–362.

Willadsen, S.M., Lehn-Jensen, H., Fehilly, C.D. & Newcomb, R. (1981). The Production of Monozygotic Twins on Preselected Parentage by Micromanipulation of Non-Surgically Collected Cow Embryos. Theriogenology 15, 23–30.

Wilmut, I. & Rowson, L.E.A. (1973). Experiments on the Low-Temperature Preservation of the Cow Embryos. Veterinary Record 92, 686–690.

Woodworth, P. (2013). Our Once and Future Planet: Restoring the World in the Climate Change Century. Chicago: University of Chicago Press.

Yamamoto, Y., Oguri, N., Tsutsumi, Y. & Hachinohe, Y. (1982). Experiments in the Freezing and Storage of Equine Embryos. Journal of Reproduction and Fertility. Supplement 32, 399–403.

10 Conservation Genetics

10.1 INTRODUCTION

Conservation genetics is usually applied and is instrumental to the conservation of biological resources. Previously, it has been too sophisticated and expensive but the situation now is very different and quite convenient. Nowadays, conservation genetics address many complicated management issues and queries. These analytical methods can further strengthen the quality of assessments made with the help of traditional methods. Accordingly, they throw light on the new information. Living organisms vary a great deal among themselves. Differences are obvious and common within species, between species, and across the ecosystems. When we talk about genetics, each species possesses a tremendous quantity of genetic information. This genetic information which each individual obtains from its parents generates its form, bodily functions, and its behavior in the given environment and ecosystem, ultimately governing reproductive success. The human genome contains about 20,000 to 30,000 genes.

Additionally, many species consist of multiple breeding populations. Each population has a different genetic variation. Hence, it interacts with other populations with different abilities. Furthermore, populations of different species coexist in different communities and territories. When combined with the environment and their own interactions, these communities give rise to an ecosystem. For the maintenance of a balanced and functioning ecosystem, regardless of its level conservation of biological diversity is highly important. During conservation of biodiversity, we need to consider both population conservation and the genetic pool.

10.1.1 Conservation Genetics

This is a subfield of population genetics and is a combination of conservation biology and genetics. Conservation genetics can be an applied or a basic science. It is a multidisciplinary subject. For its application to a target group or community, conservation genetics takes support from evolutionary taxonomy (study of family relationships), population genetics, ecology, molecular biology, and mathematical modeling. Conservation genetics mainly deals with the endangered and threatened populations. Under conservation genetics scientists help these populations by exploring the factors which brought these populations to the brink of extinction, and then work on the various steps which can be taken to reverse this trend. Previously, mathematical, evolutionary, and taxonomic aspects were given due consideration for conservation efforts, which provide an understanding of the dynamics of genes in a population. This information is then used for the management of natural resources and for the prevention of animal extinction. However, conservation biology is very much dependent on genetic diversity. Other than species diversity and ecosystem genetic diversity, conservation biology is one of the three fundamental measures of biodiversity and hence plays a significant role in what is now a really extended field.

Recently, conservation scientists have used genetic studies which have given comparatively better and more comprehensive information about biodiversity among the individuals in a population. Conservation genetics only help to conserve the right population. In this way, it does not waste valuable as well as those resources which are actually not endangered. With changing habitats, the animal population is endangered or threatened. Proper management of animal ecosystems helps in the preservation of threatened and endangered species.

DOI: 10.1201/9781003472001-10

10.2 GENETIC DIVERSITY

DNA of all the living organisms is organized in genes like letters in words. Genes control the functions of our bodies. DNA code differs from species to species, as well as within the same species, which is called genetic diversity. Genetic diversity gives color our hair, flavor to tomatoes, and shapes of beaks of birds (Minter et al., 2021). All these biological variations are attributed to genes. Genetic diversity constitutes the mean number of alleles per locus, or the percentage of polymorphic loci, ultimately heterozygosity, and expected heterozygosity. Genetic diversity is a raw material for evolution. It generates variation in various life forms, communities, and ecosystems. On earth, genetic variation give shapes to individuals, species, subspecies, populations, and ultimately the animal/plant kingdom. Genetic diversity among individuals is due to the presence of different alleles present in the gene pool of the particular community or population (Mukhopadhyay & Bhattacharjee, 2016). All the phenotypic plasticity is dependent on the genetic variability of any organism, and hence genetic diversity is very important from both the individualistic and population point of view. This phenotypic plasticity helps an individual or a population in the adaptation and evolution of that population in different environmental pressures. For the development of appropriate strategies for conservation biology as well as in many other applied fields, information about genetic diversity of that particular community or population is necessary. Genetic diversity also plays a fundamental and crucial role for the evolutionary potential of a species. Genetic diversity at the population level is a crucial focus for conservation genetics because it reduces the fitness. Accordingly, it impacts both the health and the long-term survival of populations. Juvenile mortality is the major issue. Higher juvenile mortality means reduction in immunity, which results in the culmination of population growth (Leberg, 1990). When combined, all of them contribute to the extinction risks of the population (Saccheri et al., 1998; Frankham, 2005).

In population genetics, heterozygosity measures genetic diversity. It determines the chance of a population which can survive to an environmental change. It further gives information on the effect of novel pathogens on the survival of the population and the average fitness of that population over successive generations. Further, heterozygosity has a deep connection to population genetics theory as well as to population size and hence it has a fundamental importance to conservation. As mentioned above, small populations have low genetic diversity and hence they are less heterozygous. Comparatively, therefore, these populations have lower chances of survival of the challenges of the environmental pressures (www.sciencedirect.com. Retrieved, 2023).

In a small population without gene flow, the probability of mating with close relatives becomes very high, which is very much plausible over successive generations. Consequently, this reduces the fitness of the population due to inbreeding depression. Inbred parents produce homozygous individuals, while only outbred individuals produce heterozygous offspring. When heterozygosity is low it produces less fit individuals. Low genetic diversity inhibits the chromosomes crossing over during meiosis with non-creation of new combinations of alleles on chromosomes. The average length of un-recombined tracts of chromosomes that are inherited from parents is increased. In turn, the efficacy of selection is reduced due to these lengthy chromosomes. Due to close relationship between the population size, mutation rate, and genetic diversity, immediate recognition of those populations which are on the verge of genetic loss is very important. Because only mutation and gene flow can restore loss of biodiversity, it is easily lost.

Advances in genetic and molecular technologies are the most powerful catalysts in the field of conservation. These advances lead to a wide variety of molecular methodologies which can be applied in conservation and population genetic studies. Recently, there has been intensive utilization of molecular methods in conservation biology and genetic diversity studies. These methods are primarily used as selective molecular tools to resolve the empirical questions of conservation and evolutionary concern.

10.3 IMPORTANCE OF GENETIC DIVERSITY

Genetic diversity is important because it gives species a better chance of survival. However, when populations get smaller isolated genetic diversity can be lost. Loss of genetic diversity decreases ability of the species to adapt to the environment, which ultimately affects the survival of that species (Minter et al., 2021). Low genetic diversity of many genes present in species or populations is risky in any case, which can lead to extinction of the species. Very low diversity shows that approximately all the individuals are identical. Under this scenario, if any disease or any other environmental pressure comes into the population, all the individuals in that community or population become sick and may die due to this disease. On the other hand, when high genetic diversity prevails in a certain population individuals will be stronger and capable enough to resist any environmental pressure. Hence, genetic makeup of some individuals helps them in survival and the production of viable progeny. Genetic diversity is, however, not constant but always changing. The number of variants of genes in the population will not have any effect on the survival of the individuals or population. Only those variants can contribute to future species which will survive in the next generation. The lost variants of a gene are not only difficult to recover but, in fact, recovery is totally impossible.

10.4 WHEN TO APPLY CONSERVATION GENETICS

a. *Loss of Habitat*: Different factors, e.g., habitat destruction, put a population at risk. Then scientists and conservation managers investigate these populations which are currently at risk. For example, they may study such plant population which has been disturbed due to construction of buildings at their habitat and they are no longer able to thrive. Second, they can focus their studies on disturbed population of ducks and geese due to introduction of new hunting regulations and laws. In addition to human interference, storms and diseases can also diminish animal and plant populations.
b. *Changes in population size*: Small populations are particularly sensitive to change and hence their surveillance is more critical than larger populations. Unpredictable events such as genetic mutations, environmental changes, and natural catastrophes can suddenly decrease population sizes. When the startup population of a species is small, and when there is further decline in that population, there is a sharp reduction in their genetic diversity. Small populations are not only susceptible to decreases in genetic diversity but they are also more prone to genetic drift. Moreover, when animals separate geographically and then each individual group establishes a new population from only a few individuals (founder effect), genetic diversity further declines. All these factors have critical effects on the production of new individuals and types of alleles transferred to their offspring.
c. *Genetic Drift*: With each new generation, some individuals are more viable; they survive successfully and reproduce further with the transfer of alleles to their offspring. At the same time, there are some individuals from the same progeny which are not viable, cannot survive as well as their peers, and are not capable of further production, and hence they are eliminated. With the passage of time, individual alleles do not have very much common in a population. Then the process of natural selection takes place because inherited characteristics determine and decide the survival and death of the individuals. However, some random factors like genetic drift determine survival of individuals. Due to the process of genetic drift, some alleles entirely disappear from a population. Larger populations are less prone to genetic drift than smaller ones. The presence of particular allele in a few individuals will probably not allow individuals to reproduce, which will only happen by chance. This non-production of individuals will lead to the loss of that allele.
d. *Geographical Isolation*: Even when larger populations are geographically isolated, they lose their genetic diversity. The development of new habitats creates barriers among the populations and they become geographically isolated. When a new housing colony is

established or a river changes its course, groups of individuals—which may be plants or animals—are isolated from each other and hence are divided into two groups. It is just by chance that an isolated population carries a different gene pool. For example, the establishment of a housing society separates a more delicate plant with prettier flowers to the south end, and a tall, sturdy plant with few flowers to the north end of the subdivision. Despite several efforts for their preservation, they cannot exchange their genes due to sufficient isolation. Due to this drawback, they are not able to produce more strong and advantageous plants and hence there is the possibility that they may be wiped out whenever any environmental pressure develops, which ultimately diminishes the gene pool with the decrease in the population size.

10.5 MANAGEMENT STRATEGIES

While working on the identification of endangered organisms and their subsequent management strategies, close collaboration of scientists and managers is highly important. They must work out the degree of adaptability of an organism to various temperatures, soils, and water conditions. They examine the factors like identification and characteristics of plant pollinators which influence the genetic diversity. The health and welfare of pollinating species is critical to the survival of an endangered plant species, which also needs to be examined. Finally, scientists and managers study the threats which are disintegrating the species. The common threats which risk the survival and wellbeing of the species are destruction of habitats due to human interference, climate changes, and some other miscellaneous factors. On the completion of these studies and tasks, environmental managers and scientists develop and establish a preservation plan and strategy. This preservation plan should be intelligent, comprehensive, and doable, and be effective in the preservation of animal and plant species on a wider scale.

10.5.1 Factors that Cause Extinction of the Species

Although there are several factors which can cause extinction of a species, inbreeding and inbreeding depression is at the top (Charlesworth & Charlesworth, 1987). Due to inbreeding depression, deleterious mutations accumulate (Lynch et al., 1995). This decreases heterozygosity in the population, lowering the ability to deal with change in the environment during evolutionary processes. Similar to inbreeding depression is outbreeding depression, which operates with a similar mechanism. In addition to these factors, if the population is fragmented and divided into different either related or non-related groups, they cannot maintain the desired genetic pool and hence are no longer viable in the presence of harsh climatic conditions (Vrijenhoek, 1994). Further, when there are uncertainties in the taxonomy of animals and their relationships with one another are not known, this also results in homozygosity with the production of animals with a low resistance capability to their changing environment (Haig, 1998). Normally, when the process of natural selection operates strong organisms with a good genetic pool are selected, but the opposite occurs when it is replaced by genetic drift when there is the possibility of the production of weaker organisms with low resistance capability. Moreover, when mating takes place between endemic and invasive species—and if they have dominant traits—they totally wipe out the endemic fauna as well as the gene pool, resulting in homozygosity.

10.6 MECHANISM OF CONSERVATION GENETICS

For the formulation of management strategies and choices, conservation geneticists use DNA data from an organism. Conservation scientists for any stipulated and investigative work use a well-established approach, as practiced in other scientific fields. As there are so many species of organisms which are currently endangered and threatened, it is hard for the scientists to focus on all of them.

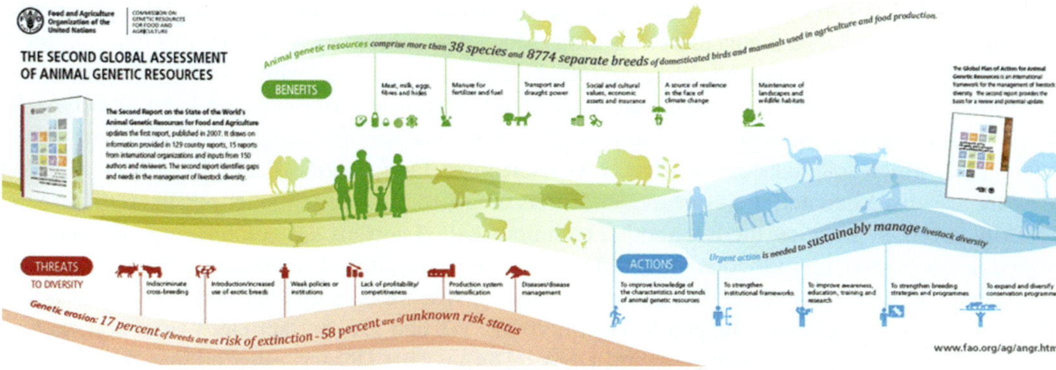

FIGURE 10.1 Management of animal genetic resources. (*Source:* www.fao.org (2007) open access.)

Hence, they define and identify certain populations and/or areas where they want to work. From these selected species or populations, they then prioritize the species or population according to its current importance status in the ecosystem. After prioritizing, they observe the selected population from different conservation angles and consider possible complications which they, might confront in initiating conservation strategies. At species level, conservation scientists first study different and known forms of selected species, with further exploration on its suspected or known relatives, along with its physical characteristics. Scientists use these characteristics to classify the species and their different forms. To examine the genetic characteristics of the organisms (DNA or protein data), scientists develop hypotheses highlighting the relationships between populations and/or species. Finally, developed hypotheses are tested for their validity and inferring of conclusions. Normally, various mathematical models are applied for the analysis of the data collected during the investigative process. These models throw light on the presence of diversity in separate populations of the species. Further, these models also tell the rate of exchange of genes among populations (gene flow) (Fig. 10.1).

10.7 TECHNIQUES USED TO ASSESS THE GENOMES OF A SPECIES

From the conservation point of view, techniques used for the assessment of genome of species are quite specific (Haig, 1995). The analysis of genome of a species can be performed in two ways. Either historic DNA is collected from the fossils and is then analyzed for genome studies or DNA is collected from the living individual for the achievement of this analysis. Currently, a variety of techniques is used to differentiate individuals from populations or vice versa. The most important of them include alloenzymes, random fragment length polymorphisms(RFLP), amplified fragment length polymorphisms (AFLP), single strand conformation polymorphism, minisatellites, microsatellites, random amplification of polymorphic DNA, DNA sequencing, and single-nucleotide polymorphisms. Within animals and plants, these different techniques focus on different and variable areas of the genomes of particular animal or plant. The specific information collected from the sample/species determines the parts of the genome to be analyzed. The status of this information further tells about the appropriate technique to be used. For example, in animals because of its high substitution rate mitochondrial DNA is useful for identifying differences between individuals. The problem therefore is its relatively small genome size and its inheritance capability only in female line. In plants, mitochondrial DNA structurally mutates at very high speed hence it is rarely used as a genetic marker. Hence, chloroplast DNA is used instead of mitochondrial DNA. There is frequent use of other sites in the genome, including major histocompatibility complex, and micro, and mini-satellites due to their high mutation rate. Sufficient information can be collected from these techniques on long-term conservation of genetic diversity, with further expounding demographic and ecological matters such as taxonomy (Haig, 1998).

Historic DNA allows geneticists to understand the changing behavior of animal species and hence, use of historic DNA is considered advantageous for genetic analysis. The results thus obtained can be extrapolated to similar species in the future. This technique has helped scientists a great deal in the study of samples and its remains are sometimes preserved in museums and caves (Wayne & Morin, 2004). All over the world, museums contain a wide range of species as a whole, their parts, or remnants, which scientists can use freely for their investigative work. Animals preserved in the caves are less disturbed and therefore evidence thus obtained provides a longer perspective (Wayne & Morin, 2004). In genetics, there is another technique called non-invasive monitoring, which relies on specific genetics of an individual under observation or study. In this technique, DNA is extracted from organic material that an individual leaves behind, when it dies and decays. Feathers of birds can be quoted as an example (Wayne & Morin, 2004). We can also extract DNA from soil, water, and air, which are called an environmental DNA (eDNA). This DNA is released into the environment from the tissue cells which organisms deposit into the environment, they degrade, and DNA is released into the environment (Barnes & Turner, 2016). In this type of deposition and release, the animal is least disrupted and provides quite useful information. This technique can provide information about the movement, diet of an individual, sex, and kinship. There are several other general techniques which can be used for the correction of those genetic factors which promote the risk of extinction, as well as complete extinction and wiping out of the particular individual or population. For example, when inbreeding is minimized and genetic variation is increased, it encourages taking multiple steps towards conservation of the genetic pool of any population or individual. There are several factors which help in reducing inbreeding. The most important of them are the increase of heterozygosity through immigration, breeding from older animals, or increasing the generational interval through cryopreservation. The purpose is ultimately increasing the effective population size through equalization of family size. Deleterious alleles arise through mutation. However, inbreeding causes prevalence of certain recessive alleles (Frankham, 1995). Deleterious mutations which arise from inbreeding or natural selection are, however, removed by purging (Frankham, 1995). In the wild loss of genetic diversity, inbreeding depression and genetic adaptation of individuals to captivity when they are enclosed is not beneficial or advantageous because it always leads to the production of weak individuals. These techniques can resolve most of the issues mentioned in the former statement, ultimately increasing heterozygosity. In addition, if we create and establish a captive environment for animals which is very much similar to the wild, this decreases the response to selection. In addition to this, fragmentation of population decreases response to selection, as well help with captivity (Woodworth et al., 2002). Different solutions applied to control the factors which make animals extinct or near extinction are overlapped and are quite hard to separate for their independent dealing in the field. For example, genes mutate and they add deleterious mutations to the population. However, deleterious mutations that result from inbreeding are more harmful, which is the focus of the most of the conservation biologists. A reduction in inbreeding, however, can easily control these deleterious mutations. So whichever technique will help in reducing inbreeding will also be helpful in decreasing the accumulation of deleterious mutations.

Particularly in defining species and sub-species of salmonids, these techniques have wide-ranging applications (Haig, 1995). Frequent hybridization in salmonids is an important issue. This issue has conservation implications as well as social, economic, and political ones. Analysis of mtDNA and allo-enzymes shows recurring hybridization in native and non-native cutthroat fish. This uncontrolled and unintentional hybridization has resulted in the decline of native cutthroat fish population. Therefore, an attempt was made to remove some hybridized population to facilitate ready breeding of native populations. This state of affairs impacts the economy of local fishermen, as well as bigger companies. For successful conservation of mammals, proper defining of species and sub-species is important too. For example, the northern white rhino and the southern white rhino are morphologically quite similar. Hence, during previous conservation practices both were defined as the same species and conservation strategies were equally applied to both of them and therefore most critical species could not receive the required level of treatment and started to decline. Recent mtDNA analyses, however, have shown that genetically both are distinct species (Groves et al., 2017). Due to this misconception in identification, there is a continuous decline in the population of northern white rhino. Now it is near extinction due to poaching and it has been realized that the assumption that it can breed freely with the southern population is totally misguided. This misunderstanding has seriously hampered the conservation effort of the white rhino population. Forensic genetic identification is the more recent approach. It can conveniently and reliably identify the species in case of poaching. Experts use DNA data bank from Wildlife to regulate poaching of animals, laundering of wildlife species, and trade of protected species (Ogden et al., 2009). When combined with other scientific disciplines, conservation genetics techniques can work more effectively in the identification and conservation of endangered and extinct species. For example, both conservation genetics and landscape genetics are normally used for the identification of corridors and population dispersal barriers. Information collected from this technique can give sufficient insight helping effective conservation management practices (Keller et al., 2015).

For a significant breakthrough, currently conservation genetics (CG) is well backed by solid theory and experimental data. To meet the suspected deficiencies and flaws, many still emerging tools have been set up with direct linkage with next generation sequencing (NGS). These tools will provide detailed and new insights in the next few years (Willi et al., 2022). Reduced-representation sequencing (e.g., RADseq) and whole-genome sequencing (WGS) are the most common techniques in NGS. These techniques are commonly used in CG for the population of pooled samples or individuals. RADseq on grasshoppers (Hoffmann et al., 2021) and individual WGS of an endangered pheasant species can be quoted as examples (Wang et al., 2021). NGS generates a massive amount of data which can test complex demographic scenarios. It links current changes with near and the more distant past. Hence, NGS can examine and detect the impact of migration and inbreeding across the genome of selected population. Recently, the genetic load of an individual or population has been comprehensively estimated and addressed because of its importance in CG (Liu et al., 2020). Eventually, potentially adaptive genetic variation and evolutionary potential (Caballero & Garcıa-Dorado, 2013) of an individual or a population is assessed with a prediction of their output and impact on CG when applied. These estimates and predictions in turn can help in the achievement of diversity estimation present at loci which are related to expressed traits (quantitative-trait loci, QTLs).

10.8 CONSEQUENCES OF THE USE OF NEW TECHNOLOGIES

However, the development of new technologies has provided tremendous convenience as well as authenticity in the appropriate identification of species for conservation of their genetic pool, but they also have several implications for the future of conservation biology. Several advancements have been observed in these techniques at the molecular level. Those that analyze mini-satellites and MHC are quite common (Haig, 1998). These molecular techniques can clarify taxonomic relationships more reliably. By determining kinship they can identify the best individuals for reintroduction to a population for recovery. This technique has far-reaching consequences. Species conserved in this way affect humans in a variety of ways such as political, economic, social, and political realms (Haig, 1998). In this context, increased genotypic diversity has been observed in the biological realm. It has helped in ecosystem recovery and recovery of the community of grasses can be quoted as an example. Under this recovery system, the community of grasses showed resistance to geese populations, which tried to disturb this community of grasses through greater genotypic diversity (Wayne & Morin, 2004). With increased diversity in the species, there is enhancement in the functioning of ecosystem. Application of new conservation genetic techniques has wider reaching effects than have been observed before. Conservation genetics is a very wide subject as well as one of the least explored. There are several challenges to conservation geneticists which must be addressed in future to fully comprehend the pros and cons of this field, as well as exploration of ignored areas.

10.9 CHALLENGES AND OPPORTUNITIES FOR CONSERVATION GENETICISTS

The first and foremost challenge to the conservation geneticists is how to classify species, subspecies, populations, and geographic races phylogenetically and how the phylogenetic diversity and uniqueness in communities and populations can be measured. Second is the identification of hybridization in natural populations and hybrids produced from these hybridization processes. Further to this, the extent as well as history of introgression between species is required to be explored. Third, conservation geneticists need to look into the genetic structure of managed and natural populations. In this context, they need to explore and identify evolutionary significant units (ESUs) and management units, which can help the conservation of certain endangered and extinct populations. Fourth, it is important for conservation geneticists to assess the level of genetic variation within species or populations with the inclusion of small and endangered populations. Further estimation of effective population size is also required to be determined. Inbreeding and outbreeding depressions significantly impact the genetic diversity of population, as well conservation strategies. Hence, there is need to measure their impact and how this impact is related to the level of heterozygosity leading to fitness of species or populations. When population is disturbed in nature, it interferes in choice of mating of individuals ultimately the reproductive strategy. It is required to explore how this sequence of events affects genetic pool and its conservation and management. Further to this it is a challenge to conservation geneticists how forensic science and techniques can be used to identify the status of endangered species and then how their illegal trade can be minimized or controlled totally to maintain their population. During captive breeding programs, practical methods for monitoring and maximizing genetic diversity need to be established. Schemes and mathematical models need to be established for the introduction of captive bred species and study of their behavior in nature when they are released in the wild. When genetically modified organisms are introduced in nature their conservation issues differ from those of already established populations. Studies need to be done about how to tackle the issues of these genetically modified organisms when introduced in nature. There are considerable interactions between environmental contaminants and organisms living in that environment. These contaminants definitely cause mutations following adaptations of organisms to the environment. It is required to study the extent of these mutation rates and impact on the behavior of organisms in the environment where they are living. The impact of industrial advancements needs to be explored to modify the further genetic conservation strategies of endangered populations.

REFERENCES

Barnes, M.A. & Turner, C.R. (2016). The Ecology of Environmental DNA and Implications for Conservation Genetics. Conservation Genetics 17(1), 1–17. https://doi.org/10.1007/s10592-015-0775-4.

Caballero, A. & Garcıa-Dorado, A. (2013). Allelic Diversity and Its Implications for the Rate of Adaptation. Genetics 195, 1373–1384.

Charlesworth, D. & Charlesworth, B. (1987). Inbreeding Depression and Its Evolutionary Consequences. Annual Review of Ecology and Systematics 18(1), 237–268. https://doi.org/10.1146/annurev.es.18.110187.001321.

FAO. (2007). The State of the World's Animal Genetic Resources for Food and Agriculture, edited by Rischkowsky, B. & Pilling, D., Rome (available at http://www.fao.org/docrep/010/a1250e/a1250e00.htm).

Frankham, R. (2005). Genetics and Extinction. Biological Conservation 126 (2), 131–140. https://doi.org/10.1016/j.biocon.2005.05.002.

Groves, C.P., Cotterill, F.P.D., Gippoliti, S., Robovský, J., Roos, C., Taylor, P.J. & Zinner, D. (2017). Species Definitions and Conservation: A Review and Case Studies from African Mammals. Conservation Genetics 18 (6), 1247–1256. https://doi.org/10.1007/s10592-017-0976-0

Haig, S.M. (1998). Molecular Contributions to Conservation. Ecology 79(2), 413–225. https://doi.org/10.1890/0012-9658(1998)079[0413:MCTC]2.0.CO;2.

Hoffmann, A. et al. (2021). An Endangered Flightless Grasshopper with Strong Genetic Structure Maintains Population Genetic Variation Despite Extensive Habitat Loss. Ecology and Evolution 11, 5364–5380.

Keller, D., Holderegger, R., van Strien, M.J. & Bolliger, J. (2015). How to Make Landscape Genetics Beneficial for Conservation Management?. Conservation Genetics 16(3), 503–512. https://doi.org/10.1007/s10592-014-0684-y.

Leberg, P.L. (1990). Influence of Genetic Variability on Population Growth: Implications for Conservation. Journal of Fish Biology 37, 193–195. https://doi.org/10.1111/j.1095-8649.1990.tb05036.x.

Liu, L. et al. (2020). Genetic Consequences of Long-Term Small Effective Population Size in the Critically Endangered Pygmy Hog. Evolutionary Applications. 14, 710–720.

Lynch, M., Conery, J. & Burger, R. (1995). Mutation Accumulation and the Extinction of Small Populations. The American Naturalist 146(4), 489–518. https://doi.org/10.1086/285812.

Minter, M., Nielsen, E., Blyth, C., Bertola, L., Kantar, M., Morales, H., Orland, C., Segelbacher, G. & Leigh, D. (2021). What Is Genetic Diversity and Why Does It Matter?. Frontiers for Young Minds 9, 656168. https://doi.org/10.3389/frym.2021.656168

Mukhopadhyay, T. & Bhattacharjee, S. (2016). Genetic Diversity: Importance and Measurements. In: Conserving Biological Diversity: A Multiscaled Approach, eds. Mir, A.H. & Bhat, N.A., Research India Publications, New Delhi, India.

Ogden, R., Dawnay, N., McEwing, R. (2009). Wildlife DNA Forensics—Bridging the Gap between Conservation Genetics and Law Enforcement. Endangered Species Research 9, 179–195. https://doi.org/10.3354/esr00144.

Saccheri, I., Kuussaari, M., Kankare, M., Vikman, P., Fortelius, W. & Hanski, I. (1998). Inbreeding and Extinction in a Butterfly Metapopulation. Nature 392 (6675), 491–494. https://doi.org/10.1038/33136.

Science Direct Topics retrieved. (2023) "Effective Population Size - An Overview | ScienceDirect Topics". www.sciencedirect.com. Retrieved 2023-02-11.

Vrijenhoek, R.C. (1994). Genetic Diversity and Fitness in Small Populations. In Conservation Genetics, eds. Loeschcke, V., Jain, S. K., Tomiuk, J., pp. 37–53. Birkhäuser, Basel. https://doi.org/10.1007/978-3-0348-8510-2_5.

Wang, P. et al.(2021) Genomic Consequences of Long-Term Population Decline in Brown Eared Pheasant. Molecular Biology and Evolution 38, 263–273.

Wayne, R. & Morin, P. (2004). Conservation Genetics in the New Molecular Age. Frontiers in Ecology and the Environment 2(2), 89–97. https://doi.org/10.1890/1540-9295(2004)002[0089:CGITNM]2.0.CO;2.

Willi, Y., Kristensen, T.N., Sgro, C.M., Weeksd, A.R., Michael Ørstedb, M. & Hoffmannd, A.A. (2022). Conservation Genetics as a Management Tool: The Five Best-Supported Paradigms to Assist the Management of Threatened Species. PNAS 119(1), 1–10.

Woodworth, L.M., Montgomery, M.E., Briscoe, D.A. & Frankham, R. (2002). Rapid Genetic Deterioration in Captive Populations: Causes and Conservation Implications. Conservation Genetics 3(3), 277–288. https://doi.org/10.1023/A:1019954801089.

11 Wildlife Conservation Laws and Policies

11.1 INTRODUCTION

Wildlife constitutes those animals which are not domesticated. Wild animals are found on each continent and in every country. One-third of the Earth's land area is forested and inhabited by wildlife.. The numbers and types of wildlife species present may vary from country to country. Wildlife includes mammals, birds, reptiles, aquatic vertebrates, and reptiles, as well as invertebrates. Lions, tigers, elephants, snakes, turtles, hawks, and butterflies are but a few examples. Other than animals, plants, fungi, and microorganisms living in their natural environment without human disturbance and interference are also component parts of wildlife. Being an integral part of our natural heritage, wildlife not only maintains our ecosystem and keeps it in balance but offers several benefits to human beings, providing aesthetic value to humans along with a major part of their food. The provision of clean air, water, shelter, and medicines are some benefits which we gain from wildlife (Forgeard, 2023). Parallel to these benefits, wildlife offers many opportunities to scientists for research. In addition to these major roles, wildlife also maintains the ecological balance, as well as natural biodiversity. This biodiversity plays a variety of roles in nature. The most important of them are seed dispersal, pollination of flowers, and control of the population of insects and other organisms when they become excessive and deleterious to human interests. This biodiversity also helps to prevent soil erosion and flooding during the rainy season.

Wildlife also plays an important role in the tourism industry. The World Travel and Tourism Council reported that wildlife tourism directly contributed US$120.1 billion to global GDP in 2018, a figure that is continuously on the rise. In the United States alone,45 million people visit wildlife refuges annually and, during their visits, people take photographs, explore rare, endangered, and threatened species, and learn about the benefits of wildlife. Their observations then lead to ideas regarding the conservation and preservation of such species. Most of the students working in wildlife launch their research projects at such facilities. The more the diversity of wildlife in a given area, the more people are attracted to that area, where they spend money on food and lodging. Tourists go to Alaska to see walruses and to South Africa and New Zealand to see penguins. The presence of lions, elephants, and giraffes in Africa's Serengeti continues to attract tourists and America's Yellowstone National Park is a draw for tourists from all over the world to see grizzly bears, bison, and moose. And yet, we still do not know the total number of species that exist in terrestrial and aquatic environments. Biodiversity is so huge that scientists claim that more than 91% of marine species still remain to be classified. This means that life on earth is highly diversified and can be translated into a single comprehensive term—"Biodiversity."

11.2 CHALLENGES TO WILDLIFE

Our natural system is very complex. Different plant and animal species interact with each other. Decline or loss of even a single species can disrupt the ecosystem balance. This imbalance has long-lasting consequences for other species found in that region. This is equally applicable to both plants and animals. However, our wildlife population is rapidly declining due to human interference and disruption. The invasion of exotic species, overhunting, pollution, illegal wildlife trade, climate

DOI: 10.1201/9781003472001-11

change, overconsumption, deforestation, and further development are the cause of this decline, which we examine below.

1. *Climate change*: With the warming of our planet, there are changes in the seasonal cycles. These variations disturb wildlife habitats which animals use for their survival and other life-related activities. Climate change is a major threat to all types of animals, from insects to polar bears. As climate is a global problem, few are unaware OF its effects, while some know what is happening around them. Climate changes affect wildlife in so many ways. Environmental change forces animals to leave their place and move to higher elevations which may be not suitable for them. Pikas and polar bears, for example, have been pushed out of their habitats to smaller areas where they can hardly survive. In such places, they cannot find anything to eat, cannot reproduce successfully, and, ultimately, they die. At their new location, these species compete with existing species for food and shelter, putting an additional stress on then. However, living not only becomes difficult for migrants but the existing species are hard-pressed for their own survival. Parallel to these effects, the oceans are warming up too. Their water chemistry has changed, and so some species have become threatened or extinct. This temperature change favored invasive species, which then spread into new areas. In the warmer months of the year, those animals which hibernate in the winter lose their protective layers of fat because they may have to live without food or water for an extended period of time. If there are abrupt changes in temperature, aquatic species such as salmon cease to spawn and become extinct. These climatic changes cause tremendous upheavals in ecosystems and lead to competition among some animals and plants due to limited resources. Therefore, a strong possibility emerges for the extinction of the entire population.
2. *Invasive species*: Sometimes human beings introduces new species into their existing ecosystem. The foreign species compete with the native species for food and space. If an invasive species become dominant, then there is no food for local fauna, ultimately wiping out all the native species. Due to this disruption, some successfully living species have become endangered or extinct.
3. *Habitat loss and fragmentation*: With the increase in human populations, there is a commensurate need for the production of food and shelter for living. This activity affects wildlife in various ways. Either it destroys or degrades the natural habitat or the wildlife population is divided into groups by fragmentation. Under this scenario, animals cannot mate freely and their gene pool is restricted, leading ultimately to weakened wildlife animals with poor resistance to changing environmental and climatic changes. Habitat loss and destruction thus reduces the space for wildlife animals, ultimately leading to the extinction of that wildlife. Habitat destruction and fragmentation causes the accumulation of species in a limited space with minimum resources like food and water. Due to their closeness and poor living conditions, the chances of disease spread increases. Because species are already very vulnerable, the population decreases at first and then perishes completely (Richard et al., 2017). Due to the COVID-19 pandemic, human behavior toward wildlife changed significantly and has further squeezed wildlife habitats, sequestering them in a very limited space with poor resources, leading to yet further extinction of wildlife.
4. *Poaching*: People hunt wildlife to fulfill their living requirements. For example, nomadic tribes in South Africa and northern Asia still use traditional hunting methods for their survival. Millions of poor people in developing countries are unable to produce enough high-quality protein from livestock and hence they depend on wildlife to fulfill their dietary demands. These unavoidable circumstances often lead to overhunting that threatens the existence of wildlife, although further growth in the population and the environment no longer allows people to depend totally on wildlife hunting for their survival and wellbeing. Due to this increased demand for food and other accessories, people turned to agriculture and other sources of food. However, hunting of wildlife still remained a popular means of

obtaining food for subsistence or trade or recreation purposes. Illegal harvesting and hunting of wildlife is also very common, including, f or example, hunting of rhino for horns, ungulates for meat and fur, and elephants for ivory. This type of hunting and poaching for commercial purposes or local usage make wildlife species vulnerable to extinction and is the major cause of the decline in the wildlife population.

Wildlife is also traded illegally. At a global level, this trade was worth US$7.23 billion back in 2013 (Share the Facts about Wildlife Trafficking, 2023). This revenue is generated against the marketing of around 100 million plants and animals (Wildlife Trade, 2023). It was estimated in 2021 that, due to this trade, species abundance declined by up to 40% and another 60% became endangered (Wildlife Trade, 2023).

This trade is very distressing for both animals and human beings. Export and exchange of animals spread zoonotic diseases to humans and result in depletion and extinction of some animals. Small animal vectors like ticks or ingestion of food and water contaminated by them can spread pathogens to humans. When the fittest species of animals are hunted or poached, then less fit organisms mate, producing weak and vulnerable animals. Due to disturbance in the sex ratio, their reproduction is also slowed and their extinction rate increases (KKIENERM, Wildlife, Forests, & Fisheries, 2023). When people earn THEIR LIVING in ecotourism, they destroy natural resources including wildlife.

5. *Pollution*: Air, soil, water, noise, and light pollution indiscriminately affect the wildlife population. Pollution changes their feeding and breeding behavior, inflicting severe stress on their reproduction, growth, and survival, which ultimately causes their decline and extinction. There are varieties of pollutants which have negative impacts on the health of wildlife. Some pollutants like pesticides are very toxic and their simple exposure can inflict a great deal of damage. When wildlife species inhale air pollutants and ingest toxic metals, they are exposed to their potential effects, which can be different for different animal species. Some may be affected more badly than others. Some of the important pollutants are listed below.

   Air pollutants come from the burning of fossil fuels or emissions from industries. Air pollutants like smog, ground-level ozone, and particulate matter decrease air quality. These pollutants directly affect wildlife health where they are living. The emission of sulfur dioxide (EPA, 2023), however, contributes to acid rain and harms both terrestrial and aquatic ecosystems.

   Heavy metals like arsenic, lead, and mercury occur at low levels in nature. However, ingestion of these metals in higher concentrations can damage animal organs, thus causing cancer. The level of their toxicity, however, depends on the type of metal and its concentration ingested by the animal. Human activities like smelting, burning fossil fuels, industrial processes, and mining contribute to the increase in heavy metals in the living environment of terrestrial as well as aquatic life.

   Toxic chemicals come from oil spills, pesticides, and industrial wastewater. These chemicals can persist in the environment for a long period of time. Their presence can cause cancer, nervous system problems, reproductive disorders, and immune system problems in animals exposed to these chemicals. Greater and wide varieties of toxic chemicals cause a wide variety of health effects in biodiversity (USEPA, 2023).

6. *Overexploitation*: Harvesting or loss of animals at a faster rate than their recovery rate is called overexploitation. Overexploitation is equally applicable to plants, reptiles, amphibians, fish, and mammals. For example in the past century when tuna and salmon (marine predatory fish) were overharvested, both their numbers and their size declined (Michael et al., 2013).
7. *Deforestation*: When forests are cut down and cleared for agriculture, logging, or for fuel, wildlife habitat is destroyed. According to one estimate, forests host more than 80% of the variety of wildlife species. When these forests are cut down and/or removed, wildlife habitats diminish, resulting in animal species vanishing. Deforestation therefore threatens wildlife and if this process continues can cause extinction of species. For Tropical forests

which host so many animal species, their deforestation is very critical to wildlife fauna. The Amazon forests, home to more biodiversity than any other biome can be quoted as an example. In some countries, wildlife conservation policies have been enacted, e.g., the Wilderness Act of 2022 can be quoted as one of the example.

8. *Ocean Acidification*: With the increase of carbon dioxide concentration in the atmosphere, there is its proportionate increase in the ocean. Carbon dioxide absorbed by the ocean goes straight down to the seabed. This excessive emission of carbon dioxide can increase the stratification of ocean water and slows down the biological pump which, in turn, decreases the level of pH in the ocean water, making it more acidic and which can ultimately destroy habitats for many coral reef habitats. This acidification is related to current as well as past extinction events(Mills, 2012).
9. *Culling*: When governments kill selective wildlife animal species it is called culling. The killing of sharks in Queensland and New South Wales (Australia), along with whales, dolphins, turtles, and some other marine life can be quoted as an example (Thom, 2015). During this shark control program, 50,000 sharks and more than 84,000 marine animals were killed (Thom, 2015). Mass killing programs of wildlife have also been practiced in the USA, where bison in Montana and swans, geese, and deer were killed in New York on a mass scale (Hadidian, 2015).

11.3 CONSERVATION OF WILDLIFE

When we attempt to conserve and protect endangered animal/plant species in the wild environment together with the habitats associated with them, the process is called wildlife conservation. Therefore, wildlife conservation refers to the practice of protecting wild species and their habitats. Protection of wildlife helps in maintaining healthy wildlife species or populations. It also restores, protects, and improves natural ecosystems. The IUCN estimates show that 42,100 wildlife species are at risk and could become extinct at any time (The IUCN Red list,, 2017). A UN report of 2019 put this figure even higher at possibly 1 million species. One estimate shows that, due to human interference, the current wildlife extinction rate is about 1,000 times greater than its normal rate of extinction (Pimm et al., 2014). Of these 42,100 species listed, 40% are amphibians, 14% are birds, and 25% are mammals (The IUCN Red List of Threatened Species, 2019). Due to limited resources, selected species are always conserved because of their ecological and economic importance, as well as their attractiveness and good aesthetic value.

It is further reported that most of the ecosystems which host endangered species are disappearing. Hence, for the conservation of perishing wildlife fauna, prominent agreements like the 1973 Convention on International Trade in Endangered Species of Wild Fauna and Flora (CITES) and the 1992 Convention on Biological Diversity (CBD) have been enacted (Stop Global destruction, 2019). The World Wildlife Fund, Conservation International, the Nature Conservancy, and the Wild Animal Health Fund are some of the non-governmental organizations which are dedicated to conservation of wildlife species.

The conservation of wildlife can only happen when we realize its importance and that we cannot live without it. Wildlife conservation's primary purpose is always maintenance of a sustainable population of wild animal species. At the same time human needs like medicine, fuel, clothing, food, or other products derived from animals or plants also need to be considered. Both humans and wildlife interact with each other and both carry intrinsic values; importance of both of them cannot be denied. Therefore, conservation of wildlife is mandatory to maintain ecological balance in the environment. However, the conservation of wildlife is becoming increasingly important due to negative human activities which are ultimately causing the extinction of some animal species. These threats are universally prevalent at the global level. The IUCN has listed orang-utans, blue whales, giant pandas, eastern lowland gorillas, Amur leopards, hawksbill turtles, and Javan rhinos as endangered species.

11.4 ENDANGERED WILDLIFE SPECIES OF PAKISTAN

11.4.1 Mammals

An endangered animal species is one going that will be extinct in the future. The term "vulnerable" is used for those species which will soon be endangered. The population of these animals is limited. The major cause of the decrease in their population is climate change and anthropogenic activities (ENVPK, 2021). The most common anthropogenic activities are killing, destruction of natural habitat, poaching, hunting, habitat fragmentation, and illegal trade, all of which are threatening the existence of these animals. Extinct species are completely wiped out and no longer available on the earth. Pakistan hosts a vast variety of animals. Among other provinces, Baluchistan and Khyber Pakhtunkhwa have unique species of animals. Most of the animals present in the four other provinces of Pakistan are under threat as follows:

1. Markhor: This is also called the screw horn goat (*Capra falconeri*) and is found in northern and central Pakistan. According to the IUCN, its status is near threatened.
2. The Indus River Dolphin (*Platanista minor*): Locally it is called Bhulan, a blind dolphin and found in the Indus River. According to IUCN status, it is endangered in Pakistan and very few individuals are found in the Indus River.
3. The Sindh Ibex (*Capra aegagrus blythi*), also called wild goat, and ibex. It is present in the Kirthar National Park in Sindh and its national status is vulnerable.
4. The Asian Black Bear (*Ursus thibetanus gedrosianus*), also called kala reech is an inhabitant of Baluchistan and according to the IUCN its status is vulnerable.
5. The Kashmir vole (*Alticola montosa*). It is predominantly present in Azad Kashmir and Khyber-Pakhtunkhwa and according to the IUCN status, it is a vulnerable species of mammals.
6. The Fishing cat (*Prionailurus viverrinus*), also called Jangli billi, is found in Sindh, Baluchistan, and Punjab. According to the IUCN its status is vulnerable.
7. The Black-tailed gazelle (*Gazella subgutturosa*), also called Goitered gazelle and locally it is also called hiran. It is commonly found in Southwestern Pakistan and according to the IUCN status it is an endangered species.
8. The Indian Wild Ass (*Equus hemionus khur*), locally called khur, is mostly found in Sindh and Baluchistan is near threatened (IUCN).
9. The Asian Leopard (*Panthera pardus*), locally called Cheetah, is commonly found in Khyber Pakhtunkhwa, Sindh, Baluchistan, Punjab, and Azad Kashmir. According to the IUCN species status, it is an endangered animal species.
10. The Himalayan Musk Deer (*Cervus canadensis hanglu*), also called White-bellied/Himalayan musk deer and Himalayan Hiran, is found in Northern Pakistan and according to the IUCN status, it is an endangered species.
11. The Smooth Coated Otter (*Lutrogale perspicillata*), locally called Oad Balao, is s vulnerable species of wildlife fauna of Pakistan. It is commonly found in KPJ, Punjab, and Sindh.
12. The Urial (*Ovis vignei*), also called Wild sheep, is commonly found in northern mountain regions of Pakistan, including the Hindu Kush and the Himalayas. It is a vulnerable species of mammals in Pakistan (IUCN).

11.4.2 Birds

For any healthy ecosystem, birds are extremely important. Pakistan hosts a wide variety of native as well as migratory birds (ENVPK, 2021) During the winter season, thousands of birds migrate from the colder regions of Siberia and Russia to Pakistan. However, an unfortunate part of Pakistan's ecosystem is that many of birds have become extinct and several others are endangered. Some birds which are currently declared endangered by the IUCN include:

1. The Houbara bustard (*Chlamydotis undulata*), also called Tilor, is commonly found in Thar Desert Sindh and is vulnerable.
2. The Indian skimmer (*Rynchops albicollis*), also called the Indian scissors-bill, is commonly found along the Indus river and is endangered.
3. The Indian Spotted Eagle (*Clanga hastate*) is a roaming and wandering bird that is also a vulnerable bird species of Pakistan.
4. The Kashmir Flycatcher (*Ficedula subrubra*) is commonly found in the Kashmir region and its current status is vulnerable.
5. The Black bellied Tern (*Sterna acuticauda*) is located along the River Indus and has been declared endangered.
6. The Asian White Backed Vulture (*Gyps bengalensis*), also called the White-rumped vulture, Oriental white-backed vulture, or Gidh is mainly found in Punjab and is now critically endangered.
7. The Cheer Pheasant (*Catreus wallichii*) is also called Chir pheasant and Wallich's pheasant. It is commonly found in Kashmir and the Himalayan region, and its current status is vulnerable.
8. The Yellow-eyed Pigeon (*Columba eversmanni*) is also called the Pale-backed pigeon and Kabootar. It is commonly found in Northeast Pakistan and the Kashmir region. Its current status is vulnerable.
9. The Saker Falcon (*Falco cherrug*) is commonly found in Northern Pakistan and its current status is endangered.
10. The Sarus Crane (*Antigone antigone*) is commonly found in the Tharparkar District and is thought to be extinct in Pakistan.
11. The Sociable Lapwing (*Vanellus gregarious*), also called a plover, is currently endangered.
12. The Western Tragopan (*Tragopan melanocephalus*), also called the Western horned tragopan, is commonly found in Pakistan and the Himalayan region. Currently its status is vulnerable in Pakistan.
13. The Indian Vulture (*Gyps indicus*), also called Gidh, is commonly found in Southern Punjab, Southwestern Balochistan, the Thar desert, Sindh, and Azad Kashmir and is currently critically endangered.

11.4.3 Reptiles

The Russian tortoise (*Testudo horsfieldii*), also commonly known as the Afghan tortoise, the Central Asian tortoise, Horsfield's tortoise, four-clawed tortoise, and the (Russian) steppe tortoise, as well as the "Four-Toed Tortoise" belongs to the family Testudinidae. It is commonly found in Central Asia from the Caspian Sea south through Iran, Pakistan, and Afghanistan. East across Kazakhstan to Xinjiang and Chinal also host this species. Currently, its status is threatened due to human activities in its native habitat (Kennerley, 2017).

11.4.4 Fish

The Banded Eagle Ray (*Aetomylaeus nichofii*) is a species of fish in the family Myliobatidae. Countries like Bangladesh, Brunei, Japan, India, Pakistan, North Korea, South Korea, Australia, Cambodia, China, Indonesia, Malaysia, Myanmar, Pakistan, Papua New Guinea, the Philippines, Singapore, Sri Lanka, Taiwan, Thailand, Vietnam, the Maldives, and possibly Mozambique host this species of fish. Its natural habitats are open seas, shallow seas, and coral reefs. However, due to loss of its habitat, currently it is threatened (Kyne et al., 2016). Another fish species which was endemic to Pakistan is bigeye tuna (*Thunnus obesus*). It belongs to the genus *Thunnus*, and family Scombridae. It is found in the open waters of all tropical and temperate oceans, but currently its status is threatened in Pakistan (Roxy et al., 2016).

11.5 WILDLIFE CONSERVATION STRATEGIES AT GLOBAL LEVEL

After realizing the accelerated loss of several wildlife species all over the world, UNEP and IUCN developed a strategy and formulated it in the form of a book for implementation (Babu, 2016). It is a go-to document for immediate action on conservation measures. The following guidelines were suggested for conservation and protection of the threatened and endangered species of the wildlife ultimately maintaining a healthy ecosystem. These suggestions are:

1. To ensure sustainable use of species and ecosystems that can be useful and supportive to communities and industries.
2. To maintain such ecosystems and ecological processes that are valuable for routine human activities. The most important of them are soil regeneration, water cleansing, nutrient cycling, etc.
3. To preserve genetic diversity of threatened and endangered animal species of the wildlife.

These strategies are not only meant for conservationists but are equally important for the government and development proponents. It is a basis for policy framework and aids commercial units to take decisions in the best interests of ecosystems. In the light of these objectives and guidelines, priority tasks can be developed for immediate conservation measures. Now, it is the responsibility of government and non-governmental organizations to ensure that key ecosystem processes are not degraded to the point where that area cannot be used for any economic activity, because ecosystems like forests, rivers, and coastal ecosystems directly contribute to our economies as well daily domestic needs. This sustainable maintenance of ecosystems demands that wildlife habitat is not fragmented, pollution is controlled in wildlife areas, and wildlife is not overexploited. If these factors are not controlled, they can lead to depletion and extinction of the species.

Those species that are, however, genetically strong can adapt and withstand the environmental stresses. Then they can transfer this genetic makeup with stronger capabilities to their offspring, which will ensure that the coming generations are better and more successfully able to survive under the harsh environmental conditions generated by human beings. Further, the sustainable usage of natural resources is suggested. It is further added that the release of pollutants should be controlled to save terrestrial and aquatic faunas. Water should be managed effectively with smart utilization of sources leaving sufficient potential for their regeneration. This can only be achieved when we understand that: (1) the planet cannot support our population continuously; (2) at some places people are forced to destroy natural resources indiscriminately to obviate their starvation and poverty. Although people know the importance of wildlife they still take up arms and kill the animals; (3) there is persistent shrinkage in resources like fisheries, meat, and forestry; (4) we need energy and related resources to meet our standard of living.

11.5.1 Wildlife Conservation and Management: National and International Concerns

During conservation measures, national governments need not ignore the short-term needs of rural communities. Implementation of conservation strategies is not possible if only long-term benefits are considered. Further to this, for the implementation of conservation measures in letter and spirit, all the stakeholders need to participate in the planning of conservation measures and strategies. At the national level strong legislation, participation, proper training, rural development, and environmental education are the keys to success in conservation.

At the international level, those environmental aspects that cross national boundaries are normally considered. Among these aspects the drafting of legislation, education of the authorities concerned, and monetary aid is important, irrespective of national borders We should care and look after the species to be conserved, not their national affiliations. Poaching and selling of animal products are the issues which also need to be addressed from the conservation and management point of view, keeping in mind international protocols and procedures. Long-term conservation planning should always be a priority, which is unfortunately always lacking at national and international levels.

11.6 WILDLIFE CONSERVATION STRETGY OF PAKISTAN

The main objective of conservation strategy is mainly to maintain the integrity of ecosystems and the health of the environment. With the increase in experience and lessons learned from the field, focus was enhanced on the interaction between the environment and development. Several criteria have been laid down for successful implementation of these conservation strategies (The International Institute for Environment and Development and IUCN). These criteria have been set out as follows: (1) conservation should not be only to maintain biodiversity or ecosystem or promotion of forest resources. Instead, the goal should be the sustainable development of the wildlife resources; (2) conservation should be a long-term approach. Instead, short-term goals should also be included in the strategy which is achievable without any political interference, either positive or negative; (3) whatever strategy is selected or developed should be based on specific and series of objectives and targets. It should be followed by the lists of the actions to be taken (Government of Pakistan, 2000). Choices must be set clearly with the setting up of desired priorities accordingly; (4) choices must be integrative among and between different sectors. To make the strategy really successful it should integrate social, economic, and ecological perspectives with the inclusion all related factors and choices. This means that a wide range of people should be involved in the development of the strategic process, which includes identification of the issues, defining objectives, preparation of a draft and its scrutiny; (5) whatever the strategy is or will be it should be adaptive to the continuous changes in economic, ecological, and political conditions. As monitoring and evaluation are the critical components of the strategic process, these two components should always therefore be kept in mind for formulating and preparing a strategic plan and its ultimate implementation. While preparing these plans there will definitely be extensive institutional changes. Without such modifications, strategic conservation and management of wildlife is impossible.

11.6.1 Principles of Wildlife Conservation Strategy of Pakistan

Conservation and management of wildlife will only be useful and long-lasting when it is sustainable. Methodology or strategy used for sustainable conservation of biodiversity should focus on the present needs of human beings. At the same time, special focus should be on the fact that it is not seriously or significantly jeopardizing the prospects for future generations (Biodiversity Action Plan For Pakistan, 2000), which means that the needs of future generations must not be ignored. The sustainable conservation and management of wildlife involves several imperatives which Pakistan is following to the letter and spirit for successful conservation and preservation of endangered and threatened species (Government of Pakistan, 2000). These principles are:

1. Rights of the future generations are safeguarded comprehensively.
2. Wherever decision-making is required, it will be comprehensive and prompt.
3. For successful accomplishment of conservation operations, current technology has been reoriented and will be in the future too when and where required. Accordingly, any strategy which is risky is avoided or at least minimized in usage and application.
4. Where possible, policy-makers have tried to minimize dependency on the scarce natural resources or animals and plants which are already near extinction, endangered, or

threatened. This does not mean that all natural resources are totally abandoned; rather, they are used and utilized only where they are essentially required.

5. Like other countries of the world, Pakistan is also a victim of a variety of pollutions, which has a direct bearing on the survival and proliferation of wildlife flora and fauna. Control of pollution irrespective of its type is the top priority of the Pakistani Government for the safeguarding of not only wildlife but also human beings.
6. It is also further asserted that wildlife populations or the human population should be in balanced form and there should be a minimum of conflict between humans and wildlife.
7. Another aspect for the successful implementation of conservation strategies is equity within the nation and between the nations. At the country level, the Pakistani Government is dedicated to achieving this task, as well deploying its earnest efforts whenever necessary for any cross-boundary activities. The purpose is to save life, whether it is within the country or pertains to adjacent countries.

Pakistan is sincerely working on this strategy to safeguard the survival and proliferation of wildlife in its natural habitat, for the human benefits where they are essentially needed. This policy also recommends the creation of a sustainable development policy think tank. This recommendation is really innovative and genuine. This activity will help to remedy the chronic lack of effective policy units within the Pakistan bureaucracy. This policy further recommends the creation of new agencies. At the same time, this policy discourages the formation of ineffective agencies and suggests that effective formations should be properly streamlined for positive and effective output.

11.7 LAWS FORMULATED FOR COMPHREHENSIVE CONSERVATION AND MANAGEMENT OF WILDLIFE

11.7.1 Powers of Authorized Officer

1. **Issuance of NOC, license, or permit**

   "As may be prescribed, government authorized officers may issue a license or permit for small game and big game hunting if requested by the hunter. A No Objection Certificate (NOC) is also issued when activities like possession, display, regulation of hunting procedures, and transportation of captured animals is involved. The officer can cancel the NOC, license, and permit whenever he deems it appropriate under Section 44 of this Act. This specifically happens when the rules and regulations of hunting are violated. If required, the same officer is authorized to establish a checkpoint for regulation of possession, hunting, display, dealing, and transportation of wild animals and their products" (Wildlife Act, 2015).
2. **"The authorized officer may search any hunting party and arrest them without issuing any prior warrant"**

   "Government authorized officers can check any suspicious person/persons. He/she can search his/her belongings and possessions. This includes the person himself, vehicle, vessel, animal package, premises, receptacle, or covering. For the scrutiny of these items he/she does not need any warrant. If an authorized officer finds any doubtful person, he can arrest and present him/her to the nearest police station within 24 hours for further proceeding under section 1 of the Act. In pursuance of this Act, an authorized person can seize and confiscate any wild animal, alive or dead, net trap, bow, arrow, firearm, snare, vessel, or anything suspected which might have used or can be used in committing such an offence. Further, the designated officer is authorized and may sell or dispose of any property seized or confiscated under section 48. However, he is also authorized to set free any animal which is alive or can auction openly as may be prescribed.

3. **Preventive measures and compounding of offence**

   An authorized officer of the Wildlife Department is fully competent by law to prevent the commission of any offence under this Act. Further, if an offence is proved he/she can a sum of money as a compensation to the loss accrued and in return can release the property confiscated under sub-section (1), as the case may be. It will follow the release of person in custody and release of confiscated property withholding any further proceedings. The sum of money received should in no case be less than 100000 rupees. The sum of money accepted as compensation under clause (a) of sub-section (1), shall in no case exceed the sum of one hundred thousand rupees. This person is further authorized to give 1/4th of the total amount collected as reward to the person as compensation who helped in detection, registration, and/or disposal of the offence under sub-section (3)."

4. **Powers of Wildlife Officer**

   According to the Code of Civil Procedure (1908) the Wildlife Officer has the same powers as vested in a Civil Court. In pursuance of this code authorized wildlife officer He can beckon or summon any suspect. Further he can compel him for the production of desired documents. If required he can receive evidence on affidavits, and at the same time can authorize a commission for the examination of witness.

5. **Exemptions to this Act**

   Authorized Officer irrespective of the prevailing clauses can request the permission of killing and capturing of any wild animal in the interest of any scientific or public purpose.

6. **Fee, schedule, hunting period can be revised**

   Authorized Officer can add or exclude from the Schedules any wild animals, subject to such conditions which go in the favor of public interest. He/she can alter or fix the hunting as well as open or close any area for hunting. He can also revise the fee for various types of licenses or permits where he/she thinks appropriate.

7. **Liability to inform about illegal hunting or any other offence**

   Any lessee or purchaser of wildlife property, Lambardar, Canal Patwari, Public Works Department Daroga, Chowkidar, Dafadar, Village Watchman, Abdar, Zilladar, Revenue Patwari, Qannungo and Tehsildar shall are bound by law to inform authorized officer about unauthorized killing, trapping, snaring, netting, or any other offence happening in his/her area under this Act. If offence is proved authorized officer can confiscate suspect property. If somebody abets anybody to commit any offence it will be punishable. Nobody under this act and rules is allowed to interfere or attempt to interfere in the performance of any functions or in the discharge of any duty.

8. **Reporting of offence and lodging the complaint**

   Except complained by authorized officer no court shall take cognizance of any offence in the meaning of the Code of Criminal Procedure, 1898 (Act No. V of 1898). For the purposes of prosecution that officer will act as witness. Further no Court inferior to First Class Judicial Magistrate shall take cognizance and trials of this case. Accordingly First Class Judicial Magistrate may try any offence punishable under this Act summarily. It is further clarified that if any person kills any wild animal by any means in the immediate defense of his own life or that of the life of any other person will not be considered offence. Hence it will be not liable to any punishment or trial. Withstanding to this if the owner of livestock or his employee kills any wild animal which is damaging his livestock or damage is suspected during grazing or when they are enclosed at night that will also be not considered as an offence. As prescribed, Government may compensate the loss of life and property damaged by wild animals. However during Trophy Hunting the meat or trophy of any wild animal killed under sub-section (1) shall be the property of Government. Hence authorized officer can auction and dispose of this animal or part of that animal in pursuance of laid down rules and regulations. Nonetheless when there is conflict on the ownership the claimant has to prove his belongings otherwise it will be considered property of the

Government. Authorization of that officer as a public servant has been explicitly described within the meaning of the Pakistan Penal Code, 1860 (XLV of 1860).

9. **Action by authorized officer will be fully protected**

"Any officer empowered under this act will free to his/her actions and activities. He/she will not face any suit, prosecution or other legal proceedings. All his/her actions or anything done or intended to be done in good faith will be considered in the best interest of Government. As authorized officer has done or doing everything in good faith and in the best interest of public service hence every police officer and member of other law enforcing agencies shall, upon request from the authorized wildlife officer will assist him in the due discharge of his duties under this Act".

(Note: A Summary of the Wildlife Act has been provided. For detailed study, consult the complete copy of Wildlife Act of KPK Pakistan 2015.)

REFERENCES

Babu, S. (2016). World Conservation Strategy: What Is It and Why It Is Important? Eco-Intelligent™ Making the World Ecologically Intelligent.

ENVPK (2021). Best Environmental Website of Pakistan.

Forgeard, V. (2023). Importance of Wildlife (with examples). Categories: Nature, Climate Change, Entertainment, Inspiration, Living, Society, Tourism, Travel.

Government of Pakistan. (2000). World Wildlife Fund for Nature, Pakistan and International Union for Conservation of Nature, and Natural Resources Pakistan. Biodiversity Action Plan for Pakistan - A Framework for Conserving our Natural Wealth. ISBN 969-8141-35-9. Imprint (PVT) Ltd., Rawalpindi Cantt. Pakistan. p. 88.

Hadidian, J. (2015). Wildlife in U.S. Cities: Managing Unwanted Animals. Animals 5(4), 1092–1113. https://doi.org/10.3390/ani5040401.

Kennerley, R. (2017). *Dryomys niethammeri*. IUCN Red List of Threatened Species. 2017: e.T40767A22223133. https://doi.org/10.2305/IUCN.UK.2017-2.RLTS.T40767A22223133.en. Retrieved 13 November 2021.

KKIENERM, Wildlife, Forests & Fisheries. (2023). Crime Module 1 Key Issues: Implications of Wildlife Trafficking. www.unodc.org. Retrieved 2023-04-13.

Kyne, P.M., Compagno, L.J.V. & Bennett, M.B. (2016). "*Aetomylaeus nichofii*". *IUCN Red List of Threatened Species*. 2016: e.T84784918A68607458. https://doi.org/10.2305/IUCN.UK.2016-1.RLTS.T84784918A68607458.en. Retrieved 16 November 2021.

Michael, C.L., Bowman, W.D. & Hacker, S.D. (2013). Ecology. 3rd Edition. Sinauer Associates, Sunderland, Massachusetts, U.S.A.

Mills, S.L. (2012). Conservation of Wildlife Populations: Demography, Genetics, and Management. Wiley.

Pimm, S.L., Jenkins, C.N., Abell, R., Brooks, T.M., Gittleman, J.L., Joppa, L.N., Raven, P.H., Roberts, C.M. & Sexton, J.O. (2014). The Biodiversity of Species and Their Rates of Extinction, Distribution, and Protection. Science 344(6187), 1246752. https://doi.org/10.1126/science.

Richard, F., Jonathan, B.D., Katherine, R., Mark, E.D.B., Michele, D.R., Charles, F.B., Robert, L.C. & Paul, S. (2017). Genetic Management of Fragmented Animal and Plant Populations. New York, NY: Oxford University Press.

Roxy, M.K., Modi, A., Murtugudde, R., Valsala, V., Panickal, S., Prasanna, K.S., Ravichandran, M., Vichi, M. & Lévy, M. (2016). A Reduction in Marine Primary Productivity Driven by Rapid Warming Over the Tropical Indian Ocean. Geophysical Research Letters 43(2). https://doi.org/10.1002/2015GL066979.

Share the Facts About Wildlife Trafficking. (2023). www.conservation.org. Retrieved 2023-04-13. MUNiSQ General Assembly.

Stop Global destruction (2019). "What is CITES?". CITES: Convention on International Trade in Endangered Species of Wild Fauna and Flora. support@stopglobaldestruction.org. Retrieved 2019-05-13.

IUCN 2025. The IUCN Red List of Threatened Species. Version 2024-2. https://www.iucnredlist.org

"The Wilderness Actl The Wilderness Society". (2022). www.wilderness.org. Retrieved 2022-05-29.

Thom, M. (2015). "Queensland's Shark Control Program Has Snagged 84,000 Animals". Action for Dolphins. Retrieved 2019-01-04.

United States Environmental Protection Agency. (2002). *Persistent organic pollutants: A global issue, a global response.* https://www.epa.gov/international-cooperation/persistent-organic-pollutants-global-issue-global-response

USEPA (2023). Effects of Acid Rain. Clean Air Markets Division, Environmental Protection Agency 1200 Pennsylvania Avenue, NW, Mail Code 6204M, Washington, DC 20460

Wildlife Act. (2015). KPK Wildlife and Biodiversity (Protection, Preservation, and Management) Act Pakistan.

Wildlife Trade. (2023). Wildlife Trade Drives Declines of Over 60% in Species Abundance, According to New Research. The University of Sheffield Western Bank Sheffield S10 2TN www.sheffield.ac.uk. 2021-02-15. Retrieved 2023-04-13.

12 Biodiversity Losses and a Way Forward

12.1 INTRODUCTION

Throughout evolutionary time, parallel to other natural processes, Earth's biological diversity contracted and expanded at the species, population, genetic, and ecosystem levels. Anthropogenic forces of environmental change significantly lost biodiversity at quite an accelerated pace. This loss occurred at both global and geological scales (Barnosky et al., 2011). There are several factors which cause this diversity loss. Important among them are extensive disruption and conversion of terrestrial, freshwater, and marine ecosystems, pollution and other forms of environmental degradation, widespread modification of hydrological systems and regimes, the global dispersal of invasive species, overharvesting of biological resources, and, finally, global climate change and its associated effects. Moreover, social and economic forces that shape the relationship between people, landscapes, and ecosystems where they inhabit can also destroy biodiversity. These factors are not new and have been present for a long time. Human communities and economies have always played a key role in biodiversity management throughout history. Human communities influenced ecosystems' composition, function, and characters of biodiversity. Some societies evolved behavioral norms and contributed to relatively sustainable and resilient environmental conditions. Some, however, could affect diversity. Human populations grew and spread geographically. Human economy expanded and intensified. All these developments put sustainability of animal diversity at risk. Recent decades have witnessed that human socioeconomic systems can only be secure for a long time if they are based on a healthy and resilient ecosystem, along with the Earth system where that ecosystem is embedded. In the early 20th century, political reforms took place. Accordingly, progressive complaints against biodiversity management emerged. At the same time, biodiversity conservation practices emerged prominently as a modern political and professional movement in the USA. Then, the policy-makers, scientists, business leaders, and citizens in the USA showed a response to the decade's widespread exploitation of resources. Accordingly, they started to adopt measures to stop these losses, encouraging the protection and utilization of natural resources rationally. With the shift in social values and increase in scientific knowledge, conservation has evolved continuously over the entire history of life. Practical approaches and preservationists marked and divided the conservation efforts. Nevertheless, they will continuously affect these efforts with broader and more integrated set of values. They recognized that the biodiversity has created a resilient ecosystem. Presently, biodiversity conservation faces the challenge of a type of evolution that can respond effectively to rapid changes in society and the environment appearing on a global scale. Making conservation more effective demands the evolution of its ethical foundations as well.

12.2 UNDERSTANDING OF SPECIES, HABITATS, AND ECOSYSTEMS AND IMPROVEMENT OF KNOWLEDGE

Biodiversity comprises a variety of species and ecosystems on Earth. It also encompasses the ecological processes going on within them, including species, ecosystems, and genetic diversity. Collectively, they are called biodiversity. Variation is an important feature of biodiversity. This variation can exist in ecosystems, between species, and, finally, within the given species.

 DOI: 10.1201/9781003472001-12

Biodiversity, therefore, represents all life forms, the natural processes, and functioning of these processes for the sustainability and wellbeing of diversity. To understand fully the implications of biodiversity, first we have to understand and comprehend the levels of the organization of biodiversity, which are:

- An **ecosystem** is constituted of microorganism communities, plants, and animals (biotic) and non-living (abiotic) elements. Both biotic and abiotic components interact as a functional unit. Characteristics of an ecosystem change when there is a change in its community members and its physical characteristics.
- **Species** represent an individual or group that can interbreed and produce fertile offspring. Therefore, species are unique genetic variation ensembles that are entirely self-generating. Their sub-species and populations can function independently and in essence they are self-replicating.
- **Genes** are the segments of DNA and are units of heredity. This DNA molecule encodes a single enzyme. This enzyme results in the synthesis of structural protein unit. Biodiversity emerges from genetic diversity. This genetic diversity continuously helps organisms participate in life processes, ultimately enabling populations to adapt to changing environments. Characteristically, biodiversity has three significant components. These are:
  i. **Composition** includes lists of the species living in an ecosystem or in that particular area. It also comprises the quality and richness of various species in that region.
  ii. **The structure** represents a pattern of that particular system and its physical organization. At the landscape level, it measures habitat complexity and other community elements. It also deals with patterns of habitats that different animal species occupy.
  iii. **Functions** deal with evolutionary and ecological processes going on in that particular area. Mycorrhizal associations, gene flow, natural disturbances, and predations are but a few examples. Soil development due to abiotic processes and hydrological cycles is also part of the ecosystem. This ecosystem recycles the nutrients for better productivity and strains water for purification. These ecosystem features are multi-scalar, incorporating both spatial and temporal dynamics. Because of their spatial and temporal dynamicity, different scales like landscapes, regions, and ecosystems can examine these attributes. Hence, characteristics of biodiversity and primary components of the ecosystem are closely related. To better understand and update our knowledge on ecosystems and biodiversity, we now study some of the ecological principles, concepts, and their applications in real life.

12.2.1 Ecological Concepts

Ecological concepts help to understand the facts about an ecosystem. A good understanding of an ecosystem paves the way for better management. The following paragraphs discuss different ecological concepts one by one.

Concept 1

Levels of biological organizations like ecosystems, landscapes, populations, communities, regions, and species make biological organizations. Life and its processes are dynamic all the time. To maintain its living status, it adapts a variety of patterns and uses various methods. Each scale or level carries an important position and needs proper functioning. For the conservation of biodiversity, the variety of scales and levels interact with each other and, hence, are interdependent. Their proper understanding and assessment facilitate conservation processes.

Concept 2

Native species. Existing animal species in a particular region/ecosystem without any immigration and emigration are called native species. For example, cedar plants and salmon originated in British Columbia (B.C.). Scotch brooms are native to that region. The Brown bullhead was introduced in B.C. from outside sources. Soon after introduction, it immediately invaded several ecosystems. Hence, this species is not native to B.C. Therefore, invasive or alien species can replace it easily. They threaten ecosystems, harming both the environment and the economy. Natural predators and diseases do not harm invasive species. They flourish without any checks and soon become deleterious and damaging. They dominate the ecosystem, changing its original species composition and how it functions. For example, when boat trailers spread Eurasian watermilfoil in freshwater ecosystems of B.C., it clogged gravel spawning beds, which salmon normally use for spawning. Higher phosphorus levels in Okanagan lakes have the same function as these birds.

Concept 3

A keystone species, ecosystem, or process. These factors *influence* ecosystems and biological communities unevenly. Their effect is more prominent where biomass is abundant and unevenly distributed. If the keystone species are lost, they broadly affect the community or ecosystem. Keystone species interact with other species in several ways. They show symbiotic dependencies such as plant-pollinator relationships, or ecosystem modification. Cavity nesting, beaver impoundments, and predation are some the examples. In several countries in the globe, wild salmon is an important keystone species. Because of their back and forth breeding movements they add marine nitrogen to freshwater and terrestrial ecosystems. They are also a source of food for many terrestrial animals, e.g., grizzly bears. These bears drag the carcasses of salmon into the forest, and add nitrogen to nitrogen-deficient forests and adjacent areas where they feed. Sea otters do the same. Otters feed on sea urchins. Sea urchins in turn feed on kelps. We can promote the production of kelp forests if we can control the excessive production of kelp forests. These kelps then provide habitat for various invertebrate species as well as for fishes, for the proper maintenance of the ecosystem presence of keystone species and their related processes carry pivotal importance.

Concept 4

Population viability/thresholds. These thresholds tell us how the various ecological processes, like disturbance, influence the survival of species/populations. With the decline of the habitat, populations/species will also decline and, if this process continues, the species may totally perish. In addition to habitat, healthy genetic variability also enhances the survival of the species. In natural ecosystems, extinction is a normal process; nevertheless, human activities have accelerated this process considerably.

Concept 5

Ecological resilience. This concept deals with the level of capability of an ecosystem to manage the stress and disturbance in and on the ecosystem and how long it takes to return to a stable state. Ecosystems are rarely at equilibrium; they are complex, adaptive, and dynamic. Most ecosystems can exist in various states. Moreover, they continuously change unpredictably in an environment. An ecosystem with sufficient tolerance capability does not collapse but instead rebuilds itself before these stages approach. Resilience in an ecosystem depends on functional diversity and response diversity. Functional diversity enhances the system's productivity, while response diversity enables the community to tolerate environmental stresses and shocks.

Concept 6

Disturbances. In ecological systems, disturbances are distinct events that can be either natural or induced by human beings. In addition to anthropogenic interferences, natural disturbances also overturn the ecosystems. Wildfires, floods, lake turnovers, droughts, wind throw, and insect and disease outbreaks are the most common ones. However, rural and urban development, timber harvesting, and road building are human-induced disturbances. Humans also induce disturbances in the aquatic environment too. Important among these are pollution, wetland drainage, damming, and water extraction from rivers and streams. Some human disturbances are long-lasting. Some of the aquatic human-caused disturbances are the construction of dams over the rivers, removal of water from the rivers and streams, drainage of wetlands, and pollution (air or aquatic). Some human-related disturbances fundamentally alter ecosystems because of their long-lasting effects. Such human modifications create difficulties in the management of ecosystems.

Connectivity/fragmentation: This factor fundamentally controls the movement of the organisms in and between the land patches, which have sufficient resources. The relevant scale of connectivity varies with each species. It depends on vulnerability to human-caused mortality, habitat requirements, and sensitivity to disturbance. Connectivity to different land patches facilitates organisms to move of their own will at the place of their choice in response to changes in the climate, seasonal cycles, and occurrence of forest fires. Fragmentation of parcels of land disrupts connectivity. Loss of connectivity results in fragmentation. Differences in landscape determine the degree and the characteristics of the natural connectivity.

Concept 7

Coarse and fine filter approach. This approach mainly deals with the management of biodiversity-inhabited landscapes and the network of protected areas. Moreover, it takes into account the type of management practices applied to the biodiversity and its habitat.

Concept 8

Risk is an inherent aspect of decision-making. Any particular action or decision can result in any damage or loss, which is a potential risk of the consequence of that particular decision or action. Risk assessment is based on the likelihood of an event which may or may not occur and its consequences. Risk assessment evaluates both these elements. Risk management assesses the risk, and weighs its magnitude to reach the best decision against the expected benefits. There is a direct relationship between risk and uncertainty.

Concept 9

Adaptive management is a systematic learning process. Adaptive management formally monitors and plans the outcomes of decisions. This ultimately helps us in the management of natural resources. Suppose we need more knowledge for decision-making. In that case, we can seek help from options like (a) "trial and error," based on the best guess. Then later on we can choose from a subset that gives better results. In the "passive adaptive" option, presumably only one model is correct. In this model, we consider multiple alternate models and then link to different policy choices. Passive adaptive management incorporates uncertainty in a dynamic system providing effective management. Passive adaptive management provides citizens, scientists, managers, and other stakeholders with more opportunities. They learn a great deal from this phenomenon with maximum benefits. There is a close relationship between adaptive management and concepts of risks and uncertainty. Output from these phenomena helps continuously in the decision-making process.

Concept 10

Ecosystem-based management (EBM). This manages human activities in an adaptive way. This approach relies on place and considers intrinsic ecosystem value vs. its value to humans.

Concept

Protected area. Protected areas, including parks, sanctuaries, etc., are often considered the essential and foremost component of a coarse filter approach applied for conservation. In addition to conservation, protected areas can play several other roles pertaining to conservation. Protected areas are used to finely filter the desired species. For example, this approach is applicable to the protection of a significant landform or a population of a rare species. This concept also provides connectivity and can also serve as a benchmark. In addition to the above applications, we can apply this concept to education and research opportunities.

12.3 EXTENSION OF BIODIVERSITY SCIENCE: AN ACQUISITION OF KNOWLEDGE

For conservation-oriented decisions, scientists must undertake well-designed studies on the abundance, diversity, and distribution of different taxa. Understanding of further biodiversity patterns is important to identify and assess the effects of physical, climatic, biological, and human-related variables on the distribution and abundance of different taxa, communities, and populations. In this connection, we need to conduct a systematic survey of reptile biodiversity to explore their association with different habitats in light of various aspects of ecology. In addition to reptiles, there is a need to estimate the abundance of birds such as raptors and inland-wetlands birds. Therefore, for proper management of biodiversity understanding the dynamics of these diverse wildlife assemblages in the country is crucial. It is also crucial to describe the composition of the mammalian predator assemblages across different sites in the country. The knowledge gained in this way can help in driving the colonization processes, consequently extirpating the mammal populations. All these activities will generate and maintain biodiversity databases. These databases in turn will support proper management actions and decision-making regarding the protected areas.

Applied research should be our priority. Following its applications we then monitor the effects of human activities on different taxa. For example, arthropods show different feeding behaviors during their life stages. They act as detritivores, herbivores, and predators. Hence, at various levels of the trophic web, arthropods play a crucial role in nutrient recycling and energy flux (Ayal, 2007). Many arthropod species are very loyal to their habitat and hence when there is any change in the surrounding environment they respond immediately (Martínez, 2013). Disturbance in any ecosystem changes abundance of the animals, richness, and species composition. We can express this variation at a taxonomic and functional level. For example, during habitat loss there is disturbance in the richness and birds, reptiles, and small mammals. This happens in all the regions of the country. There is a need to explore the consequences of this loss and its assessment accordingly. Human populations mostly settle and recreational activities are established along sandy areas and coastal habitats. These activities affect the animal population both in abundance and distribution specifically their impact is more significant on reptile population found in that particular area. The development of the region following human establishment increases the road traffic load. Road accidents frequently kill middle-sized mammals such as foxes, cats, maras, grisons, and armadillos. Even birds like tinamous and various reptiles cannot escape from such accidents. Therefore, it is important to assess the intensity of road killings and devise ways to prevent such accidents from happening again. In addition, higher mammals predate on smaller ones. Estimation of these animals' deaths due to this predation effect is also important.

The exact quantification of these deaths can help plan actions to mitigate these actions to save lives. On the basis of these observations, we need to explore and devise effective, non-lethal methods to control and guard predators. Dogs and donkeys may be quoted as examples (Novaro et al., 2016). Internationally, Pakistan has renowned marine and coastal biodiversity in addition to diverse and attractive terrestrial wildlife. Due to rich wildlife fauna, we can develop responsible, low-impact tourism activities. Therefore, it is required to work on specific actions to promote responsible, sustainable human development activities. These activities should be based on sound scientific knowledge ensuring that they are not endangering biodiversity conservation needs. Only by following such measures can conservation policies be implemented in their true essence.

12.4 ESTABLISHMENT OF QUARANTINE PRACTICES TO MINIMIZE THE OCCURRENCE AND IMPACT OF INVASIVE AND DESTRUCTIVE PESTS, DISEASES, AND WEEDS

Cross-border movements of animals and their products spread transboundary animal diseases, such as FMD. This problem is more severe in countries without adequate implementation of laws and regulations governing these movements (Di Nardo et al., 2011). In Southeast Asia, transport of live animals, either tamed or from the wild, is the main factor facilitating the spread of different diseases (Gleeson, 2002). Distributions of viral strains that have been isolated and collected from these areas witness the pathways used for the trade of livestock or other animals transported legally or illegally (Di Nardo et al., 2011). Poor regulations of extensive land borders further facilitate the unchecked movement of livestock.

When infected animals interact, they spread diseases, mainly through an aerosol transmission or environmental contamination. The movement of the animals from one place to another facilitates contact between infected and susceptible animals and plays a pivotal role in spreading diseases. At local levels, animals may interact with each other at water-drinking sources or pastures when grazing, which then become the source of the spread of diseases (Naing Oo., 2010).

12.4.1 Drivers of Animal Movement

Price or money is the major driver of animal movements at the national or international level. Whenever there is disparity in demand and supply, there are always price differentials. If demand for one product increases at one place due to its low supply, its price will rise automatically, and then the products and animals are attracted from other places. Hence, price or money is the major driver instigating the movement of animals.

Other factors that promote animal movements are the preparation of ornaments and the consumption of animals at religious or other festivals. This increases the consumption levels of animals, with depletion of stock in one place. Animals are imported from different areas to fill this gap. Noteworthy among them are Islamic religious festivals, Chinese New Year, a religious festival in southern Thailand, and Tet in Vietnam. To facilitate the transportation of commodities among the countries, the process of the development of corridors is expanding rapidly (ADB, 2014; Hong Kong Trade Development Council, 2015). This activity will improve existing infrastructure, allowing rapid movement of animals over longer distances. Nevertheless, in most cases, traders commit this movement illegally. They use more minor roads and prohibited border crossings. This limits the utility of these corridors. Until and unless the official movement of these animals is properly regulated, these corridors are purposeless.

12.4.2 Quarantine and Inspection

In this practice, all the animals are critically inspected and strictly isolated to control the spread of diseases. Quarantine measures are applied to several animal groups to introduce them to a new area (farm, village, etc.), or if they are moved among the countries, or if an animal is infected or any animal housed with infected animals and accompanying infected animals during transportation.

12.4.3 Establishment of Quarantine Systems National Level

Quarantine systems prevent and control the spread of animal diseases into and within the country. Moreover, these systems check in and out movement of animals. Geering et al. (1999) have suggested the strict use of quarantine programs to control the international borders. These measures can help in preventing the smuggling and uncontrolled entry of animals, animal products, and other potentially dangerous goods. Scientists and experts should issue proper certificates authenticating sound animal health. This is possible only if they launch such program which can ensure pre- and post-quarantine measures. These programs should not encourage the smuggling of animals, which leads these programs toward counterproductive efforts. Hence, there is a need to negotiate quarantine conditions to be fully agreed between exporting and importing countries to import animals and their products safely. These conditions require the examination of animal health and its certification, and quarantine before export. On the arrival of animals in the host country, it is also necessary to inspect the animal and quarantine it before its release or transfer to an enclosure or housing.

12.4.4 Cross-border Movement of Livestock Needs Quarantine/import Regulations

When any country imports any animal, tamed or wild, the importing country may execute risk analysis and work on the impact of different risk mitigation measures. Applying quarantine restrictions to the effects of these restrictions on illegal traders should also be considered.

12.4.5 Location and Establishment of Quarantine Labs

The Veterinary Authority controls the quarantine laboratory works. In this facility, animals are kept in isolation and cannot interact with each other directly or indirectly. The purpose is to ensure the non-transmission of any specific pathogen to the local and wild fauna. If considered appropriate and essential, animals are also tested and treated.

The place and methodology used for quarantine should be decided mutually by both exporting and importing countries with the assurance that quarantine facilities decided are fit for the purpose. Certification from the exporting country should be so comprehensive and authenticated that it should be acceptable to the importing or transit countries. The government can run the quarantine stations, or any private organization can run them with the government's permission. Nevertheless, whoever operates it should have the prior approval of the governments of both the exporting and importing countries.

12.4.6 Role of Quarantine

Quarantine will play an integral role in case of an outbreak of any disease. Combined with other measures, it prevents the mixing of infected species with others and helps in testing the disease and its treatment if required.

12.5 MANAGEMENT PRACTICES OF BIODIVERSITY AND ENGAGEMENT INDUSTRY LANDHOLDERS

If we look at the global level, biodiversity conservation depends heavily on protected areas. Protected areas have helped biodiversity to halt its decline and ensure the survival and existence of biodiversity in the future. These protected areas may contain either private or public land and have a powerful role in the promotion of conservation of nature. Functionally, these protected areas facilitate in situ large-scale conservation of both flora and fauna (Bishop et al., 2004). Due to limited land available (13.9% of the total global land area) they contain a small fraction of global biodiversity. Hence, we need to consider sufficient measures for conservation.

Further, these areas need effective improvement because they are susceptible to human degradation (Chape et al., 2003). This conservation measure also carries demerits. The land is limited. Therefore, it is hard to convert every tract of land into formal recognized protected areas. Hence, we restrain from the implementation of conservation measures (Figgis, 2004). Presently, private land cannot fulfill all of the conservation requirements. However, if land is available it can maintain connectivity with other protected areas, helping in the extension of species and their habitats (Clough, 2000). For example, 73.8% of the total land encompassed by national parks in Great Britain belongs to private people. The private sector in Costa Rica contributes up to 45% of the total land of biological reserves.

Similarly, the private sector has contributed a minimum of 14 million hectares of land for wildlife management in South Africa (Chacon, 2005). Therefore, the focus should be on the biodiversity-rich areas instead of the formally recognized protected areas, regardless of whether the land belongs to the private or the public (Knight, 1999). Nevertheless, acquisition of land ownership and its integration in conservation planning and management is a complicated process due to the interrelationship of social and economic traits (Mascia, 2003). Nonetheless, negative consequences appeared when the top-down approach was applied to conserving biodiversity on private land. Private landowners are reluctant to participate in such conservation strategies where they get or are offered benefits (Grodzińska-Jurczak & Cent, 2011). Hence, globally, for conservation of nature the exploration of land from the private sector should be based on participatory site selection approaches and legally modulated for financial incentives to owners (Doremu, 2003). These efforts started much earlier in developed countries. Unfortunately, an adequate system with legal and government support does not exist in most developing countries. Hence, the promotion of the conservation of biodiversity on private land still lags far behind that practiced in developing countries (Figgis, 2004). Nonetheless, we need a new vocabulary parallel to the IUCN protected areas system. These conservation security strategies should be in line with important facets of conservation objectives means conservation of biodiversity on private land.

12.5.1 Acquisition of Private Land and Promotion of Biodiversity: Role of External Interventions

If we want to use private land for the conservation of nature, then there are several policy options that we can use. They can be regulatory prohibitions. Second, the government can directly acquire land from private farmers with some incentives. However, for the successful implementation and finalization of any decision on conservation policy, the public consultation component must be included (Ostermann, 1998). Some strategies for the conservation of nature, however, have specific goals. Unlike the earlier strategies, there are some other conservation strategies. Working objectives of these strategies are quite broad. Biodiversity conservation of these strategies is a secondary objective (Adetunji, 2013). The existing biodiversity conservation options can be voluntary, where the landowner has no say in decision-making.

Nonetheless, it can also be voluntary, where the landowner has complete power to say yes or no. However, options are open, either for their individual use or their adoption in combination form.

12.5.2 Involuntary Strategies

In this strategy, government agencies are fully competent. They can use prescriptions or prohibitions. There is minimal participation of landowners in the decision-making process of management of biodiversity or usage of land donated by these landowners.

12.5.3 Compulsory Displacement of the Land Owner and Total Acquisition of Land Required for the Conservation of Nature

Landowners need to be consulted frequently when we want to start the establishment of protected areas for the conservation of nature. The experts acquired the land compulsorily, and the facility was erected according to its requirement, including its volume and prescribed area (Stroup, 1997). This option, however, was not applicable in developing countries with limited land holdings. Nonetheless, irrespective of developing or developed countries, wherever conflict emerged between the wildlife population and local inhabitants, people were displaced. The land was acquired, and nature was preserved and protected according to its requirements (Brockington & Schmidt-Soltau, 2004).

12.5.4 Imposed Restrictions/regulations

This strategy does not permit the consultation of private landowners. After acquisition of the private land it is declared a protected area or part of the protected area for conservation of biodiversity. The authorities then start to impose restrictions on the use of land acquired from private landowners. These restrictions hamper the pace and progress of developmental activities, which in turn has a negative impact on the conservation of ecosystems, species, and habitats. Although strategy does not disturb or displace the landowners, confrontation does develop on property rights. Both parties also need help to utilize this land. This strategy fits very well on those lands which are located within the areas which are strictly protected. In such situations, restrictions and regulations apply to public land equally as they do to private land (Eli, 2003). Such imposed restrictions are of no use and are currently seldom visible in developed countries which are devoid of compensation, easements, or contracts. On a smaller scale, this is also practiced in the Habitat Conservation Plan in the USA. Here, no specific rules and regulations exist for the local use of land or for dividing into zones for any option. This practice is more prevalent in underdeveloped countries or in those in a transition state. Because these countries cannot purchase all the land that falls under protected areas due to their financial disability, neither are they in a position to pay compensation to landowners (Eli, 2003).

12.5.5 Voluntary Strategies and Tools for the Use of Private Land

There are several voluntary strategies to acquire and use private land for the conservation of biodiversity. Despite their context specificity, these strategies can be successfully adapted and used for different sites and regions. In this scenario, the government and other agencies support the land acquisition mechanism and incentives to be proposed for the private landowner. However, these, activities can not bind the farmer to handing over the land for conservation purposes. The farmer is still independent, and the decision rests with the farmer on what to do. North America, Australia, Africa, many countries in Latin America, and most of the large private conservation organizations purchase private land and themselves manage all the conservation efforts. Sometimes, these organizations acquire the land from the private sector and donate it to government agencies for conservation purposes (Eli, 2003). Whatever the way of working, their primary goal is biodiversity conservation. For this purpose, they take up considerably large tracts of land with better spatial and temporal protections.

12.5.6 Formal and Informal Status of Private Reserves

Sometimes, a parcel of land is set aside legally for the protection of nature. Although it is still under private ownership, both parties can benefit equally from it. Such areas are called private reserves, which can be either wildlife or game reserves or ranches. Whatever the position is, they will be used for the protection of biodiversity. Keeping within sustainable limits, game reserves and ranches are used for trophy hunting. These protected areas may be either formal or informal. This means that government bestows its legal status based on existing and current ecological and technical criteria. If the landowners agree, then it can be used for various activities for protection of nature without declaring its status. However, this depends upon the provisions available in the country. The size and nature of these land reserves depend on the objectives of the organization that is responsible for the protection of nature in that area/region. Accordingly, they vary in size, tenure of the land and its management regime, and methodology (Krug, 2001). Private reserves and game reserves are considered the most popular conservation strategies in the country. It does not matter whether an individual possesses it and other investors are the part of it. Due to their effective usage, normally such reserves have very rich mega-fauna. Judicial and sustainable management of these areas can generate large incomes from game hunting, wildlife viewing, eco-tourism, and safaris. In the absence of other options, these reserves can do a great deal for biodiversity conservation. Nonetheless, conservation options lose their enthusiasm if economic benefits directly contribute to conservation and maintenance of wildlife habitat (Lindsey et al., 2006). In the African continent, traditionally such private game reserves, used for the management of wildlife, are very common. For example, Namibia, South Africa, Kenya, Zimbabwe, and Tanzania have many private reserves used for the conservation of wildlife and which are continuously gaining popularity in these areas (Krug, 2001). Namibia has approximately 150 and South Africa has 1,000 game reserves with successful conservation activities (Krug, 2001). For example, in Australia large private companies with the support of National Reserve System programs establish private wildlife sanctuaries and private protected areas for conservation of nature (Figgis, 2004). The U.K. has also taken up a similar initiative under the National Nature Reserves program. This program—with the approval of statutory conservation bodies—declares the private land as a protected area. Hence, these larger private organizations take hold of protected areas from establishment to management instead of individuals (Reid, 2011).

12.5.7 Conservation Covenants, Easements and Restrictions Thereof

An easement literally means acquisition of land for a specific purpose from the private sector. Fundamentally, a conservation easement is a voluntary activity. Nonetheless it is a legal binding and agreement grantor (a landowner) and grantee (an organization such as a land trust or a government agency). Though it is new concept but it has gained lot of popularity in conservation strategy despite others which are currently under use. These are becoming increasingly expensive and intrusive. Moreover, they generate conflicts over property rights. Hence, for conservation of nature conservation easements offer a more effective and less expensive tool than involuntary approaches (Main, 1999). In this strategy, the landowner relinquishes some rights over the land, but nonetheless becomes an economic beneficiary. Mainly the private landowner sells developmental rights and avails tax relief (Bernstein & Mitchell, 2005). Easements can be successfully and favorably applied to lands with historical, cultural, or scenic value. Applications of easements can be further extended to agricultural lands and private forests if required and considered worthy of such use (TNC, 2011). Covenants or conservation easements started in the USA in 1950. Now they are used effectively with considerable success in Australia, the U.K., and Africa, as well as across Latin America (Leva, 2002).

12.5.8 Conservation Contracts/programs as Incentives to Private Landowners

GOT Offer of incentives to landowners attract the private land owners to voluntarily apply conservation measures. For successful implementation of this program we need to lay down well-defined conservation goals. These goals should be ecological sound and at the same time acceptable to landowners. Such programs help in the restoration of quality of land benefitting landowners economically and limit those activities which limit the impacts of those factors which impact biodiversity negatively (George, 2002). The voluntary programs are of different types and very diverse in forms with probably the largest number is coming from USA. Instead of getting payment from government landowner through conservation contracts can share the cost, and technical assistance. This working strategy can be another incentive to private landowner too. If private landowner is lending forests, wetlands, and/or agriculture land and wildlife species worth lot then the landholder can be offered special reward for initiating the conservation measures (Clough, 2000).

12.5.9 Conservation Networks

Presently, landowners are looking for money and other monetory benefits. They have integrated all the conservation activities with economic value. Hence, several associations/networks/organizations have emerged. These organizations disseminate the required and useful information among landowners. Landowners can adjudge the most applicable conservation strategy favorable for them. These organizations therefore bridge the gap between private landowners and conservation agencies. A few examples of such networks are the Private Landowner Network and the Cooperative Conservation America in the U.S.

12.5.10 Mixed Strategies

For successful implementation of conservation measures and outcomes, traditional approaches are considered and executed. The significance of these strategies is enhanced when voluntary bottom-up strategies combine with top-down prescription. If the protected areas have patches of private land then this approach works very well. In the presence of government restrictions, this is the only viable option of land use for conservation measures. Mitigation banking and transfer of development rights (TDRs) can be quoted as examples for mixed strategies. Local governments take up TDRs (complex market-based instruments). They promote the transfer of development rights. Then they sell the particular right from ecologically sensitive areas (sending areas) and higher development potential areas (receiving areas) (Daniels, 1998). Mitigation banking, on the other hand, enhances, restores, and/or or preserves wetlands.

12.6 REVERSAL OF THE CRITICALLY THREATENED SPECIES AND ECOLOGICAL COMMUNITIES BY PLANNING AND IMPLEMENTATION OF MANAGEMENT REGIMES

12.6.1 From the Field to the IUCN Red List

At the global level, currently available data—whether on species, sub-species, or population—is used for the assessment of species. Whatever reliable sources of information are available—whether from the present or from the past—assessing scientists use them for the assessment of species. For the study of sub-species, both species and population assessment is accomplished. Red List Authorities depute at least two qualified reviewers to evaluate the submitted assessments.

12.6.2 Approaches to Recovery of Species

12.6.2.1 Science

Knowledge is a key to most, if not all, problems. If we engage scientists in our decisions and all our decisions are based on the evidence, then we can confidently claim that there is a possibility that our decisions will be successful. If successful decisions, when we monitor the results and outcomes of the decisions we can learn what type of works we need to undertake and what type of interventions we need to adapt to obtain the maximum possible best outcomes. This scientific base grows further when unique knowledge of the environment is collected and is then trickled down to generations to come; this gives different dimensions to the continuously developing and growing basis of this knowledge. At global level, some governments spend millions of dollars on the species recovery plan, e.g., the Australian Government dedicated A$30 million on a scientist-based approach meant for recovery of the species.

Continuous changes are taking place and so, accordingly, are their impacts on the living world. Science should also address these changes, as well as the impacts they have on the environment and its inhabitants. In addition to existing environmental stresses, climate changes inflict threats to biodiversity. However, it depends on geographic extent and magnitude and speed of these changes. Climatic changes can also exacerbate other threats like the breakup of unexpected fires. When climate changes, it induces changes in distribution of species and in composition and function of biological communities. Some governments are already doing a great deal of work on biodiversity conservation. Others need to invest in the development of adaptation research capacity. This adaptation research should be capable of producing practical, hands-on tools and information for local decision-makers, including a threatened species recovery plan. To achieve this approach, broadly speaking we rely on the scientific community, which is funded by states, territories, universities, private institutions, and philanthropic (charitable) sources. This is required to build the evidence for underpinning proposed or current action to be undertaken.

12.6.2.2 Action

Science and scientists provide us with the information necessary about what to do. The next step is to take proper action for remedial measures for recovery of threatened or near-threatened species. For any actions for protection and recovery of threatened species, scientists prioritize resources and efforts. All these resources and efforts should be further backed with hard and measurable targets. Governments are and should invest sufficient amounts of their budgetary allocations on initiatives like the Green Army, tree plantation, and landscape programs. Such programs are very helpful for communities and regional natural resource management (NRM) organizations. All of them then can effectively take practical actions in different or all parts of the country to protect and recover threatened species.

When state and territory governments with statutory obligations are involved in biodiversity conservation measures, they necessarily take effective recovery actions. These recovery actions are not taken for fun. It is obligatory for community groups, scientists, and non-government organizations to exhaust their dedicated efforts and, where necessary, take local decisions to achieve successful outcomes for the protection and recovery of threatened species.

12.6.2.3 Partnership

As we all are beneficiaries of the natural environment, we all are responsible for its conservation. To develop and maintain biodiversity sustainably is a shared opportunity and responsibility for all of us. When it is left to governments alone, it may end up as a failure. For successful implementation of biodiversity conservation and management programs we need community coordination, support from scientists and other researchers, non-government organizations (big and small), state and territory governments, regional NRM organizations, local land care groups, local government, and the business sector. Partnership with these organizations and groups can ensure the effectiveness and sustainability of our actions. Until and unless we work together, we cannot prioritize and maximize our resources following protection of threatened species.

12.6.3 Effective Planning to Recover Species

Under national environmental law, whatever species are declared or listed as threatened require conservation advice and, accordingly placement, of a recovery plan. Prepared with the collaboration of stakeholders, the key planning documents help to initiate the necessary actions for stopping the decline of species and also the recovery of those species which are threatened right now. These documents provide the detailed actions which are well supported by appropriate evidence for the protection of threatened species. Prioritization of efforts and effective government programs are fundamental for the protection of threatened species. Effective recovery teams are mandatory for the facilitation of the successful implementation and effective recovery planning. These teams will hold hands with interested states and territories, scientists, private organizations, community groups, and land managers. This cooperative approach with these stakeholders strengthens recovery planning and implementation across state borders.

12.6.3.1 Regulatory Protection

Ecological communities and threatened species present within them are considered matters of national environmental significance. Hence, such matters and projects impact the threatened species. Therefore, any program that deals with the conservation of threatened species demands rigorous assessment under national environment law. This assessment process takes into account all project proponents to obviate any impacts on threatened species. Sometimes these projects require counterbalancing. This activity helps to implement specified actions in species recovery plans. Funding for facial tumor disease in Tasmanian devils can be quoted as an example of these offsets. Some countries like Australia are committed to deliver and have established a one-stop shop for environmental approvals. This activity under national environmental law will accredit state planning systems for the creation of single environmental assessment and approval process.

12.7 EXTENSION AND PROMOTION OF KNOWLEDGE ABOUT MANAGEMENT OF INVASIVE ANIMAL SPECIES

Invasive alien species are species that are not native to that particular region or country. These can threaten the environment, indigenous fauna, their health, production, and economic output. They are considered the second greatest threat to biodiversity at the global level. There are several ways in which alien species threaten our natural world, as well as our wellbeing. Gradual development in globalization results in climate changes that instigate more challenges for the management of invasive species. According to the Millennium Ecosystem Assessment, invasive species and climate change irreversibly damage the functioning of ecosystems and human well-being (Hassan et al., 2005). Invasive species impact each and every sector of life. They affect biodiversity conservation efforts and the sustainability of biodiversity. They further affect fisheries, forestry, biosafety, trade, health, and agricultural productivity, and proper storage and preservation of crops and their products. In the United States, it has been estimated that invasive species cost US$137 billion annually (Pimentel et al., 2000; Carpenter et al., 2005). Damage caused by invasive species is a global problem and its management requires immediate international action. Taxonomic support contributes a great deal in classification and identification of species and ultimately management of the invasive species. Taxonomy (basic biological knowledge) when combined with evolving technologies and tools effectively prevents the activities of invasive species, their early detection, eradication, and ultimately their control (CBD, 2001a). Despite several applications of taxonomy, it is still not clear how far the scope of taxonomy can help in the management and control of invasive species.

Invasive species are universally scattered among human activities. Accordingly, similar work is going on in different fields. Damage, issues of invasive species, and their management are similar to other environmental issues across the planet. Important among them are plant pests and pathogens, the spread of disease and disease vectors, biosecurity, and weed control in agriculture crops. Nevertheless, later issues have emerged, including distinct terminologies and frameworks required for their management and regulation. It is in our hour of need that we are seeking proper regulatory mechanisms and technical cooperation across these different areas, while at the same time accounting for their individual needs and the constraints they are facing. Taxonomy, for example, needs integration of ecology and biocontrol for its effective functioning. Other than these two disciplines, close cooperation of the user communities is needed, irrespective of whether they be from biodiversity conservation, biosecurity, agriculture, health, or trade.

12.7.1 Assessment of the Invasive Species

Assessment of any invasive species uses reliable and documentary sources for the characterization of the taxonomy of the species under consideration. Assessment of the invasive species is always executed with the objective that those personnel and/or institutions and organizations who are dealing with the management of the invasive species are in receipt of sufficient information to handle those invasive species which are threatening our existing population soon after their emergence. Assessment of invasive species is undertaken with the expectation that taxonomic sectors predict, detect, monitor, prevent, and control invasive species in a much better way.

To date, the importance of taxonomy in the identification and management of invasive species has not been well recognized. Accordingly, institutional support for this field is surprisingly weak and fragmented. The Davis Declaration (Fell, 2001) has presented an overview on taxonomy and invasive species which is the only document which can be used for this purpose, although its circulation is very limited. The Davis Declaration further stresses the information hubs which can propose solutions for the delivery of taxonomic support and expertise.

The Convention on Biological Diversity (CBD) says that there is an urgent need to assess the level of invasive species at regional, national, and global levels for the implementation of its Global Taxonomy Initiative (GTI). This assessment focuses on the benefits which taxonomy offers in the form of keys which are identifications, distributional data, and species diagnoses. Under the CBD, whatever policy and strategy will be developed will be based on the assessment of invasive species. Unavailability of this data hinders the implementation of CBD programs. The identity of the beneficiaries of this assessment, the target organizations, and the stakeholders in the assessment are listed below.

- **Beneficiaries**: These are the principal agencies and authorities which are specifically involved in quarantine and management of parks. Plant protection organizations, who have the responsibility of detecting and controlling threats posed by invasive species, are also beneficiaries of this assessment. It has been further observed that these authorities often complain about the inadequate services and support from taxonomy for the successful accomplishment of their work.
- **Target organizations**: These organizations are meant to devise any policy for museums, collections, and other taxonomic institutions. It is applicable only to interesting institutions these organizations can only get benefit from relevant research, which includes species information, identification tools, and other outputs. Such tools and information can also be helpful and beneficial for other stakeholders. Taxonomic research, training, product development, and capacity-building at national level will serve as resource for policy setup. It will also serve as a resource for setting any policy at national level relevant to. It will also help in the advancement and technical cooperation at global level to the initiatives taken by various nations for the management of invasive species bridging the gap which people experience due to weaknesses and fragmented capacity of the taxonomy.
- **Stakeholders:** Both nationally and internationally there are several stake holders of assessment of invasive species. These stakeholders are policy makers as well as decision makers. Moreover, those stakeholders who deal with the impacts of invasive species and their proper management are also included in this list. Examples of important stakeholders are the CBD, the International Plant Protection Convention (IPPC), Sanitary and Phytosanitary Agreement of the World Trade Organization (WTO–SPS), and the Ballast Water Convention. This number is, however, not enough to deal with such environmental issues. A great deal more is required for the efficient management of invasive species and their effects on the environment. Such stakeholders should take a complete hold of all the activities at different levels covering professional as well as other boundaries. These activities include the establishment of legal and regulatory authorities. These organizations raise

awareness among farmers about pest controlling insects following setting an international strategy for their effective involvement in invasive pest control. All such activities seek help and support from taxonomy, although it might not be immediately apparent.

12.7.2 Invasive Species and Their Global Importance

The Millennium Ecosystem Assessment (MEA) team is considered to be mainly responsible for the care of our planet earth. The MEA team claims that, among others, the introduction of invasive species directly affects the ecosystem processes and wellbeing of human beings. The gradual growth in trade following the introduction of invasive species has considerably increased their impact on most of the ecosystems (Millennium Ecosystem Assessment, 2005a). Of particular concern is the association of shipping and the impact of invasive species on marine ecosystems, species, and water quality in which invasive species inhabit. There are several factors which attract and facilitate the proliferation of invasive species in new environment, and a few of them are discussed below.

12.7.3 Habitat Change

Habitat fragmentation and disturbance is one of the leading factors which attract invasive species to come and propagate. Disturbed ecosystems are more susceptible to intrusion of invasive animal and plant species, including human pathogens (Millennium Ecosystem Assessment, 2005b). Other than fragmentation, other causes of ecosystem disturbance and changes in climate create well-documented and well-established opportunities for other types of invasive organisms, e.g., weeds. For example, in 2005, in southern parts of the United States, after Hurricane Katrina *Triadica sebifera*, the Chinese tallow tree, established itself over widely affected coastal areas (Sheikh, 2005).

12.7.4 Invasive Alien Species and Islands

Invasive species have dramatically affected the different islands. It has been observed that extinction rates and rates of population decline are higher on islands when compared with continents (Millennium Ecosystem Assessment, 2005b). Unfortunately, knowledge about their actual status is limited. British Overseas Territories (Varnham, 2006) (most of which are islands) have recently been comprehensively surveyed for species distribution. The information collected about occurrence of species was somewhat variable. It seems that the nature of the information obtained was very much dependent on the date and time of the survey conducted.

12.7.5 Biosecurity

At a global level, biosecurity is increasingly gaining in importance. Biosecurity concerns also arise from national species which pose a threat to human and animal health, although they are necessarily not alien. To combat the harmful effects of invasive species at the national level, support from taxonomy is sought for their proper identification and culmination. Similarly, at the international level several efforts have been exhausted to prevent the unchecked transmission of harmful pathogens and diseases that might affect and infect humans and animals. Both at national and international levels, taxonomic research institutes, specialist laboratories, and agricultural research centers collaboratively help in the diagnosis and identification of such toxins and/or pathogens.

12.7.6 Policy Context

Currently at the global level there are more than 40 international policy instruments. However, there are also some policy-related agreements which all deal with invasive species-related issues (McNeely et al., 2001). Among them, the Convention on Biological Diversity (CBD) has comprehensive and much wider scope. Unlike the CBD, the WTO–SPS and the IPPC have more limited scope. Nonetheless, as trade instruments, both politically and economically they are more significant. The Ballast Water Convention has recently been recognized as an instrument to address the major component of species which have invaded the marine environment, along with international shipping.

12.7.7 Research and Monitoring

In order to address the issue of invasive species appropriately demands adequate development of knowledge bases. States are therefore required to undertake appropriate research on invasive species and monitor alien invasive species as convenient and suitable with the inclusion of baseline taxonomic study of biodiversity. Accordingly, monitoring is very important for preliminary detection of those species which have invaded recently. Monitoring should be general as well as specific. Involvement of other sectors like local communities should be mandatory. During research, invasive alien species should be thoroughly identified. Reliable documentation of the information collected is as important as their identification. This information should cover the following facets. The most important of them is the history of its origin, ecology of invasion, time period of invasion, conditions, and place/habitat of invasion. Next, we need to know the impacts of these invasive species at the ecosystem, species, and genetic levels. In addition to these impacts, we need to monitor social and economic impacts of the invasive species. No less important is monitoring the proliferation of invasive species, and their spread over time.

12.7.8 International Plant Protection Convention (IPPC)

The IPPC came into existence in 1950. Since then, national and regional plant protection organizations have maintained its objectives, including the prevention and spread of pests and pathogens to agriculture and wild plants.

12.7.9 World Trade Organization Sanitary and Phytosanitary Agreement (WTO–SPS)

Unlike other conventions, the SPS Agreement takes up particular issues. These issues necessarily pertain to plants, animals, and human beings. The agreement also lays down rules and regulations for the coordination of different policies. Sanitary and phytosanitary measures collect products coming from disease-free areas. The SPS inspects different products from different areas. When and where required, it treats them or suggests treatments for them. Experts use international standards to improve the operational efficiency of the SPS agreement. Scientists specifically use IPPC standards for plants and the World Organizations uses it for Animal Health. Implementation of the number of standards requires support from taxonomy, hence exporting countries always need access to taxonomists if they really want successful trade in biological and agricultural commodities. However, importing countries require taxonomic information to predict potential pest distribution, which is an integral part of the risk analysis process of the imported commodity.

12.7.10 Ballast Water Convention

The International Convention for the Control and Management of Ships' Ballast Water and Sediments controls and manages the unlawful transport of animals from one nation to another. The control and management of ships' ballast water and sediments' tries to prevent and minimize the effect of harmful organisms ultimately eliminating the transfer of unwanted pathogens and aquatic organisms. Management and regulation of importation of animals under "Ballast Water and Sediments" does not match with international law. The convention has set up the GloBallast Program with support from the United Nations Development Programme and the Global Environment Facility

(Jassal, 2018). With the support of these programs, developing countries can effectively implement the measures to regulate and control the introduction of foreign marine species.

12.7.11 Role of Taxonomy on the Management of Invasive Species: The Nature of Taxonomy

According to one estimate there are about 15 million total species. Taxonomy has the main role in naming and classification of various animals. Out of 15 million species, to date approximately 1.78 million different species have been identified. Such a huge number of species in total and pending to be named demonstrates that the task of the taxonomists and taxonomy is considerable. Invasive species need to be known and identified and for the identification of many of them helping aids are not available. As mentioned above, that naming of the species may be in the past still means that the species concerned needs to be known, which is the fundamental task of the taxonomy. Taxonomy has contributed a great deal in the management of invasive alien species. It has named them accurately, updated those which are well understood and well accepted, and used and applied them internationally. The name of any species is the key which enables sharing of information pertaining to invasive or potentially invasive species. Management of invasive species is an international and inter-sectoral endeavor. Therefore, it depends heavily on the access to relevant information from different countries and different disciplines nationally or internationally. Therefore, that information is of fundamental importance for effective management of invasive species. It has been observed that there is an annual change in the given names of the species @ 1%; Fishbase, one of the most developed species databases available, can be quoted as an example well justifies this claim. Therefore, it is important to use names of animals which are currently in scientific use. Moreover, there needs to be a convenient linkage to other names used in the past. If the same species has different names between countries and regions this can become problematic. In this way, the species become misidentified and accordingly management actions can be either delayed or misdirected. Further, different countries use different names for animals, which can create confusion and complicate the decision-making. Improper decisions can then lead to incorrect action against threatened species (Anon, 2005). Therefore, appropriate naming is fundamental to the effective management of invasive alien species.

12.8 PROMOTION OF KNOWLEDGE ABOUT CONSERVATION OF GENETIC BIODIVERSITY

Animal genetic resources encompass species, breeds, and strains of all animal species which have cultural, economic, and scientific values for human beings. These benefits and values may be in the form of food or any kind of agricultural production for the present population, as well as for those who will come. Across the globe during the last several thousand years about 40 animal species have been fully or partially tamed for the benefits of human beings. Thirty percent of the total 7,616 tamed livestock breeds are at risk of extinction. Developing countries with marginal production environments hold the major share of these breeds. Buffalo, horses, chickens, pigs, goats, sheep, and cattle are the main species reared on a mass scale. In some areas of the globe, camels, donkeys, elephants, reindeer, and rabbits are used for human livelihood. Chickens, sheep, and cattle are uniformly distributed all over the world, while pigs and goats are less common and are found in selected areas of the world. In developing countries, conservation of animal genetic resources (AnGR) is a vital component of conservation and distribution of biodiversity. They are responsible for subsistence and nourishment of about 70% of the poor rural population of the world. If we categorize this 70% of the population, then 170 million are landless livestock keepers, 686 million mixed farmers, and 194 million pastoralists. The wise use of genetic diversity can help to improve food security in developing countries because these resources are rapidly depleting in developing countries due to various tangible reasons which need to be conserved at any cost.

12.8.1 The Status of Animal Genetic Resources and Programs for Their Conservation

Usually for the conservation of animal genetic resources ex situ and in-situ conservation programs are in common use. These methods are more useful if employed away from the original production sites of the animals under study. When conservation of animal genetic resources takes place in their original environment it is called in-situ conservation. In-situ breeding programs, captive breeding of animals in the zoo, and cryopreservation of genetic material in the form of semen, oocytes, embryos, and DNA are very common. The role of natural selection therefore diminishes in ex-situ conservation. Due to the fewer number of animals they do not represent the full diversity of the breed and hence they are isolated from the wider gene pool. When animal genetic resources are conserved in the form of gene bank, it does not save the historical and ecological values nor does it possess the socioeconomic role of the breed. Other than limited genetic diversity, there are several other problems which are associated with it. Availability of required infrastructure, continuous funding, and provision of adequate land are the major challenges which must be faced in the execution of such projects. Nonetheless, the scarcity of land is the major issue due to shortage of land for other purposes in developing countries. Ex-situ and in-situ are generally approaches to maintain animal genetic resources. They are universal and can be conveniently used at different tiers, including at the regional, national, and global levels. Because developing countries focus on quantity of production to meet the daily needs of their people and do not care about quality, conservation of genetic resources is not their priority. Although these methods are repeatedly in use for the improvement of the genetic resources of few breeds of buffalo, cattle, goat, and sheep, but utility of these programs varies from region to region. Among African countries, Kenya and Ethiopia and some others have started genetic resource conservation of local and exotic cattle breeds including Sahiwal cattle. Ten South American countries have also started programs to conserve the genetic resources of milk, beef, and mutton-producing animals, but progress of these programs in all of the above-mentioned countries is not encouraging. Out of the ten countries mentioned, nine claim that in addition to dairy animals they have also started genetic resource conservation programs for other animals too, although they are variable in numbers. To date, improvement work on one bovine species, seven sheep, four camelids, three horse breeds, four goats, and three pig species has been successfully accomplished. In Argentina, however, honeybee gene banks have also been established. Southeast Asian countries, including Bangladesh, Bhutan, India, Nepal, Pakistan, and Sri Lanka, have now started several genetic resource conservation programs. These programs include cattle, buffalo, goats, and sheep breeds for assessment and further improvement. It is really unfortunate that in these countries there is no functional program on the conservation of genetic resources of these animals in which farmers can receive support or fully participate. The use of available nucleus herds for multiplication of indigenous breeds is common practice in existing centers. Such species are threatened and always inclined to inbreeding. Weaknesses of institutions and organizations and their allied infrastructure do not support sustainable breeding programs. There are several constraints which hinder the successful execution of these programs. The lack of adequate funding, lack of basic facilities, and shortage of trained and skilled manpower in animal breeding are the most important ones. Some others, although less important, also constrain the success of such programs, which are the insufficient participation of farmers, weak breeding policies, poor definition of breeding objectives, and weak interactions and linkages between and within different institutions.

12.8.2 Animal Genetic Resources in the Developing Countries and Ways and Means for Their Maintenance

Policy-makers and major stakeholders in the livestock sector do not have enough knowledge about the status of animal genetic resources. These stakeholders do not know how to conserve these genetic resources and how to use them sustainably for the benefit of human beings. It is urgent to take various steps and devise an appropriate and effective strategy for the conservation of animal genetics to the desired level and standard. First, there is the need to characterize the different breeds of animals for their conservation value. Second, it is desired to know their relatives in the wild. Then comes what strategy we should adapt for conservation of genetic resources for farm animals. It is a little different from the wild setup because it demands that conservation strategies should be integrated with technical, economical, sociological, and political parameters. It is worth mentioning here that we cannot effectively manage genetic resources unless and until we do have comprehensive knowledge about the characteristics of the animal breeds. This information includes data on the production environment, geographical distribution, population size and structure, and status of genetic diversity within and between breeds. Whenever any program on the conservation of genetic resources of animals is intended to be initiated, it should be considered from the livestock keepers' perspective, because livestock keepers are the principal end users of diversity in animal genetic resources. There is an urgent need for recognition of shrinking diversity and enhanced future production demands efficient long-term strategies and efficient and purposeful animal breeding and rearing schemes. To execute these breeding programs successfully, one should have the knowledge of what is actually produced and how is it marketed. Moreover, there is also a need to ponder the characteristics of the breeds, and the socioeconomic and cultural values for the formulation of adequate breeding objectives.

This does not pertain only to the keeping of the animals but it also demands amalgamation of the knowledge that farmers have and how they rear and manage their animals to maintain domestic animal diversity. Nevertheless, whatever knowledge exists is adequately documented and characterized and hence is not well-appreciated. Due to ignorance of wealth of knowledge and rather than going behind Western designs and structures, most of the genetic diversity conservation programs in developing tropical countries have failed. Speaking practically, indigenous knowledge on the species is of paramount importance for conservation of genetic resources. For example, farmers from West Africa, and from East Africa have been breeding helminth-resistant Red Maasai sheep and trypan tolerant N'Dama cattle for centuries. Similarly, breeders of developing countries breed indigenous buffalo, goat, camel, and pig and they always try to maintain those animal breeds which can resist drought, are longer living, disease resistant, and have the ability to survive on quality feeds. Knowledge developed during these activities serves as the basis for developing strategies for sustainable exploitation of genetic resources. This knowledge further helps the continued maintenance of these genetic resources. Encouragement to the local breeders is another incentive in maintaining their local herds and their promotion. In this way, scientists can manage the required animal genetic resources comfortably and successfully in their respective ecological contexts. Rural economic opportunities go hand-in-hand with the conservation of domestic animal diversity. Lack of availability of supportive infrastructure required for conservation and genetic improvement programs is another critical issue. For successful accomplishment of any breeding program, genetic evaluation systems and appropriate and reliable functional facilities are mandatory. There should be a comprehensive system of recording the data produced. It has been observed that in low to medium level explorative facilities for conservation and genetics functional infrastructure is either totally missing or not fully developed to support ongoing breeding activities. The nomads, farmers, and indigenous communities cannot maintain the existing farm animal genetic diversity if they do not develop a properly working strategy. This strategy should facilitate the marketing of products from local breeds. At national level, this strategy can raise awareness about the value of genetic resources among consumers community. It should also describe and characterize local breeds in a better way with proper documentation for future usage. In addition to above contributions, this strategy should provide veterinary services to the local livestock when and where required. These strategic mechanisms should not end here; rather, strategies need to be developed how to use land rights. Development of mechanisms for controlling unnecessary and/or diseased animal exports is also a crucial part of this strategical chain of activities. Further to this mechanism, there is a need to devise ways and means through education and training for the capacity building of the farmers and local communities. Competent staff, long-term financial support, efficient institutions and organizations, supportive policies, and strong links among these components are also important for initiation and execution of these strategies.

At the global level, conservation of about 7,616 livestock breeds is both technically and financially impossible. There are a number of factors which contribute in the prioritization of the breeds for conservation, and important among them are: (1) the level of endangeredness of that species; (2) divergence level of that breed; and (3) specific adaptations. Other than the above factors breed merit, genetic uniqueness, uniqueness of that particular trait, cultural value of that particular breed, risk of breed extinction, and economic importance of that particular trait can also play a crucial role in the prioritization of breeds for conservation. In the very beginning, the policy formulated should prescribe types of breeds and number of breeds which need to be conserved and they will be managed when they are passing through the process of conservation. Differences exist in the species horizon of each species. Therefore, planning for conservation of genetic resources need to be designed for at least 50 years ahead from the perspective of horizon of species.

12.9 MANAGEMENT OF HABITAT LOSS AND ITS EFFECTS ON SURVIVAL OF ANIMAL SPECIES

The two most important factors that play a part in the current extinction of species are fragmentation and degradation of natural habitats (Groombridge, 1992). Loss and fragmentation of habitat reduce population sizes and then due to poor planning and future apprehension demographic factors increase the probability of extinction of species (Burkey, 1995). Although many studies show the effects of fragmentation but in most cases they are confounded.

Fragmentation has a negative effect, including a decrease in size of habitats following isolation of these patches (Celada et al., 1994; Hunter et al., 1995), ultimately resulting in large-scale habitat loss. If there is habitat loss but the number of patches does not change then whatever the effect will be from fragmentation will only be due to habitat loss. Removal of patches from the landscape further isolates the remaining patches, ultimately resulting in further decrease in fragmentation due to fewer number of patches (Ripple et al., 1991). If an area of the habitat does not change, then fragmentation either has no effect on survival of the population (Middleton & Merriam, 1983; Hamel et al., 1993) or affects population survival negatively (Irlandi, 1994; Dytham, 1995). Fragmentation and habitat loss work together, and therefore it is hard to differentiate which factor has the greater effect on the extinction of species. If fragmentation is important in some areas, then losses due to fragmentation should be mitigated with the assurance that the remaining part of the habitat is not fragmented. If we assume that the losses due to fragmentation are trivial compared to total loss, reducing fragmentation can mitigate their effects but is risky for the overall conservation strategy of the biodiversity (Danielson, 1994; Kareiva & Wennergren, 1995).

12.9.1 Some Possibilities

To maintain biodiversity under a changing climate we need to change our priorities and active and adaptive management of existing resources. In some places, active management will protect biodiversity from human interference, while in other places process of conservation of biodiversity takes in other measures like avoiding the interferences in ecosystem processes and species status. Later options seem stronger and more practical than those currently available. In either case, we need to actively consider the values of biodiversity in the context of climate change, competitive usage of land, and marine resources. In this connection, we have to watch continuously what is happening around us. We should work on the predictions how the ecosystem will respond to the changing climate in combination with other factors and/or modifiers which change the dynamic interactions between species and functions of ecosystems. It is important to comprehend that whatever actions are taken, understanding of the predictions of such changes will not be complete and will be a long way away from perfection. In this scenario, whatever management action will be taken it should fall into a framework which is flexible and can conveniently be adapted when required. Further, we need to establish environmental thresholds to ascertain the limits for society so society is achieving the targets of sustainability, while keeping itself within these threshold limits. Therefore, it is required that when economic and policy mechanisms are formulated they should be based on four principal criteria, which are:

1. **Long-term environmental effectiveness:** This criteria emphasizes the closeness of development plans regarding climatic changes following conservation of biodiversity and climate policies to the established thresholds. Consistency is therefore a basis to judge the success of the environmental effectiveness of any policy mix. For example, if any policy focuses on a single component with the effort to reduce the losses, biodiversity and ecosystem services will not be so effective. Nevertheless, if someone considers all the sectors for biodiversity conservation/protection, then this policy will be more effective.
2. **Equity consideration:** In framing actions for the protection of biodiversity and ecosystems within the environmental limits, we need to consider their effects on economic agents and natural systems in the context of future projections and predictions (ex ante) (Beg et al., 2002).
3. **Cost effectiveness:** The management process used to manage ecosystems for the sake of conservation of biodiversity offers a cost-effective method to protect ecosystems, environments, species, population, and societies from climate changes. This protection can be effective only when this management process strengthens hardness and resilience in those societies and ecosystems which are vulnerable and threatened, respectively. If the diversified ecosystems are protected and managed sustainably, they offer very cost-effective solutions to deal with climatic conditions in mitigation of its effects and adaptation to climatic changes may be either abrupt or slow. Such mechanisms fully value ecosystems and the services they are providing. Fundamentally, it is important that sustainability of an ecosystem (exceeding thresholds) should not be comprised at the cost of economic activities. Therefore, to achieve short-term economic goals and long-term sustainability, there is a need to keep a proper balance within policies.
4. **Institutional compatibility of the policy combinations:** It is desired that there should be better integration, monitoring, and regulation of policies. Political and economic differences can sometimes break this relationship and it does not happen like this all the time. If a combination of policies is compatible, it will be better able to address the environmental issues. It is worth mentioning that society depends on biodiversity in one way or another. Biodiversity supports societies by providing life-supporting goods and services. The policy devised and economic activity affects environmental quality, maintaining it within safe threshold limits. Therefore, to move it further the formulation and evaluation of economic activities and policies should be the first principle.

12.9.2 Ecosystem-based Adaptation (EBA)

This approach helps vulnerable communities (which are currently facing threats from climate change) in building resistance in their ecosystems and livelihoods (IUCN, 2009). EBA can also help in the confiscation and collection of carbon dioxide and offer social, cultural, and economic benefits to communities. Healthy ecosystems defend the negative effects of climate change, providing opportunities for sustainable economic prosperity of the country and its inhabitants.

EBA is an adaptation response to climate change. This adaptation achieves multiple goals: (1) it sustains biodiversity and provides ecosystem services to support human beings for its wellbeing; and (2) it enhances the ecosystem's ability to sequester carbon dioxide and mitigate the climate changes that negatively impact biodiversity. EBA encompasses local, national, regional, and global strategies that operate at multiple levels. EBA harmoniously facilitates society, vulnerable communities, and the environment for their convenient adaptation to climate change. The UNEP is currently is in the process of developing an EBA program. When we establish good ecosystem management practices and then further integrate these practices into local, regional, national, and global climate change strategies and action plans, then we can successfully in implement these strategies. The EBA program keeps in view the needs of the country. It delivers specific products and services to support EBA programs. The following three main components play a pivotal role in this whole process. These are: (1) support from knowledge: for adaptation and their economic value, this component analyzes ecosystem services. On the basis of knowledge, this conveys these results to decision-makers. Then these experts design and implement priority EBA policies and projects; (2) skill development and its demonstration: specifically in developing countries, for the implementation of EBA options, this component helps to create the enabling conditions. These options include capacity building, technology development and diffusion, and piloting and demonstration. The technologies developed during this process are accessible for restoration of biodiversity and ecosystems. Moreover, by adapting and utilizing the ecological engineering they can easily adapt to the changing environment at the diffusion stage, soon after their development, ultimately reducing the risk for disaster; (3) integration of EBA options into National Adaptation Plans. This component integrates EBA into national adaptation plans. Institutional structures mobilize different stakeholders, thus establishing a coordinated and integrated approach. This approach emerges through modification of the frameworks and processes used to develop or revise policies, programs, and projects.

12.9.3 Biodiversity as the Fundamental Unit of Economic Activity

Our dominant economic model does not see any economic benefits in nature, which is the main symptom and the root cause of loss of biodiversity. Our thinking values those things for which we pay. However, nature provides us with clean air to breathe, freshwater to drink, and for its utilization for other purposes, including the fertility of soil, control and management of floods, control of droughts, and climate stability. These services are not traded in the market and we do not have to pay anything for them; rather, they are provided free by Mother Nature. The majority of researchers agree that our current economic global model is not sustainable. If economics were to become part of our solution, then it could reverse our declining ecosystems and loss of biodiversity. We have to work now on that part of economics which recognizes the value of the services nature is providing us. Then we need to set a stage for a new green economy to assess its cost. If the benefits of biodiversity previously forgotten are recognized, they become worth preserving. Many examples have been set where payment is made against each ecosystem service provided. Here, people who are custodians of that service mange it and interact with it in one way or another. They receive payments against the benefits that an ecosystem is providing to the wider community, for example: (1) in Costa Rica, those farmers which maintain their forests in low earning pastures are paid for these efforts. This is in addition to their benefits which they get from land and water; (2) in India, if any person restores ecosystems and extracts water he is paid and he has to pay respectively for either of any of these activities; (3) the cities of San Francisco and New York receive their freshwater supply from reservoirs and lake watersheds. These water sources are surrounded by well-managed forests; (4) in Indonesia, the converse is true. Here, the people of Lombok Island (Indonesia) pays US$0.60 per household. Managing organizations use this money for the sustainable management of their watershed forests. In this way, they maintain vital water supplies to the local communities.

12.9.4 The Way Forward

Due to biodiversity and ecosystem services, many poor people can successfully adapt to a variety of strategies required for their living. Ecosystem services help the poor and also mitigate the negative effects of climate change. Enough is done in this way yet it is hard to meet all these objectives. We can say that our decisions are based on full understanding of environmental-economic trade-offs, as well as good science. Whatever the solutions are proposed to address climatic changes; they should not damage biodiversity and ecosystem services otherwise they promote non-uniformity and aggravate poverty. A similar type of decision needs to be taken on a global scale without compromising environmental protection. Such decisions demand some basic structural changes in the current economic models of the world. Under this scenario, consumption of resources is the primary driver which degrades the environment, induces biodiversity losses, and will ultimately result in unsustainable societies. Rather, we need to develop such models which fully value the environment and, in this context, reverse those market failures which have occurred due to the existing models. These newly developed economic models have the ability to maintain a balance in the reproductive capacity along with provision of essential services to all the ecosystems of the world. At the same time, these models should modify and rebuild the ecosystems to uniformly meet the basic needs of all components of human society. Such models are required to nourish individual and global collective responsibility and, at the same time, facilitating that all the resources are shared among the needy in an equitable manner. Bilateral organizations, governments, the private sector, and individuals should search as well as try to integrate available information, thinking, and decision-making in such an admirable way and with an assurance that initiatives taken by the Minimum Development Goals (MDGs), the United Nations Framework Convention on Climate Change (UNFCCC), and the CBD, which help them in the achievement of their objectives are well supported. Governments should recognize the interactions and synergies present between the objectives that must support both top-down and bottom-up initiatives, ultimately encouraging the desired change. Similarly, businesses and communities should take the maximum economic advantage gained from an ecosystems-based approach. It is hard to develop synergistic solutions regarding biodiversity conservation unless and until issues including biodiversity loss, climate change, and poverty are addressed in an integrative way.

12.9.5 Recommendations

There is a need to strengthen the CBD for the optimization of benefits from the conservation of biodiversity. This optimization can mitigate climate change effects and help society to adapt to the changing environment successfully. The UNFCCC is then required to lay its foundations and build its structure at the progress and development made at CoP15 (Conference of the Parties), which will ultimately recognize the role ecosystems are playing in the regulation of our climate. In this context, there is a need to maximize the synergies in the protection of environment. These synergies will not only face and address the changes in the climate but also for those regarding poverty alleviation, and sustainability in food and water security. Therefore, there is a need for greater integration between the CBD and the UNFCCC. If adaptation to climatic change is based on ecosystems, it should receive international recognition and heavy funding which will ensure that this has a place in the core strategic planning for sustainability of an economic development. These new economic models are equally good for government and businesses, and for the creation of both private opportunity and public goods. These models will be fully supported to form new partnerships between businesses, citizens, and their governments.

REFERENCES

ADB. (2014). *Economic corridor development for inclusive Asian regional integration: modelling approach to economic corridors.* Asian Development Bank (ADB), Mandaluyong City. http://www.adb.org/publications/economic-corridor-development-inclusive-asian-regional-integration-modeling-approach [accessed 13th December, 2015].

Adetunji, J. (2013). Private land is an important piece of conservation jigsaw. Conversation UK.

Anon. (2005). Ivory-bill takes Perch on Money Tree. Birder's World, December, 18.

Ayal, Y. (2007). Trophic Structure and the Role of Predation in Shaping Hot Desert Communities. Journal of Arid Environments 68, 171–187.

Barnosky, A.D., Matzke, N., Tomiya, S., Wogan, G.O., Swartz, B., Quental, T.B., Marshall, C., McGuire, J.L., Lindsey, E.L., Maguire, K.C., Mersey, B. & Ferrer, E.A. (2011). Has the Earth's Sixth Mass Extinction Already Arrived? Nature 471(7336), 51–57.

Beg, N. et al. (2002). Linkages between Climate Change Sustainable Development. Climate Policy 2, 129–144.

Bernstein, J. & Mitchell, B.A. (2005). Land Trusts, Private Reserves and Conservation Easements in the USA. Parks: Protected Areas Programme – Private Protected Areas 15(2), 48–60.

Bishop, K., Dudley, N., Phillips, A. & Stolton, S. (2004). Speaking a Common Language: The Uses and Performance of the IUCN System of Management Categories for Protected Areas. Cardiff University, IUCN, UNEP and WCMC. http://cmsdata.iucn.org/downloads/speaking_common_language.pdf

Brockington, D. & Schmidt-Soltau, K. (2004). The Social and Environmental Impacts of Wilderness and Development. Oryx 38(2), © 2004 FFI, Oryx, 38(2), 140–142 DOI: 10.1017/S0030605304000250

Burkey, T.V. (1995). Extinction Rates in Archipelagos: Implications for Populations in Fragmented Habitats. Conservation Biology 9, 527–541.

Celada, C., Bogliani, G., Gariboldi, A. & Maracci, A. (1994). Breeding Bird Communities in Fragmented Wetlands. Biology Conservation 69, 177–183.

Chacon, C.M. (2005). Fostering Conservation of Key Priority Sites and Rural Development in Central America: The Role of Private Protected Areas. Parks: Protected Areas Programme – Private Protected Areas 15(2), 39–47.

Chape, S., Blyth, S., Fish, L. & Fox, P.M. eds. (2003). United Nations List of Protected Areas. Cambridge, Spalding.

Clough, P. (2000). Encouraging Private Biodiversity: Incentives for Biodiversity Conservation on Private Land. Report to the Treasury. Wellington, New Zealand Institute of Economic Research.

Daniels, T. (1998). "Purchase of Development Rights, Agricultural Preservation and Other Land-Use Policy Tools – The Pennsylvania Experience." http://ageconsearch.umn.edu/bitstream/17934/1/ar980034.pdf

Danielson, B.J. (1994). A Justification for the Conservation of Large Reserves. In: Principles of Conservation Biology, eds. Meffe, G.K. & Carroll, C.R., pp. 276–277. Sinauer, Sunderland, Mass.

Di Nardo, A., Knowles, N.J. & Paton, D.J. (2011). Combining Livestock Trade Patterns with Phylogenetics to Help Understand the Spread of Foot and Mouth Disease in Sub-Saharan Africa, the Middle East and South-East Asia. Revue Scientifique et Technique de L'Office International Des Epizooties 30(1), 63–85.

Doremus, H. (2003). A Policy Portfolio Approach to Biodiversity Protection on Private Lands". Environmental Science & Policy 6, 217–232.

Dytham, C. (1995). The Effect of Habitat Destruction Pattern on Species Persistence: A Cellular Model. Oikos 74, 340–344.

ELI (Environmental Law Institute). (2003). Legal Tools and Incentives for Private Lands in Latin America: Building Models for Success. Washington DC, Environmental Law Institute. https://cmsdata.iucn.org/downloads/spn___legal_tools_and_incentives_for_private_lands_conservation.pdf

Fell, A. (2001). Davis Declaration calls for Action on Invasive Species. Andy Fell, Research news (emphasis: biological and physical sciences, and engineering), 530-752-4533, ahfell@ucdavis.edu

Figgis, P. (2004). Conservation on Private Lands: the Australian Experience, 1–34. Cambridge, IUCN.

Geering, W.A., Roeder, P.L. & Obi, T.U. (1999). Manual on the Preparation of National Animal Disease Emergency Preparedness Plans. FAO, Rome. Available at: https://www.fao.org/4/x2096e/X2096E.htm

George, S. (2002). "Conservation in America: State Government Incentives for Habitat Conservation – A Status Report." Defenders of Wildlife, USA. https://defenders.org/sites/default/files/publications/conservation_in_america_state_profiles.pdf

Gleeson, L.J. (2002). A Review of the Status of Foot and Mouth Disease in South-East Asia and Approaches to Control and Eradication. Revue Scientifique et Technique de L'Office International Des Epizooties 21(3), 45–475.

Grodzińska-Jurczak, M. & Cent, J. (2011). Expansion of Nature Conservation Areas: Problems with Natura 2000 Implementation in Poland? Environmental Management 47(1), 11–27.

Groombridge, B., Editor. (1992). Global Biodiversity: State of the Earth's Living Resources. Chapman and Hall, New York, N.Y. 585, 40–46.

Hamel, P.B., Smith, W.P. & Wahl, J.W. (1993). Wintering Bird Populations of Fragmented Forest Habitat in the Central Basin, Tennessee. Biological. Conservation 66(2), 107–115.

Hong Kong Trade Development Council. (2015). *The Belt and Road Initiative*, Hong Kong Trade Development Council: Research. [Online] Hong Kong Trade Development Council. Available at: http://china-trade-research.hktdc.com/business-news/article/One-Belt-One-Road/The-Belt-and-Road-Initiative/obor/en/1/1X000000/1X0A36B7.htm [accessed 13th December, 2015].

Hunter, J.E., Gutierrez, R.J. & Franklin, A.B. (1995). Habitat Configuration Around Spotted Owl Sites in Northwestern California. Condor 97, 684–693.

Irlandi, E.A. (1994). Large- and Small-Scale Effects of Habitat Structure on Rates of Predation: How Percent Coverage of Seagrass Affects Rates of Predation and Siphon Nipping on an Infaunal Bivalve. Oecologia 98, 176–183.

IUCN (2009). Position Paper, UNFCCC Bangkok meeting, 28th Sept–9th Nov 2009.

Jassal, R. (2018). Ballast Water Management: What We Need to Know and How to Comply. MySea Time.

Kareiva, P. & Wennergren, U. (1995). Connecting Landscape Patterns to Ecosystem and Population Processes. Nature 373, 299–302.

Knight, R.L. (1999). Private Lands: The Neglected Geography. Conservation Biology 13, 223–224.

Krug, W. (2001). Private Supply of Protected Land in Southern Africa: A Review of Markets, Approaches, Barriers and Issues. World Bank/OECD International Workshop on Market Creation for Biodiversity Products and Services. Paris, January 25–26.

Leva, C.E.D. (2002). The Conservation of Nature and Natural Resources through Legal and Market-Based Instruments. Review of European Community & International Environmental Law 11(1), 84–95.

Lindsey, P.A., Alexander, R., Frank, L.G., Mathieson, A. & Romanch, A.A. (2006). Potential of Trophy Hunting to Create Incentives for Wildlife Conservation in Africa Where Alternative Wildlife-Based Land Uses May Not Be Viable. Animal Conservation 9, 283–291.

Main, M.B. (1999). Evaluating Costs of Conservation. Conservation Biology 13(6), 1262–1272.

Martínez, F.J. (2013). Estructura de las comunidades de artrópodos epigeos en ambientes representativos de Península Valdés. Dissertation, Universidad Nacional de la Patagonia San Juan Bosco, Argentina.

Mascia, M.B. (2003). Conservation and the Social Science. Editorial. Conservation Biology 17(3), 649–650.

Millennium Ecosystem Assessment (MA) (2005a and 2005b). Initiated in 2001. Assessment of the Consequences of Ecosystem Change for Human Well-Being and the Scientific Basis for Action to Enhance the Conservation and Sustainable Use of Those Systems and a State-of-the-Art Scientific Appraisal of the Condition and Trends in the World's Ecosystems and the Services. Restore, Conserve and Enhance the Sustainable Use of Ecosystems.

Naing Oo, K. (2010) Epidemiological Study to Support the Establishment of a Progressive Zoning Approach for the Control of Foot and Mouth Disease in Myanmar. Murdoch University, Western Australia. PhD thesis. Available at: http://researchrepository.murdoch.edu.au/2974/ [accessed 21st February, 2016].

Novaro, A.J., González, A. & Pailacura, O. et al. (2016). Manejo del conflicto entre carnívoros y ganadería en Patagonia utilizando perros mestizos protectores de ganado. Mastozoología Neotropical 24(1), 47–58.

Ostermann, O. (1998). The Need for Management of Nature Conservation Sites Designated Under Natura 2000. Journal of Applied Ecology 35, 968–973.

Reid, C.T. (2011). The Privatisation of Biodiversity? Possible New Approaches to NatureConservation Law in the UK. Journal of Environmental Law 23(2), 203–231.

Ripple, W.J., Bradshaw, G.A. & Spies, T.A. (1991). Measuring Forest Landscape Patterns in the Cascade Range of Oregon, USA. Biology Conservation 57, 73–88.

Stroup, R.L. (1997). The Economics of Compensating Property Owners. Contemporary Economic Policy 15(4), 55–65.

TNC (The Nature Conservancy, USA). (2011). Private Lands Conservation. http://www.nature.org/about-us/private-lands-conservation/index.htm

Index

For Product Safety Concerns and Information please contact our EU representative GPSR@taylorandfrancis.com Taylor & Francis Verlag GmbH, Kaufingerstraße 24, 80331 München, Germany